Advanced Inorganic Chemistry

Advanced Inorganic Chemistry

Editor: Bridget Kent

NY RESEARCH
P R E S S

New York

Published by NY Research Press
118-35 Queens Blvd., Suite 400,
Forest Hills, NY 11375, USA
www.nyresearchpress.com

Advanced Inorganic Chemistry
Edited by Bridget Kent

International Standard Book Number: 978-1-63238-667-0 (Hardback)

Cataloging-in-Publication Data

Advanced inorganic chemistry / edited by Bridget Kent.
    p. cm.
Includes bibliographical references and index.
ISBN 978-1-63238-667-0
1. Chemistry, Inorganic. 2. Chemistry. I. Kent, Bridget.
QD151.3 .A38 201
546--dc23

Contents

Preface

The study of inorganic compounds is within the domain of inorganic chemistry. Some of the branches of study in inorganic chemistry are descriptive inorganic chemistry, theoretical inorganic chemistry, thermodynamics and inorganic chemistry, mechanistic inorganic chemistry and synthetic inorganic chemistry. Each of these branches focuses on different aspects of the study of inorganic compounds. Descriptive inorganic chemistry deals with the classification of compounds based on their properties. Theoretical inorganic chemistry involves various semi-quantitative and semi-empirical approaches such as molecular orbital theory, ligand field theory and approximations methodologies like density functional theory. This book presents the complex subject of inorganic chemistry in the most comprehensible and easy to understand language. It is a valuable compilation of topics, ranging from the basic to the most complex advancements in this field. The extensive content of this book provides the readers with a thorough understanding of the subject.

This book has been the outcome of endless efforts put in by authors and researchers on various issues and topics within the field. The book is a comprehensive collection of significant researches that are addressed in a variety of chapters. It will surely enhance the knowledge of the field among readers across the globe.

It gives us an immense pleasure to thank our researchers and authors for their efforts to submit their piece of writing before the deadlines. Finally in the end, I would like to thank my family and colleagues who have been a great source of inspiration and support.

Editor

Olivine-Based Blended Compounds as Positive Electrodes for Lithium Batteries

Christian M. Julien [1],*, Alain Mauger [2], Julie Trottier [3], Karim Zaghib [3], Pierre Hovington [3] and Henri Groult [1]

[1]   PHENIX, UMR 8234, *Sorbonne Universités*, Univ. Pierre et Marie Curie, Paris-6, 4 Place Jussieu, 75005 Paris, France; henri.groult@upmc.fr
[2]   IMPMC, *Sorbonne Universités*, Univ. Pierre et Marie Curie, Paris-6, 4 Place Jussieu, 75005 Paris, France; alain.mauger@impmc.jussieu.fr
[3]   Energy Storage and Conversion, Research Institute of Hydro-Québec, Varennes, QC J3X 1S1, Canada; Trottier.Julie@ireq.ca (J.T.); zaghib.karim@ireq.ca (K.Z.); Hovington.pierre@ireq.ca (P.H.)
*   Correspondence: christian.julien@upmc.fr

Academic Editor: Tom Nilges

Abstract: Blended cathode materials made by mixing $LiFePO_4$ (LFP) with $LiMnPO_4$ (LMP) or $LiNi_{1/3}Mn_{1/3}Co_{1/3}O_2$ (NMC) that exhibit either high specific energy and high rate capability were investigated. The layered blend LMP–LFP and the physically mixed blend NMC–LFP are evaluated in terms of particle morphology and electrochemical performance. Results indicate that the LMP–LFP (66:33) blend has a better discharge rate than the $LiMn_{1-y}Fe_yPO_4$ with the same composition ($y = 0.33$), and NMC–LFP (70:30) delivers a remarkable stable capacity over 125 cycles. Finally, *in situ* voltage measurement methods were applied for the evaluation of the phase evolution of blended cathodes and gradual changes in cell behavior upon cycling. We also discuss through these examples the promising development of blends as future electrodes for new generations of Li-ion batteries.

Keywords: blend; insertion compounds; olivine; layered materials; Li-ion batteries

---

1. Introduction

Recent advances to develop highly effective electrode materials for Li-ion batteries (LIBs) derived from composites or blended architectures are new technological approaches to designing high-energy and high-power density storage systems [1,2]. The blend electrode is an active material formed by a mixture of two or more distinct lithium insertion compounds that possess percolating frameworks for ions and electrons to achieve better balanced electrochemical performance than that of an individual compound. We can consider four configurations: a complete physical mixture, a segregated blend, a layered-type blend and an integrated blend cathode (Figure 1). Sometimes, the notion of a blend is restricted to the case of physical mixture, but the notion in the present work is extended to all the configurations, including another one, the core-shell structure, where one of the components is the core and the other one is the shell of a particle. For materials currently used as positive electrodes in lithium batteries, the different crystal chemistries are examined from the basic reaction driven by a charge transfer from the guest Li^+ ions to the conduction band of the host compound. Thus, electron-donating species of a blend electrode composed of two components—$(M^{n1}X)_1$ and $(M^{n2}X)_2$ (where M is a transition metal with oxidation state n_i, and X is an anion) host components—can give rise to a reversible reaction classically represented by the following scheme:

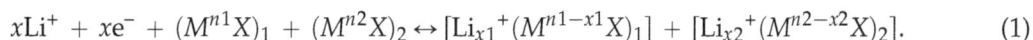

$$x\text{Li}^+ + xe^- + (M^{n1}X)_1 + (M^{n2}X)_2 \leftrightarrow [\text{Li}_{x1}^+(M^{n1-x1}X)_1] + [\text{Li}_{x2}^+(M^{n2-x2}X)_2]. \qquad (1)$$

In the usual case, x_i is the molar lithium insertion fraction, and $[Li_{xi}^+(M^{n,i-x,i}X)_i]$ is the final product. The concept of using two or more electrochemically active materials in a blend matrix is not new; since 1976, Margalit suggested applying $xAg_3PO_4 \cdot (1-x)Ag_2CrO_4$ to a Li primary battery for pacemakers and watches [3]. This concept has been extended to numerous cathode materials to avoid their individual drawbacks [4–7]. Pynenburg *et al.* [4] patented a rechargeable lithium cell composed of a physical mixture of spinel $Li_xMn_2O_4$ and one of the lithiated metal oxides from the Li_xMO_2 group (M = Ni, Co), wherein $0 < x \leqslant 2$. For instance, pure $LiMn_2O_4$ has a better rate capability than pure $LiNi_{0.8}Co_{0.15}Al_{0.05}O_2$, but, at a low rate, the specific energy of $LiNi_{0.8}Co_{0.15}Al_{0.05}O_2$ is higher than that of $LiMn_2O_4$, so that mixing the two elements should lead to an optimized compromise [5,7]. Other effects not necessarily expected have been observed. Numata *et al.* [5] showed the advantages of blending $Li_{1+x}Mn_{2-x}O_4 \cdot LiNi_{0.8}Co_{0.15}Al_{0.05}O_2$, which decreased the Mn dissolution from the spinel into the electrolyte. Meanwhile, however, it increased the irreversible capacity loss in the first cycle and deteriorated the rate capability. Albertus *et al.* [7] adopted a mathematical model to multiple types of active materials in a single electrode to treat both direct (galvanostatic) and alternating (impedance) currents and concluded that combining a sloped-potential system ($LiNi_{0.8}Co_{0.15}Al_{0.05}O_2$) with a flat-potential system ($LiMn_2O_4$) may assist in electrode state-of-charge (SOC) determination. In 2011, LIBs (composed of 288 Li-ion cells) using a blended cathode formed by the mixture $LiMn_2O_4 \cdot LiNi_xMn_yCo_{1-x-y}O_2$ were developed by LG Chem. Ltd. and applied to power the GM Chevy-Volt electric vehicles [8]. In 2006, Kim *et al.* [6] studied the $0.5LiCoO_2 \cdot 0.5LiNi_{1/3}Mn_{1/3}Co_{1/3}O_2$ (LCO–NMC)-mixed cathode exhibiting a stable cycleability maintained at *ca.* 163 mAh·g^{-1}.

Depending on the nature and the combination of materials, the blending found different advantages (but not all of them at the same time), such as (i) the inhibition of overheating effects during overcharge conditions; (ii) a reduction in irreversible capacity decay; (iii) an improvement of cycling life; (iv) a rise of the discharge C-rate; and (v) endurance of high temperature and thermal stability [9–12].

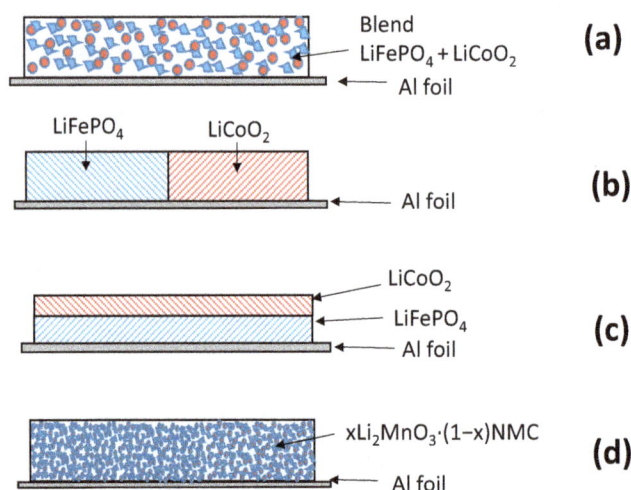

Figure 1. Scheme of the four different configurations of blended cathodes: (**a**) completely intermixed, (**b**) segregated, (**c**) layered, and (**d**) integrated cathode.

The literature of olivine-based blended active cathode materials is rather sparse. The good thermal stability of olivine-structured compounds $LiMPO_4$ (M = Fe, Mn) makes them attractive cathode materials, but the relatively flat voltage plateau at 3.45 V *vs.* Li^0/Li^+ is a limitation, which yields difficult the SOC evaluation. This drawback can be overcome using a blend of olivine with a layered material. Various mixtures have been tested such as the blend of $LiFePO_4$ (LFP) with Li-rich NMC [13,14], which exhibits improved power capability, thermal stability, and easy SOC monitoring. In another approach, a blend of $LiFePO_4$ with $Li_3V_2(PO_4)_3$ (LVP) delivers lower capacity and has a voltage profile consisting of multiple plateaus due to the LVP component [15]. Uniform blends of micron-sized $LiMn_2O_4$ with nano-sized $LiFePO_4$ (1:1 mass ratio) prepared by ball milling showed that LFP powders not only cover the $LiMn_2O_4$ particles but also effectively fill the cavity of the spinel network [16]. Recently, blend electrodes were prepared by mixing the olivine $LiFe_{0.3}Mn_{0.7}PO_4$ (LFMP) and $LiMn_{1.9}Al_{0.1}O_4$ spinel (LMO) in order to obtain a composite electrode, combining the high capacity of LFMP and the rate capability of the spinel [17].

In this paper, we discuss the architecture, the structural properties, and the electrochemical performance of two $LiFePO_4$-based blend systems as positive electrodes for lithium batteries: (i) the blend fabricated by $LiFePO_4$ with $LiMnPO_4$ (LMP) olivine structures, and (ii) the blend formed by the $LiFePO_4$ olivine structure with the $LiNi_{0.33}Mn_{0.33}Co_{0.33}O_2$-layered compound. In both systems, the use of carbon-coated LFP results in an increase in electrode conductivity and rate capability. Thus, we show that blending may be a good solution to avoid individual drawbacks of components.

2. Results

2.1. LiMnPO$_4$–LiFePO$_4$ (LMP–LFP) Blended Electrode

To overcome the difficulty of carbon coating on $LiMnPO_4$ (LMP) olivine particles, it is possible to prepare a $LiMnPO_4$–$LiFePO_4$ (LMP–LFP) composite, in which the carbon-coated LFP component encapsulates the core of $LiMnPO_4$ particles to take advantage of the catalytic reaction of Fe with C. The STEM images of several particles representative of the sample are shown in Figure 2. Figure 2a displays well-crystallized flaky grains several tens to 100 nm in size. Some of the grains have with many dislocations. A typical EDX map (Figure 2b) shows the two components of blended particles: LMP (in red) in the center, surrounded with a LFP layer (in green). Elemental analysis indicates the LMP–LFP (66:33) composition. The Fe content is minimal at the center and maximal at the surface, while the opposite holds true for the Mn content. The HRTEM image (Figure 2c) shows the lattice fringes with the sharp interface between $LiMnPO_4$ and $LiFePO_4$ and the continuous carbon-coating layer (~2–3 nm thick). To summarize the results illustrated by the structural analysis, LFP forms a continuous layer that partly covers most of the LMP particles, but it also forms few small $LiFePO_4$ particles distributed sparsely (Figure 2). The strain at the interface between the two materials is accommodated by extended defects such as dislocations and grain boundaries; on the other hand, the LFP layer is well crystallized as evidenced by the lattice fringes in the HRTEM image (Figure 2c). This result suggests that LFP is deposited by some epitaxial effect on the well-crystallized surface of LMP, since the free surface of LFP at 600 °C is disordered [18].

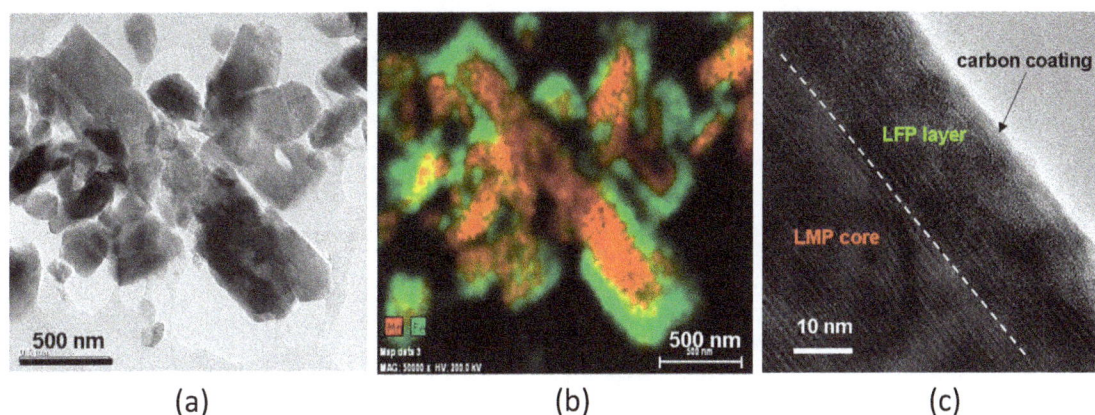

(a) (b) (c)

Figure 2. (**a**) STEM image of the $LiMnPO_4$–$LiFePO_4$ (LMP–LFP) blend particle showing the different layers; (**b**) The typical EDX map of LMP (in red)–LFP (in green) particle; (**c**) HRTEM image showing the lattice fringes with the sharp interface between $LiMnPO_4$ and $LiFePO_4$ and the continuous, carbon layer.

The electrochemical features of the LMP–LFP (66:33) blend is illustrated in Figure 3, showing the charge–discharge profiles *vs.* specific capacity of the Li//blend coin cell for the first two cycles.

Electrochemical tests were carried out at C/12 in the potential range 2.0–4.5 V $vs.$ Li^0/Li^+. The upper flat part corresponds to the limit of 4.5 V imposed on the potential to protect the electrolyte. The first plateau observed at 3.5 V is characteristic of the Fe^{2+}/Fe^{3+} redox potential $vs.$ Li in $LiFePO_4$. The next plateau at 4.0 V is characteristic of the Mn^{2+}/Mn^{3+} potential in $LiMnPO_4$, so that both components efficiently contribute to the electrochemical properties. We find that the specific capacity of the first charge is 144 $mAh \cdot g^{-1}$ (charge transfer of $x = 0.86e^-$) followed by the discharge/charge ratio close to unity at the second cycle. Using such a blended electrode, the electrochemical properties are improved with respect to the carbon-coated $LiMn_{2/3}Fe_{1/3}PO_4$ solid solution with a comparable Fe/Mn ratio, as shown in Figure 4. The Peukert plots show that, at 10C, the delivered capacity of the blend is 65.5 $mAh \cdot g^{-1}$ against 23.5 $mAh \cdot g^{-1}$ for $LiMn_{2/3}Fe_{1/3}PO_4$. Therefore, the use of a $LiFePO_4$ as a buffer layer between the high-density cathode element (like $LiMnPO_4$) and the carbon layer opens a new route to improve the performance of the olivine family as the active element of the cathode for Li-ion batteries. This result is attributable to the fact that the LFP shell allows for the coating of the composite, while it is difficult to deposit the carbon at the surface of LMP at a temperature small enough to avoid its decomposition ($\approx$650 °C).

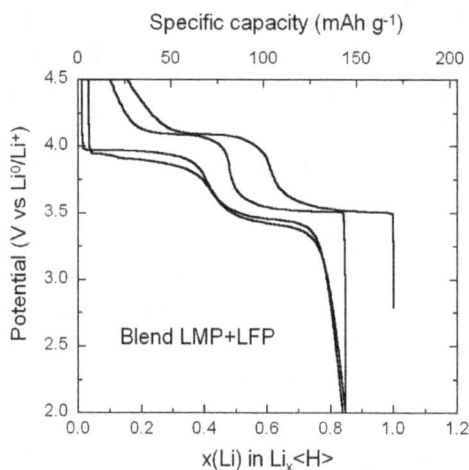

Figure 3. Charge-discharge profiles of Li coin cell with a LMP–LFP (66:33) blend positive electrode for the 1st and 2nd cycle. Electrochemical tests were carried out at C/12 in the potential range 2.0–4.5 V $vs.$ Li^0/Li^+.

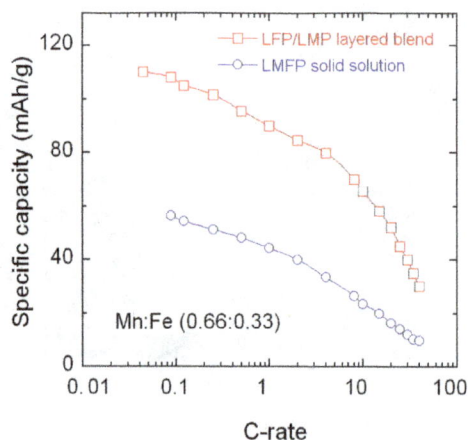

Figure 4. Peukert plot of the LMP–LFP blend compared with the cathode of the same composition formed by the $LiMn_{1-y}Fe_yPO_4$ ($y = 0.33$) solid solution.

2.2. LiNi$_{1/3}$Mn$_{1/3}$Co$_{1/3}$O$_2$–LiFePO$_4$ (NMC–LFP) Blended Electrode

A blended cathode formed by the LiNi$_{1/3}$Mn$_{1/3}$Co$_{1/3}$O$_2$–LiFePO$_4$ (NMC–LFP) mixture exhibits excellent rate capability by using nano-sized LFP particles and delivers higher capacity, *i.e.*, 180 mAh·g^{-1}, due to the NMC component. Consequently, a Li-ion battery including NMC–LFP physical blending provides a good solution that achieves both energy and power capability and exhibits improved electrochemical features with thermal stability similar to LFP. In the SEM images of the NMC–LFP (70:30) blend shown in Figure 5, one discerns the two distinct components of this blend. The cathode material prepared by ultra-microtomy exhibits large spherically shaped NMC particles (secondary particles ~10 μm in size, having a very close size distribution) and small LFP nano-sized particles (<0.5 μm), but no LFP coating on NMC is observed. This could be due to the mechanical force applied during the microtomy. Figure 6 shows the TEM images and EDX maps of NMC–LFP blends with different compositions: (a) (70:30) with Fe in blue, Mn in violet and (b) (80:20) with Fe in red, Mn in green. The differences in morphology and particle size between the two samples are evidenced. Higher SEM magnification confirms that the co-grounded blend sample possesses a better coverage of NMC with LFP than the (80:20) sample.

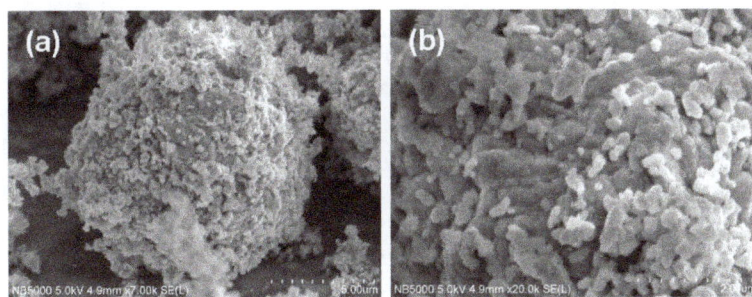

Figure 5. SEM images of the LiNi$_{1/3}$Mn$_{1/3}$Co$_{1/3}$O$_2$–LiFePO$_4$ (NMC–LFP) (70:30) blend powders showing (**a**) big NMC particle surrounded by nano-sized LFP and (**b**) a detail of the blend.

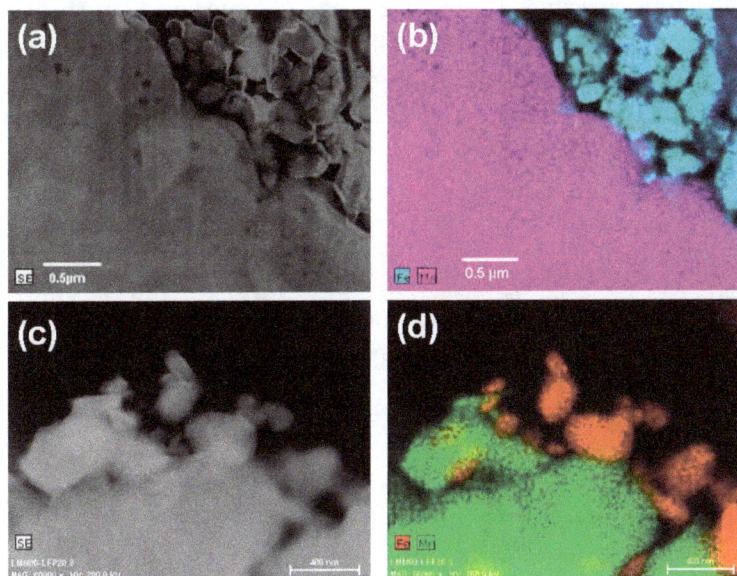

Figure 6. TEM images and EDX maps of NMC–LFP blends with different compositions. (**a,b**) 70:30 (Fe in blue, Mn in violet) and (**c,d**) 80:20 (Fe in red, Mn in green).

The structural properties of the NMC–LFP blends were studied by Raman spectroscopy as shown in Figure 7. A typical Raman spectrum displays three sets of bands: (i) the vibrational features of

NMC with Ni–O mode at 471 cm^{-1} and the broad band at *ca.* 600 cm^{-1} attributed to the Co–O/Mn–O stretching modes [19]; (ii) the symmetric stretching vibration of PO_4^{3-} molecular units of $LiFePO_4$ at around 950 cm^{-1}; and (iii) two broad Raman bands of the carbon surrounding the LFP particles centered at 1336 and 1607 cm^{-1} assigned to the D- and G-bands, respectively. Note that the high intensity of the latter bands is due to the strong Raman scattering efficiency of carbon [20].

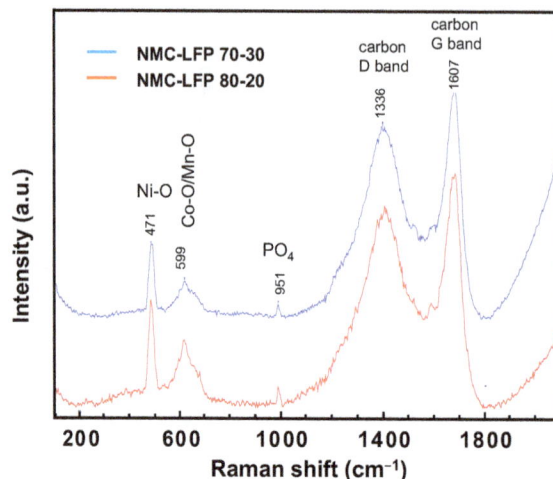

Figure 7. Raman spectra of NMC–LFP blends. Spectra were recorded using a 514.5 nm laser line excitation.

Figure 8 shows the discharge profiles (5th cycle) as a function of the C-rate for a Li-coin cell with the NMC–LFP (70:30) blend as positive electrode. Electrochemical tests were carried out in the potential range 2.0–4.2 V *vs.* Li^0/Li^+. The shape of the curves shows combined features of LFP with the potential plateau at 3.35 V and the sloppy profile of the NMC component from 3.4 to 4.2 V (see discussion below). This behavior suggests the independent charge–discharge profiles of the cathode constituents. The discharge capacity is 148 mAh·g^{-1} at C/12 rate and 42 mAh·g^{-1} at 10C. Note that, as some Li^+ ions are not removed from NMC at 4.2-V cut-off voltages, this causes a lower specific capacity. The electrochemical performance of a Li-coin cell with the NMC–LFP (70:30) blend positive electrode as a function of the number of cycles at 1C rate (Figure 9) shows that this cathode material exhibits very stable cycleability after 125 cycles.

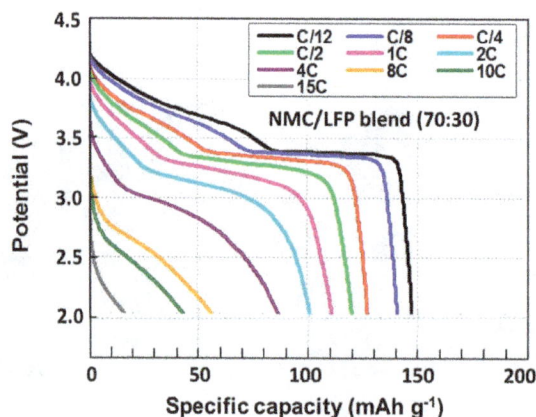

Figure 8. Discharge profiles (5th cycle) as a function of the C-rate of Li-coin cell with a NMC–LFP (70:30) blend positive electrode. Electrochemical tests were carried out in the potential range 2.0–4.2 V *vs.* Li^0/Li^+.

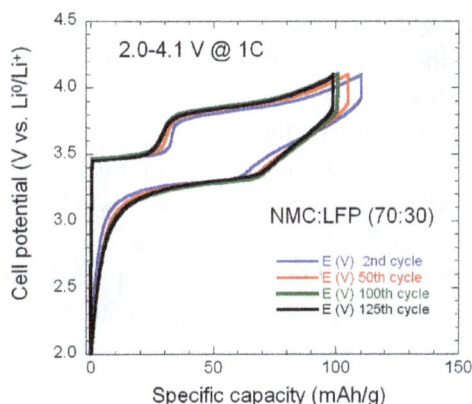

Figure 9. Electrochemical performance of a Li coin cell with a NMC–LFP (70:30) blend positive electrode as a function cycle.

The Peukert plots of the NMC–LFP (70:30) blended electrode cycled at different cut-off voltages are shown in Figure 10. These results indicate that (i) the 4.8-V cut-off provides the higher specific capacity of 180 mAh·g^{-1} at a C/10 rate, but shows the lowest performance at high rates; (ii) higher capacity was obtained when the voltage cut-off was higher than 4.1, 4.2, and 4.3 V; and (iii) for the 4.1-V cut-off, the capacity fade is less pronounced, *i.e.*, 0.12 mAh·g^{-1} per cycle. The impedance measurements, after formation cycles, show an increase in the charge transfer resistance at a voltage cut-off of 4.8 V, and a comparable result is observed with 4.1 and 4.3 V cut-off. As pointed out by Imachi *et al.* [12], who investigated a double-layered LiCoO$_2$·LiFePO$_4$ blended cathode, the advantage of blending LFP consists in the creation a resistive barrier that improves the tolerance against overcharging the electrochemical cell.

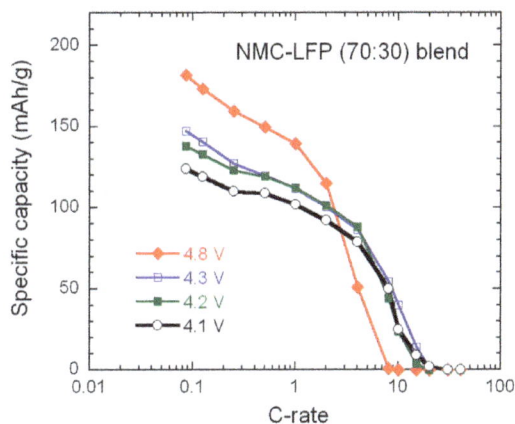

Figure 10. Specific capacity of the NMC–LFP (70:30) blended electrode as a function of C-rate in lithium cells cycled at different cut-off voltages.

3. Phase Evolution

In situ voltage measurement methods have been applied for the evaluation of the phase evolution of blended cathodes and gradual changes in cell behavior upon cycling. Using the so-called incremental capacity (IC) with respect to cell potential, dQ/d$V = f(V)$, and the differential voltage curve dV/d$Q = f(Q)$ could provide such information, since voltage plateaus in the V–Q profile is identifiable by dQ/dV peaks on IC curves [21,22].

Figure 11 presents the incremental capacity *vs.* potential for the Li coin cell with the NMC–LFP (70:30) blend cathode. The peaks in Figure 11 should only relate to the positive electrode material since the negative electrode is lithium metal. The typical IC curves of the blend clearly display two sets of redox peaks, each one related to the NMC or LFP component. The IC of the blend exhibits a sharp peak at 3.47 V corresponding to the voltage plateau of the charge profile Q vs. V due to the Fe^{2+}/Fe^{3+} oxidation in the LFP component followed by a broad feature at 3.85 V corresponding to the Ni^{2+}/Ni^{4+} oxidation in the NMC component. The reduction processes occurred at 3.65 and 3.32 V, respectively. The IC curves show that the decrease in the area of the LFP peak at 3.47 V on charge, along with the slight evolution of the peak at 3.32 V on discharge, while the redox NMC peaks are almost unchanged after 125 cycles. These results reveal the possible mechanisms involved in the blend material that could be associated with the strain–stress field interaction in the particles due to the mechanical milling preparation with the net effect of shortening the 3.47-V plateau. Thus, changes in the dQ/dV curves can help to identify ageing of electrode materials [23].

Another typical test to characterize the power capability of a cell at different state of charge (SOC) is the plot of the area specific resistance (ASR) vs. the cell potential defined by the Equation (2):

$$\text{ASR} = \frac{V_{oc} - V_{cell}}{I} A, \tag{2}$$

where V_{oc} is the open-circuit voltage, I the current, and A is the electrode area. As presented in Figure 12, the ASR of the NMC cell drops sharply for potential $V_{cell} < 3.5$ V, which is an inherent electrochemical feature of layered oxides, while the ASR of the NMC–LFP blend increases sharply at lower potential, ca. 3.25 V. These results show that the blend cathode can deliver high power at low SOC due to the 3.5-V plateau of the discharge curve of LFP.

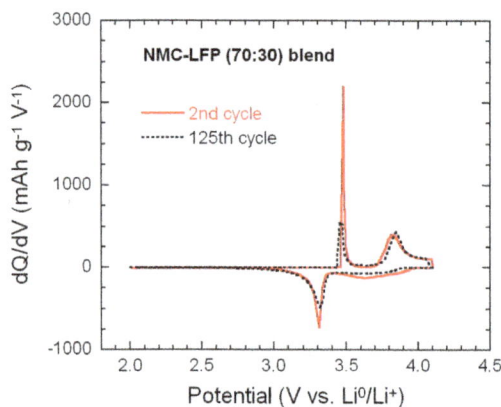

Figure 11. Incremental capacity dQ/dV of the NMC–LFP (70:30) blended electrode vs. potential as a function of the cycle number.

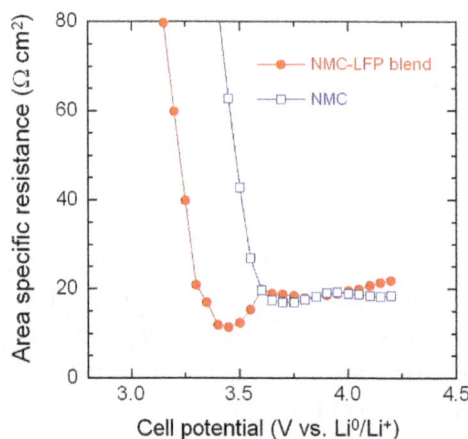

Figure 12. Plot of the area specific resistance of NMC and NMC–LFP blend as a function of the cell potential showing the increase in area specific resistance (ASR) at low state of charge (SOC).

4. Discussion

The results we report here enable the evaluation of the blends as new cathodes in a given geometry. This remains to be determined and chosen to optimize the electrochemical performance. In the case of the LMP–LFP blend, the core-shell geometry must be chosen because the purpose is to have a conductive carbon coating layer that could be deposited more efficiently on LFP than on LMP. In the case of NMC–LFP, only the case of mixed powders was investigated. To our knowledge, the only work where the effect of the geometrical distribution of the powders has been investigated is the comparison between the electrochemical properties of cells prepared with positive electrodes made of $Li[Li_{0.17}Mn_{0.58}Ni_{0.25}]O_2$–LFP blend in the completely intermixed, segregated, and layered geometries (see Figure 1) [13]. The discharge profiles are reported in Figure 13 for the case of an equal ratio of the two powders at different C-rates after few initial cycles devoted to the stabilization of the cell and solid-electrolyte interface. The characteristic sloping discharge curve starting above 4 V can be attributed to the $Li[Li_{0.17}Mn_{0.58}Ni_{0.25}]O_2$ part, and the relatively level portion of the curve occurring at less than 3.4 V can be attributed to the carbon-coated $LiFePO_4$. Clearly, the performance of the three electrodes is different. The low-rate capability of $Li[Li_{0.17}Mn_{0.58}Ni_{0.25}]O_2$ lowered overall electrode specific energy when the materials were mixed intimately or layered on top of each other in a single electrode. The best geometry for this blend is thus the segregated case, because it is the only configuration where the current can be driven through the LFP part at a high discharge rate. Otherwise, the active particles of both components must be held at the same potential. On the other hand, if $Li[Li_{0.17}Mn_{0.58}Ni_{0.25}]O_2$ is replaced by $LiCoO_2$, a material with a higher rate capability than $Li[Li_{0.17}Mn_{0.58}Ni_{0.25}]O_2$, the layered cell's performance will greatly improve. Therefore, more systematic studies of blends as a function of the different geometries should be investigated in the near future.

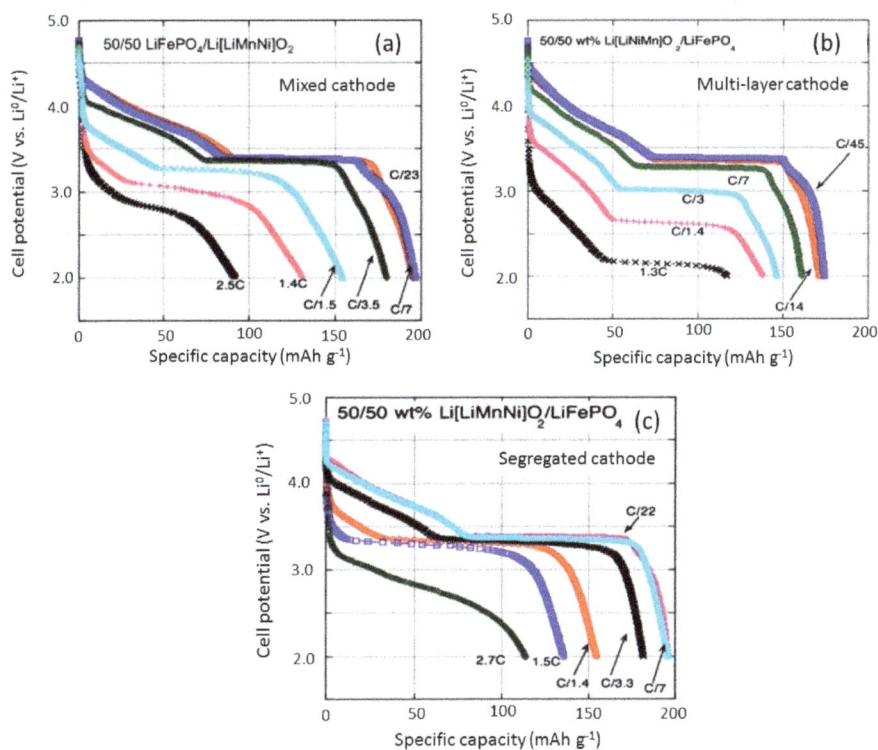

Figure 13. Discharge curves of $Li[Li_{0.17}Mn_{0.58}Ni_{0.25}]O_2$–LFP blended cathodes in different geometries [13]. (**a**) physically mixed cathode; (**b**) multilayer cathode and (**c**) segregated cathode.

Another field of investigation that should be developed is the optimization of the mixed powders. In practice, it is very difficult to mix powders. The phenomenon of de-mixing has been well-known for a long time [24]. In practice, the sizes of the particles of the components of the blend are different, in which case "size-segregation" may occur. This phenomenon is made possible by the gaps that open up around particles when they are shaken, allowing percolation to occur. Large particles tend to move toward less dense regions of smaller particles, so that large objects can rise to the top of vertically shaken layers, a well-known effect in granular materials. In addition, the particles can also spontaneously form patterns in the process [25]. The properties of mixed powders then depend on different parameters: the shape and size of the particles of the two blend elements, the packing structure of the powder mixtures in the scale of particle contacts (coordination number of each powder, defined as the number of particles surrounding and having contact with a reference particle), and the distribution of the contact chains, which, in the present case, also determines the electrical resistance of the blend. So far, the development of powder dynamics has been investigated in the framework of vibrations and flow. In the context of blends for electrodes, the dynamics is associated with the volume change of the particles induced by the lithium insertion of or de-insertion upon cycling, which is a new area of research to be developed in the near future.

5. Materials and Methods

5.1. Synthesis and Characterizations

The synthesis procedure of the LMP–LFP composite consists of three steps: (i) $LiMnPO_4$ olivine was synthesized by the hydrothermal method assisted with ascorbic acid ($1.3 \, g \cdot L^{-1}$) and 3% Denka black using $LiOH \cdot H_2O$, $MnSO_4 \cdot H_2O$, and H_3PO_4 (Li:Mn:P molar ratio 3:1:1) as raw materials; (ii) As-prepared LMP powders were added to the LFP precursor solution made by the dissolution of $LiOH \cdot H_2O$, $FeSO_4 \cdot 7H_2O$, and $(NH_4)_2HPO_4$ in distilled water. The reactive medium thus obtained is poured into a polytetrafluoroethylene (PTFE) vessel, put into a stainless steel chamber (Parr, Volume = 325 mL), and heated at 220 °C for 7 h. After cooling at room temperature, the powder is obtained by filtration, washed three times in distilled water, and dried at 90 °C during 12 h in N_2 atmosphere. At the end, 15.1 g of LMP–LFP blended material was obtained; (iii) The carbon-coating of the composite was obtained by mixing the product of step (ii) with a solution of lactose, the ratio lactose/composite being 1/10. This product was heated at 400 °C for one hour, then heated at 600 °C for 3 h [26].

$LiNi_{0.33}Mn_{0.33}Co_{0.33}O_2$-layered powders were prepared using the co-precipitation method described elsewhere [27] by a hydroxide route, for which transition-metal hydroxide and lithium carbonate were starting materials. Using a lithium excess $\eta = Li/M$ ratio of 1.05 the sample shows low cationic mixing $Ni^{2+}(3b) < 2\%$. The composite was obtained by physical mixing of NMC with carbon-coated LFP using a ball milling technique with different weights (70:30) and (20:80) ratios (NMC:LFP).

The characterization of the samples was done using a Hitachi HD-2700C dedicated STEM with a CEOS aberration corrector and a cold field-emission gun equipped with a newly designed silicon-drift detector (SDD) from Bruker (East Milton, ON, Canada), which offers over 10 times the solid angle of conventional energy-dispersive X-ray Si(Li) detectors. This dedicated STEM has demonstrated an image resolution of 78 pm in high angle annular dark field (HAADF).

5.2. Electrochemical Tests

Electrochemical tests were performed with composite electrode materials prepared by mixing 80% (w/w) of the active material, 10% (w/w) super C65 carbon (TIMCAL, Willebroek, Belgium), and 10% (w/w) polyvinylidene fluoride (Solef PVdF 6020 binder, Solvay, Brussels, Belgium) in N-methyl-2-pyrrolidone (NMP, Sigma-Aldrich, St. Louis, MO, USA) to produce a slurry. A 90-μm-thick film was obtained by coating this slurry on aluminum foil that gives an electrode loading ~8 $mg \cdot cm^{-2}$. This film was dried overnight at 80 °C, and then punched out with a diameter of 1.4 cm. The cell was assembled as coin-type cells in an argon-filled glove box using lithium foil (Alfa Aesar, Ward Hill,

MA, USA) as an anode, LP30 (1 mol·L^{-1} LiPF$_6$ in (1:1) ethylene carbonate (EC)—dimethyl carbonate (DMC) as an electrolyte, and glass microfiber filters (Whatmann$^®$-GF/D 70 mm Ø, Sigma-Aldrich) as a separator. VMP3 multi-channel potentiostat (Bio-Logic, Claix, France) was used to electrochemically characterize the electrodes at 25 °C in the voltage range of 1.5–4.8 V *vs.* Li^0/Li$^+$ in galvanostatic tests.

6. Conclusions

In this paper, we studied two blended cathode materials based on a mixture of nano-structured particles of LiFePO$_4$ olivine with another micro-structured insertion compounds such as LiMnPO$_4$ olivine and LiNi$_{1/3}$Mn$_{1/3}$Co$_{1/3}$O$_2$-layered compound. To overcome the difficulty of carbon coating on LiMnPO$_4$ olivine particles, the preparation of the LMP–LFP (66:33) composite was realized. Using such a blended electrode, the electrochemical properties are improved with respect to the carbon-coated LiMn$_{2/3}$Fe$_{1/3}$PO$_4$ solid solution with a comparable Fe/Mn ratio. The blended cathode formed by the NMC–LFP mixture exhibits excellent rate capability by using nano-sized LFP particles and delivers a higher capacity, *i.e.*, 180 mAh·g^{-1}, due to the NMC component. Consequently, Li-ion batteries including NMC–LFP physical blending provide a good solution that achieves both energy and power capability and exhibits improved electrochemical features with thermal stability similar to LFP.

In situ voltage measurement methods were applied for the evaluation of the phase evolution of blended cathodes and gradual changes in cell behavior upon cycling using the so-called incremental capacity (IC) with respect to cell potential, $dQ/dV = f(V)$. Another typical test to characterize the power capability of a cell at different states of charge (SOCs) is the plot of the area specific resistance (ASR) *vs.* the cell potential. Both methods have demonstrated that blending cathode materials with LiFePO$_4$ nano-particles improves their electrochemical performance. We hope that the electrochemistry will lean on the progress that has been made in mathematics to model non-linear dynamics of powder mixing and will benefit from the powder technology sustained by their application in different fields of the industry in the near future, aiming to optimize the blends to be used as new electrodes of Li-ion batteries.

Acknowledgments: This study was supported by the Laboratoire Physicochimie des Electrolytes et Nanosystems Interfaciaux, Université Pierre et Marie Curie, Paris.

Author Contributions: Christian M. Julien and Alain Mauger conceived and wrote the paper, Karim Zaghib designed the experiments; Julie Trottier, Pierre Hovington and Henri Groult assisted characterization of the materials.

References

1. Chikkannanavar, S.B.; Bernardi, D.M.; Liu, L. A review of blended cathode materials for use in Li-ion batteries. *J. Power Sources* **2014**, *248*, 91–100. [CrossRef]
2. Zaghib, K.; Trudeau, M.; Guerfi, A.; Trottier, J.; Mauger, A.; Julien, C.M. New advanced cathode material: LiMnPO$_4$ encapsulated with LiFePO$_4$. *J. Power Sources* **2012**, *204*, 177–181. [CrossRef]
3. Margalit, N. Non-Aqueous Primary Battery Having a Blended Cathode Active Material. U.S. Patent 3,981,748, 21 September 1976.
4. Pynenburg, R.; Barker, J. Cathode-Active Material Blends of Li$_x$Mn$_2$O$_4$. U.S. Patent 5,429,890, 4 July 1995.
5. Numata, T.; Amemiya, C.; Kumeuchi, T.; Shirakata, M.; Yonezawa, M. Advantages of blending LiNi$_{0.8}$Co$_{0.2}$O$_2$ into Li$_{1+x}$Mn$_{2-x}$O$_4$ cathodes. *J. Power Sources* **2001**, *97–98*, 358–360. [CrossRef]
6. Kim, H.S.; Kim, S.I.; Kim, W.S. A study on electrochemical characteristics of LiCoO$_2$/LiNi$_{1/3}$Mn$_{1/3}$Co$_{1/3}$O$_2$ mixed cathode for Li secondary battery. *Electrochim. Acta* **2006**, *52*, 1457–1461. [CrossRef]
7. Albertus, P.; Christensen, J.; Newman, J. Experiments on and modeling of positive electrode with multiple active materials for lithium-ion batteries. *J. Electrochem. Soc.* **2009**, *156*, A606–A618. [CrossRef]
8. GM's new battery chemistry? It's already in Chevy Volt. Available online: http://www.popsci.com/cars/article/2011-01/gm%e2%80%99s-new-battery-chemistry-it%e2%80%99s-already-chevy-volt (accessed on 7 January 2011).

9. Tran, H.Y.; Täubert, C.; Fleischhammer, M.; Axmann, P.; Küppers, L.; Wohlfahrt-Mehrens, M. $LiMn_2O_4$ spinel/$LiNi_{0.8}Co_{0.15}Al_{0.05}O_{0.2}$ blends as cathode materials for lithium-ion batteries. *J. Electrochem. Soc.* **2011**, *158*, A556–A561. [CrossRef]

10. Gao, J.; Manthiram, A. Eliminating the irreversible capacity loss of high capacity layered $Li[Li_{0.2}Ni_{0.13}Mn_{0.54}Co_{0.13}]O_2$ cathode by blending with other lithium insertion hosts. *J. Power Sources* **2009**, *191*, 644–647. [CrossRef]

11. Kitao, H.; Fujihara, T.; Takeda, K.; Nakanishi, N.; Nohma, T. High-temperature storage performance of Li-ion batteries using a mixture of Li–Mn spinel and Li–Ni–Co–Mn oxide as a positive electrode material. *Electrochem. Solid State Lett.* **2005**, *8*, A87–A90. [CrossRef]

12. Imachi, N.; Takano, Y.; Fujimoto, H.; Kida, Y.; Fujitami, S. Layered cathode for improving safety of Li-ion batteries. *J. Electrochem. Soc.* **2007**, *154*, A412–A416. [CrossRef]

13. Whitacre, J.F.; Zaghib, K.; West, W.C.; Ratnakumar, B.V. Dual active material composite cathode structures for Li-ion batteries. *J. Power Sources* **2008**, *177*, 528–536. [CrossRef]

14. Gallagher, K.G.; Kang, S.H.; Park, S.U.; Han, S.Y. $xLi_2MnO_3 \cdot (1-x)LiMO_2$ blended with $LiFePO_4$ to achieve high energy density and pulse power capability. *J. Power Sources* **2011**, *196*, 9702–9707. [CrossRef]

15. Zheng, J.C.; Li, X.; Wang, Z.X.; Niu, S.S.; Liu, D.; Wu, L.; Li, L.J.; Li, J.H.; Guo, H.J. Novel synthesis of $LiFePO_4$–$Li_3V_2(PO_4)_3$ composite cathode material by aqueous precipitation and lithiation. *J. Power Sources* **2010**, *195*, 2935–2938. [CrossRef]

16. Qiu, C.; Liu, L.; Du, F.; Yang, X.; Wang, C.; Chen, G.; Wei, Y. Electrochemical performance of $LiMn_2O_4$/$LiFePO_4$ blend cathodes for lithium ion batteries. *Chem. Res. Chin. Univ.* **2015**, *31*, 270–275. [CrossRef]

17. Wohlfahrt-Mehrens, M.; Klein, A.; Axmann, P. Blend Performance of $LiMn_{0.7}Fe_{0.3}PO_4$–$LiMn_{1.9}Al_{0.1}O_4$ electrodes: Properties beyond physical mixtures? In Proceedings of the 18th Int. Meeting on Lithium Batteries, Chicago, IL, USA, 19–24 June 2016.

18. Trudeau, M.L.; Laul, D.; Veillette, R.; Serventi, A.M.; Zaghib, K.; Mauger, A.; Julien, C.M. *In-situ* HRTEM synthesis observation of nanostructured $LiFePO_4$. *J. Power Sources* **2011**, *196*, 7383–7394. [CrossRef]

19. Julien, C. Local cationic environment in lithium nickel–cobalt oxides used as cathode materials for lithium batteries. *Solid State Ion.* **2000**, *136–137*, 887–896. [CrossRef]

20. Julien, C.M.; Zaghib, K.; Mauger, A.; Massot, M.; Ait-Salah, A.; Selmane, M.; Gendron, F. Characterization of the carbon-coating onto $LiFePO_4$ particles used in lithium batteries. *J. Appl. Phys.* **2006**, *100*, 63511. [CrossRef]

21. Dubarry, M.; Svoboda, V.; Hwu, R.; Liaw, B.Y. Incremental capacity analysis and close-to-equilibrium OCV measurements to quantify capacity fade in commercial rechargeable lithium batteries. *Electrochem. Solid State Lett.* **2006**, *9*, A454–A457. [CrossRef]

22. Weng, C.; Cui, Y.; Sun, J.; Peng, H. On-board state of health monitoring of lithium-ion batteries using incremental capacity analysis with support vector regression. *J. Power Sources* **2013**, *235*, 36–44. [CrossRef]

23. Dubarry, M.; Liaw, B.Y.; Chen, M.S.; Chyan, S.S.; Han, K.C.; Sie, W.T.; Wu, S.H. Identifying battery aging mechanisms in large format Li ion cells. *J. Power Sources* **2011**, *196*, 3420–3425. [CrossRef]

24. Shinbrot, T.; Muzio, F.J. Noise to order. *Nature* **2001**, *410*, 251–258. [CrossRef] [PubMed]

25. Mullin, T. Coarsening of self-organized clusters in binary mixtures of particles. *Phys. Rev. Lett.* **2000**, *84*, 4741–4745. [CrossRef] [PubMed]

26. Vediappan, K.; Guerfi, A.; Gariépy, V.; Demopoulos, G.P.; Hovington, P.; Trottier, J.; Mauger, A.; Zaghib, K.; Julien, C.M. Stirring effect in hydrothermal synthesis of C–$LiFePO_4$. *J. Power Sources* **2014**, *266*, 99–106. [CrossRef]

27. Zhang, X.; Jiang, W.J.; Mauger, A.; Li, Q.; Gendron, F.; Julien, C.M. Minimization of the cation mixing in $Li_{1+x}(NMC)_{1-x}O_2$ as cathode material. *J. Power Sources* **2010**, *195*, 1292–1301. [CrossRef]

2

Monoanionic Tin Oligomers Featuring Sn–Sn or Sn–Pb Bonds: Synthesis and Characterization of a Tris(Triheteroarylstannyl)Stannate and -Plumbate

Kornelia Zeckert

Institute of Inorganic Chemistry, University of Leipzig, Johannisallee 29, D-04103 Leipzig, Germany; zeckert@uni-leipzig.de

Academic Editor: Axel Klein

Abstract: The reaction of the lithium tris(2-pyridyl)stannate $[LiSn(2\text{-}py^{6OtBu})_3]$ ($py^{6OtBu} = C_5H_3N\text{-}6\text{-}OtBu$), **1**, with the element(II) amides $E\{N(SiMe_3)_2\}_2$ (E = Sn, Pb) afforded complexes $[LiE\{Sn(2\text{-}py^{6OtBu})_3\}_3]$ for E = Sn (**2**) and E = Pb (**3**), which reveal three Sn–E bonds each. Compounds **2** and **3** have been characterized by solution NMR spectroscopy and X-ray crystallographic studies. Large $^1J(^{119}Sn\text{-}^{119/117}Sn)$ as well as $^1J(^{207}Pb\text{-}^{119/117}Sn)$ coupling constants confirm their structural integrity in solution. However, contrary to **2**, complex **3** slowly disintegrates in solution to give elemental lead and the hexaheteroarylditin $[Sn(2\text{-}py^{6OtBu})_3]_2$ (**4**).

Keywords: tin; lead; catenation; pyridyl ligands

1. Introduction

The synthesis and characterization of catenated heavier group 14 element compounds have attracted attention in recent years [1–5]. However, contrary to silicon and germanium, there are limitations for tin and lead associated with the significant decrease in element–element bond energy. Hence, homonuclear as well as heteronuclear molecules with E–E bonds become less stable when E represents tin and or lead. Moreover, within this class of compounds, discrete branched oligomers with more than one E–E bond are rare compared with their linear analogs [6–10]. It has been shown that reactions of lead(II) halides with Grignard reagents involved disproportionation processes leading to plumbates of type $[MgBr(THF)_5][Pb(PbPh_3)_3]$ [11] and $[Mg(THF)_3(\mu\text{-}Cl)_3Mg(THF)_3][Pb(PbBp_3)_3]$ (Bp = 4-Biphenyl) [12]. The formation of closely related anions $[Sn(SnR_3)_3]^-$ (R = Me or Ph) was found as part of decomposition of corresponding stannyl anions in the presence of distannanes [13–16]. Further rare investigations included the reduction reaction of $SnPh_4$ with elemental lithium in liquid ammonia to give $[Li(NH_3)_4][Sn(SnPh_3)_3]$ [17] or reactions of the organotin chlorides $SnPh_2Cl_2$ as well as $SnMe_3Cl$ with lanthanoid metals, yielding complexes $[Yb(DME)_3Cl]_2[Sn(SnPh_3)_3]_2$ [18] and $[Ln(THF)_4\{Sn(SnMe_3)_3\}_2]$ (Ln = Sm, Yb) [19]. Complete structural characterizations, however, are missing due to the very low solubility and sensitivity of all compounds reported.

Recently, conventional salt metathesis reactions have been employed for synthesis of neutral distannylstannylene species $Sn(SnAr_3)_2$ (Ar = $C_6H_3\text{-}2,6\text{-}(O^iPr)_2$) [20] and $Sn(SnPh_2Ar^*)_2$ ($Ar^* = C_6H_3\text{-}2,6\text{-}(Trip)_2$, Trip = $C_6H_2\text{-}2,4,6\text{-}^iPr_3$) [21], bearing a divalent tin atom bonded to two triarylstannyl groups. In both cases, the catenation was implemented by a stannyl lithium derivative $LiSnAr_3$ or $LiSnPh_2Ar^*$, respectively. Our studies on related triheteroaryltin(II) compounds revealed a diverse reactivity. Thereby, it was found that compounds $[Li(THF)Sn(2\text{-}py^R)_3]$ ($py^R = C_5H_3N\text{-}5\text{-}Me$ or $C_5H_3N\text{-}3\text{-}Me$), in which the lithium cation is well embedded within the tris(pyridyl)stannate core, can either act as a two-electron donor leading to a tin–element bond formation [22,23] or may be used for salt metathesis reactions with the lone pair of electrons at the tin(II) atom remaining available [24].

Continuing these studies, it is now shown that the reaction between the lithium tris(2-pyridyl)stannate [LiSn(2-py^{6OtBu})$_3$] (py^{6OtBu} = C$_5$H$_3$N-6-OtBu), **1**, with group 14 element(II) amides E{N(SiMe$_3$)$_2$}$_2$ (E = Sn or Pb) lead to the straight formation of the lithium salts [LiSn{Sn(2-py^{6OtBu})$_3$}$_3$] (**2**) and [LiPb{Sn(2-py^{6OtBu})$_3$}$_3$] (**3**) each with the lithium cation intramolecularly coordinated by the novel tris(stannyl)metallate enabled through the unique donor functionality of the complex anion [Sn{Sn(2-py^{6OtBu})$_3$}$_3$]$^-$ as well as [Pb{Sn(2-py^{6OtBu})$_3$}$_3$]$^-$ itself.

2. Results

Compounds **2** and **3** have been prepared in reactions of the lithium stannate **1** upon addition of tin(II) or lead(II) amide E{N(SiMe$_3$)$_2$}$_2$ (E = Sn, Pb) in a 3:1 stoichiometry (Scheme 1). The initial yellow THF solution of **1** gradually changed to a reddish-brownish solution of **2** and a green solution of **3**. The complexes were obtained as the only products in good isolated yields as colorless (**2**) as well as yellow (**3**) crystalline material upon work up and re-crystallization from toluene or benzene. It is likely that neutral E{Sn(2-py^{6OtBu})$_3$}$_2$ was initially generated. However, this molecule, which could neither be isolated nor detected, may have then reacted further with another equiv of **1** to give the lithium complexes **2** or **3**, respectively, by virtue of a donor–acceptor interaction between two low valent group 14 element species. This may be explained with the less steric requirement of the pyridyl ligand as well as the pronounced two-electron donor ability of **1** comparable to other tris(2-pyridyl)stannates [22–24].

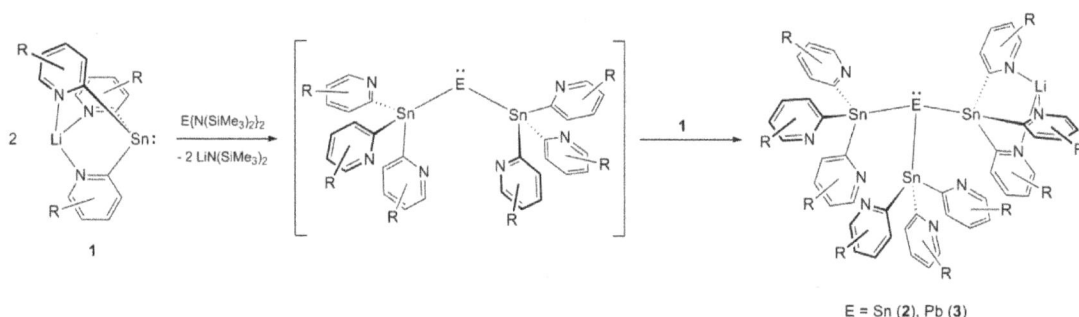

Scheme 1. Illustration of the possible reaction pathway to give compounds **2** and **3**; with py^R = py^{6OtBu} = C$_5$H$_3$N-6-OtBu.

Indeed, the molecular structures of complexes **2** and **3** (Figure 1, Table 1) imply the donation of a lithium stannate fragment to the central element(II) atom of a E{Sn(2-py^{6OtBu})$_3$}$_2$ molecule due to the intramolecular *N*-coordination of the lithium cation by only one tris(2-pyridyl)stannyl unit in **2** and **3** as found for previously reported tris(2-pyridyl)stannate complexes [22–25]. Thus, compounds **2** and **3** can be described as zwitter-ionic complexes with a Li···Sn separation of 3.339(14) Å in **2** as well as 3.321(10) Å in **3**.

In complex **2**, the central Sn atom adopts a trigonal pyramidal geometry with Sn–Sn(1)–Sn angles of 92.52(2), 93.22(2), and 96.13(2)°. The Sn–Sn bond lengths range from 2.8127(7) to 2.8217(6) Å and lie well within the previously known range (2.811–2.836 Å) for complexes of the [Sn(SnPh$_3$)$_3$]$^-$ anion [16–19] but are considerably shorter than those found for the distannylstannylenes Sn(SnAr$_3$)$_2$ (Ar = C$_6$H$_3$-2,6-(O^{i}Pr)$_2$) (2.86 Å) [20] as well as Sn(SnPh$_2$Ar*)$_2$ (Ar* = C$_6$H$_3$-2,6-(Trip)$_2$, Trip = C$_6$H$_2$-2,4,6-^{i}Pr$_3$) (2.96 Å) [21]). The geometry at Sn(2), whose interligand angles vary from 93.8(2)° to 132.5(2)°, is strikingly distorted from idealized tetrahedral geometry. This can be attributed to the binding of the lithium cation by the three pyridyl groups of this stannyl unit contrary to atoms Sn(3) and Sn(4).

Compound **3** crystallized as two structurally similar, but crystallographically independent molecules. Therefore, only one of the crystallographically independent molecules is shown in Figure 1. The molecular structure of **3** has many similarities to that of **2**. In **3**, the lead atom may also be described as trigonal pyramidal with the lone pair of electrons located at the apex. Accordingly, the Sn–Pb–Sn angles are relatively small with values ranging from 90.85(1)° to 94.49(1)°. Similar structural features can be found in the previously reported tris(plumbyl)plumbate anions, where angles around the central lead atom average to 93.0° [11] and 94.1° [12]. The Pb–Sn bond lengths amount to 2.8743(3) and 2.8919(4) Å, and may be compared with 2.7844(4) Å in Pb(Ar)SnAr$_3$ (Ar = C$_6$H$_3$-2,6-(O^{i}Pr)$_2$) and 2.9283(4) Å in Pb(SnAr$_3$)$_2$ [26].

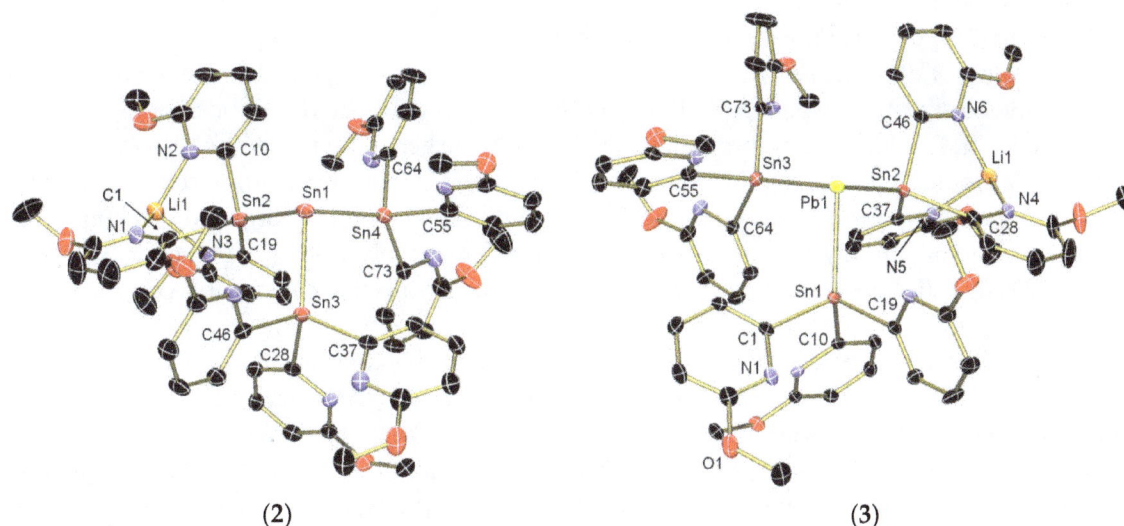

(2)   (3)

Figure 1. Molecular structures of [LiSn{Sn(2-py^{6OtBu})$_3$}$_3$] (**2**) and [LiPb{Sn(2-py^{6OtBu})$_3$}$_3$] (**3**) with 30% probability ellipsoids. Hydrogen atoms and methyl groups are omitted for clarity.

Table 1. Selected bond lengths (Å) and bond angles (°) of compounds **2** and **3**.

	2		3
Sn(1)–Sn(2)	2.8127(7)	Pb(1)–Sn(1)	2.8743(4)
Sn(1)–Sn(3)	2.8181(6)	Pb(1)–Sn(2)	2.8919(4)
Sn(1)–Sn(4)	2.8217(6)	Pb(1)–Sn(3)	2.8818(4)
Sn(2)···Li(1)	3.339(14)	Sn(2)···Li(1)	3.321(10)
Sn(2)–C$_{ipso}$	2.168(7), 2.179(7), 2.184(8)	Sn(1)–C$_{ipso}$	2.154(5), 2.167(6), 2.182(5)
Sn(3)–C$_{ipso}$	2.161(6), 2.165(7), 2.183(6)	Sn(2)–C$_{ipso}$	2.172(6), 2.183(6), 2.191(6)
Sn(4)–C$_{ipso}$	2.163(6), 2.167(6), 2.172(7)	Sn(3)–C$_{ipso}$	2.167(6), 2.169(5), 2.177(5)
Li(1)–N	2.024(13)–2.038(15)	Li(1)–N	1.996(11)–2.030(11)
Sn(2)–Sn(1)–Sn(3)	92.52(2)	Sn(1)–Pb(1)–Sn(2)	90.90(1)
Sn(2)–Sn(1)–Sn(4)	93.22(2)	Sn(1)–Pb(1)–Sn(3)	94.49(1)
Sn(3)–Sn(1)–Sn(4)	96.13(2)	Sn(2)–Pb(1)–Sn(3)	90.85(1)
C$_{ipso}$–Sn(2)–C$_{ipso}$	93.8(2), 98.9(3), 99.8(3)	C$_{ipso}$–Sn(1)–C$_{ipso}$	101.2(2), 103.2(2), 104.8(2)
C$_{ipso}$–Sn(3)–C$_{ipso}$	100.8(2), 102.8(2), 104.7(2)	C$_{ipso}$–Sn(2)–C$_{ipso}$	92.4(2), 98.6(2), 100.0(2)
C$_{ipso}$–Sn(4)–C$_{ipso}$	100.4(2), 101.2(3), 103.6(2)	C$_{ipso}$–Sn(3)–C$_{ipso}$	101.8(2), 103.3(2), 104.3(2)
C$_{ipso}$–Sn(2)–Sn(1)	112.1(2), 114.0(2), 132.5(2)	C$_{ipso}$–Sn(1)–Pb(1)	109.2(1), 112.5(1), 123.8(1)
C$_{ipso}$–Sn(3)–Sn(1)	109.2(2),112.6(2), 124.5(1)	C$_{ipso}$–Sn(2)–Pb(1)	111.8(2), 113.7(2), 134.1(2)
C$_{ipso}$–Sn(4)–Sn(1)	106.0(2), 113.0(2), 129.7(2)	C$_{ipso}$–Sn(3)–Pb(1)	104.6(1), 107.9(1), 132.0(1)

Solution NMR spectra of both complexes in d_8-THF reveal a symmetric arrangement within the [Sn{Sn(2-py^{6OtBu})$_3$}$_3$]$^-$ or [Pb{Sn(2-py^{6OtBu})$_3$}$_3$]$^-$ skeleton, indicated by three chemical equivalent stannyl moieties with ^{119}Sn NMR room temperature resonances at −151 ppm (**2**) and −180 ppm (**3**). The ^{119}Sn–$^{119/117}$Sn coupling constant in compound **2** is 7050 Hz, and may be compared with scalar coupling constants found for stannylstannylene species (6240–9400 Hz) [20,27–31]. The chemical shift of the bridging tin atom in **2** appears at −1064 ppm. This value is consistent with reported data for LiSn(SnMe$_3$)$_3$ (δ = −1062 ppm) [14,32], but reveals a significant upfield shift by *ca.* 870 ppm compared to the three-coordinate tin(II) atom in **1** (δ = −191 ppm). The ^{207}Pb NMR spectrum of **3** is characterized by one signal at −1643 ppm and a ^1J(^{207}Pb–$^{119/117}$Sn) coupling of 14900 Hz, which is smaller than observed for the hitherto isolated stannylplumbylenes (19971–22560 Hz) [26], but larger than determined for stannylplumbanes, e.g., $Ph_3SnPbPh_3$ (3469 Hz) [33,34]. The chemical shift value of the three-coordinate lead atom in **3**, however, is strikingly deviating from low field ^{207}Pb resonances usually observed for triaryl plumbates, e.g., at +1063 ppm for $LiPbPh_3$ [35] or +1020 ppm for the related $[LiPb(2-py^{6OtBu})_3]$ [36]. Thus, in **3**, the central lead atom is significantly more shielded according to a smaller contribution to the paramagnetic contribution term. In addition, fast fluctuation of the lithium counterion in **2** or **3** can be attributed to a upfield shifting of the ^7Li NMR resonances to +0.68 ppm (**2**) and +1.08 ppm (**3**) in comparison to the parent lithium stannate **1** (δ = +2.12 ppm).

Compound **3** decomposes when stored over time in solution, yielding elemental lead, the lithiumstannate **1** and the hexaorganoditin $[Sn(2-py^{6OtBu})_3]_2$ (**4**). The decomposition, which was monitored by NMR spectroscopy, proceeds smoothly with the generation of less soluble crystalline **4**. Single crystals, suitable for X-ray diffraction analysis (see Supplementary Material) were isolated from the NMR tube. Compound **4** ranks into the group of hitherto known hexaorganoditin compounds as a result of an apparently inexorable cleavage of the comparably weak Sn–Pb bond. Similar to $[Sn(2-py^{6Me})_3]_2$ [37], **4** provides further functionalities with prospective coordination characteristics due to the N-heteroaromatic ligand system.

Finally, the results obtained suggest the formation of the tris(stannyl)metallate complexes $[LiSn\{E(2-py^{6OtBu})_3\}_3]$ **2** and **3** to be favored over neutral bis(stannyl) derivatives. This may be rationalized with the following: (i) the less steric requirement of the organic substituent used, allowing further attack on E; (ii) the rapid reaction of electron rich **1**; and (iii) a stabilization of the anionic oligomers by the intramolecular coordination of the counterions. Thus, a straight route in synthesis of heavier group 14 element catenates containing more than one homonuclear or heteronuclear element–element bonds has been presented. Solution studies on **3**, however, underline the weakness of the Pb–Sn bonds in this molecule.

3. Experimental

All manipulations were carried out using standard Schlenk techniques under an atmosphere of dry nitrogen. THF and benzene have been distilled over sodium and stored over potassium mirror. Deuterated THF was stored over activated 4 Å molecular sieves. $Sn\{N(SiMe_3)_2\}_2$ and $Pb\{N(SiMe_3)_2\}_2$ were prepared according to published procedures [38].

NMR spectra have been recorded on a Bruker Avance DRX 400 spectrometer (Bruker BioSpin GmbH, Rheinstetten, Germany). The chemical shifts (δ) are quoted in ppm and are relative to tetramethylsilane (TMS). CHN analyses were carried out on a Vario EL III CHNS instrument (Elementar Analysensysteme GmbH, Hanau, Germany).

3.1. Syntheses

$[LiSn(2-py^{6OtBu})_3]$ (1): $SnCl_2$ (0.76 g, 4 mmol) was added to a solution of $Li(2-py^{6OtBu})$ (1.88 g, 12 mmol) (generated from LiPh and $2-Br-C_5H_3N-6-OtBu$) in 30 mL of diethyl ether at 0 °C. The reaction mixture was stirred overnight at room temperature. After filtration, the residue was dissolved in toluene. Further filtration from LiCl, the removal of toluene, and the re-crystallization from THF gave yellow **1** in a total of 1.75 g (76%). Characterization: Elemental analysis (%) calcd. for $C_{27}H_{36}N_3O_3SnLi$, C_4H_8O: C, 57.43; H, 6.84; N, 6.48. Found: C, 57.31; H, 6.87; N, 6.32. ^1H NMR (400.1 MHz, 298 K, d_8-THF): δ = 1.43 (s, 9H, CMe_3), 6.51 (d, 8 Hz, py), 7.29 (t, 8 Hz, py), 7.40 ppm (d, 8 Hz, py); $^{13}C\{^1H\}$ NMR (100.6 MHz, 298 K, d_8-THF): δ = 28.5 (CMe_3), 78.3 (CMe_3), 109.9 (py), 130.0 ($^2J(^{13}C-^{119/117}Sn)$ = 109 Hz, py), 135.0 ($^3J(^{13}C-^{119/117}Sn)$ = 19 Hz, py), 161.7 (py), 194.5 ppm ($^1J(^{13}C-^{119/117}Sn)$ = 345 Hz, C_{ipso}); ^7Li NMR (155.5 MHz, 298 K, d_8-THF) δ = 2.12 ppm; $^{119}Sn\{^1H\}$ NMR (149.2 MHz, 298 K, d_8-THF): δ = −191 ppm. ^1H NMR (400.1 MHz, 298 K, C_6D_6): δ = 1.30 (s, 9H, CMe_3), 6.32 (d, 8 Hz, py), 7.01 (t, 8 Hz, py), 8.15 (d, 8 Hz, py); $^{13}C\{^1H\}$ NMR (100.6 MHz, 298 K, C_6D_6): δ = 27.9 (CMe_3), 77.8 (CMe_3), 111.2 (py), 131.1 ($^2J(^{13}C-^{119/117}Sn)$ = 134.8 Hz, py), 134.8 ($^3J(^{13}C-^{119/117}Sn)$ = 25 Hz, py), 160.4 (py), 191.9 ppm ($^1J(^{13}C-^{119/117}Sn)$ = 360 Hz, C_{ipso}); ^7Li NMR (155.5 MHz, 298 K, C_6D_6): δ = 3.78 ppm; $^{119}Sn\{^1H\}$ NMR (149.2 MHz, 298 K, C_6D_6): δ = −201 ppm.

[LiSn{Sn(2-py^{6OtBu})$_3$}$_3$] (2): To a solution of **1** (1.10 g, 19 mmol) in 10 mL of THF, Sn{N(SiMe$_3$)$_2$}$_2$ (0.28 g, 6.3 mmol) was added at room temperature. The greenish solution was stirred for 1 h, and the solvent was then completely removed *in vacuo*. Crystallization from the concentrated toluene solution afforded colorless crystals of **2**. Yield: 0.73 g (63%). Characterization: Elemental analysis (%) calcd. for C$_{81}$H$_{108}$N$_9$O$_9$Sn$_4$Li: C, 53.06; H, 5.94; N, 6.88. Found: C, 52.67; H, 6.02; N, 6.85. ^{1}H NMR (400.1 MHz, 298 K, d_8-THF): δ = 1.20 (s, 9H, CMe$_3$), 6.15 (d, 8 Hz, py), 6.78–6.84 ppm (m, py); ^{13}C{^{1}H} NMR (100.6 MHz, 298 K, d_8-THF): δ = 23.7 (C*Me*$_3$), 76.0 (*C*Me$_3$), 107.3 (py), 125.7 (^2J(^{13}C–$^{119/117}$Sn) = 96 Hz, py), 133.5 (^3J(^{13}C–$^{119/117}$Sn) = 36 Hz, py), 160.9 (^3J(^{13}C–$^{119/117}$Sn) = 46 Hz, py), 174.8 ppm (^1J(^{13}C–^{119}Sn) = 285 Hz, ^1J(^{13}C–^{117}Sn) = 271 Hz, ^2J(^{13}C–$^{119/117}$Sn) = 20 Hz, C$_{ipso}$); ^{7}Li NMR (155.5 MHz, 298 K, d_8-THF): δ = 0.68 ppm; ^{119}Sn{^{1}H} NMR (149.2 MHz, 298 K, d_8-THF): δ = −151 (^1J(^{119}Sn–$^{119/117}$Sn) = 7050 Hz, ^2J(^{119}Sn–$^{119/117}$Sn) = 890 Hz, Sn(py)$_3$), −1064 ppm (*Sn*Sn$_3$).

[LiPb{Sn(2-py^{6OtBu})$_3$}$_3$] (3): Crystalline **3** was prepared from **1** and Pb{N(SiMe$_3$)$_2$}$_2$ in a 3:1 stoichiometry as described above for the synthesis of **2**. Concentration of the benzene solution and storage at 8 °C overnight gave yellow **3** in 84% yield. Characterization: Elemental analysis (%) calcd. for C$_{81}$H$_{108}$N$_9$O$_9$Sn$_3$PbLi: C, 50.62; H, 5.66; N, 6.56. Found: C, 49.77; H, 5.62; N, 6.34. ^{1}H NMR (400.1 MHz, 298 K, d_8-THF): δ = 1.27 (s, 9H, CMe$_3$), 6.20 (d, 8 Hz, py), 6.86–6.91 ppm (m, py); ^{13}C{^{1}H} NMR (100.6 MHz, 298 K, d_8-THF): δ = 29.3 (C*Me*$_3$), 78.7 (*C*Me$_3$), 110.1 (py), 128.4 (^2J(^{13}C–$^{119/117}$Sn) = 95 Hz, py), 136.1 (py), 163.8 (^3J(^{13}C–$^{119/117}$Sn) = 41 Hz, py), 180.0 ppm (^1J(^{13}C–$^{119/117}$Sn) = 214 Hz, ^2J(^{13}C–^{207}Pb) = 78 Hz, C$_{ipso}$); ^{7}Li NMR (155.5 MHz, 298 K, d_8-THF) δ = 1.08 ppm. ^{119}Sn{^{1}H} NMR (149.2 MHz, 298 K, d_8-THF): δ = −180 ppm (^1J(^{119}Sn–^{207}Pb) = 14700 Hz); ^{207}Pb{^{1}H} NMR (83.6 MHz, 298K, d_8-THF): δ = −1643 ppm (^1J(^{207}Pb–$^{119/117}$Sn) = 14,900 Hz). ^{1}H NMR (400.1 MHz, 298 K, C$_6$D$_6$): δ = 1.41 (s, 9H, CMe$_3$), 6.42 (d, 8Hz, py), 7.00 (t, 8 Hz, py), 7.35 ppm (d, 8Hz, py); ^{13}C{^{1}H} NMR (100.6 MHz, 298 K, C$_6$D$_6$): δ = 28.7 (C*Me*$_3$), 78.6 (*C*Me$_3$), 111.0 (py), 136.1 (py), 162.9 (py), 178.4 ppm (C$_{ipso}$); ^{7}Li NMR (155.5 MHz, 298 K, C$_6$D$_6$): δ = 1.20 ppm. ^{119}Sn{^{1}H} NMR (149.2 MHz, 298K, C$_6$D$_6$): δ = −161 ppm; ^{207}Pb{^{1}H} NMR (83.6 MHz, 298K, C$_6$D$_6$): δ = −1738 ppm (^1J(^{207}Pb–$^{119/117}$Sn) = 14,620 Hz).

[Sn(2-py^{6OtBu})$_3$]$_2$ (4): A NMR solution of **3** in C$_6$D$_6$ was set aside at room temperature. The color of the solution gradually changed from green to pale yellow. The formation of elemental lead was observed. After several days, the colorless crystals of **4** were grown from solution. In the ^{119}Sn{^{1}H} NMR spectrum finally only one signal at −201 ppm was detected, according to free lithium stannate **1**. Isolated crystals of **4** were dissolved in d_8-THF for characterization by NMR spectroscopy. Characterization: Elemental analysis (%) calcd. for C$_{54}$H$_{72}$N$_6$O$_6$Sn$_2$: C, 56.96; H, 6.37; N, 7.38. Found: C, 56.92; H, 6.32; N, 7.28. ^{1}H NMR (400.1 MHz, 298 K, d_8-THF): δ = 1.20 (s, 9H, CMe$_3$), 6.37 (d, 8Hz, py), 7.03 (t, 8 Hz, py), 7.27 ppm (d, 8Hz, py); ^{119}Sn{^{1}H} NMR (149.2 MHz, 298K, d_8-THF): δ = −223 ppm. No ^{119}Sn–^{117}Sn coupling could be detected.

3.2. X-ray Crystallography

Crystals of **2**, **3**, and **4** were removed from a Schlenk tube under a nitrogen atmosphere and covered with a layer of hydrocarbon oil. A suitable crystal was attached to a glass fiber and quickly placed in a low-temperature N$_2$ stream.

The crystallographic data were collected on a Xcalibur-S diffractometer (Rigaku Oxford Diffraction, Kent, UK) with MoKα radiation (λ = 0.71073 Å), solved by direct methods and refined by full-matrix least squares on F^2 (SHELX97) [39,40] using all data. All non-hydrogen atoms are anisotropic, with H-atoms included in calculated positions (riding model). CCDC1467318 (**2**), CCDC1467319 (**3**), and CCDC1467320 (**4**) contain the supplementary crystallographic data that can be obtained free of charge from The Cambridge Crystallographic Data Centre via www.ccdc.cam.ac.uk/structures. The crystallographic data for compound **2** and **3** are given in Table 2.

Table 2. Crystallographic data for compounds **2** and **3**.

	2	3
Formula	$C_{81}H_{108}LiN_9O_9Sn_4$	$C_{81}H_{108}LiN_9O_9PbSn_3$
M_r (g·mol^{-1})	1833.46	1921.96
Cryst system	triclinic	triclinic
Space group	$P\bar{1}$	$P\bar{1}$
a (Å)	16.2901(9)	17.2766(4)
b (Å)	17.2775(8)	19.4489(6)
c (Å)	19.5167(8)	28.6285(7)
α (°)	103.329(4)	88.874(2)
β (°)	101.969(4)	85.776(2)
γ (°)	118.018(5)	65.256(3)
V (Å^3)	4388.4(4)	8711.9(4)
Z	2	4
F(000)	1864	3856
T (K)	−180(2)	−130(2)
ρ_{calcd} (g cm^3)	1.388	1.465
μ (mm^{-1})	1.181	2.833
Reflns collected	31672	74704
Reflns unique	16052	41501
R_{int}	0.0620	0.0391
Final R_1 ($I > 2\sigma(I)$)	0.0574	0.0531
Final wR_2 (F^2) (all data)	0.1172	0.1050

Acknowledgments: The author thanks the *Deutsche Forschungsgemeinschaft* (DFG) for funding (Grant: ZE804/3-1) and Reike Clauss for preparative support.

References

1. Davies, A.G. Recent advances in the chemistry of the organotin hydrides. *J. Chem. Res.* **2006**, *2006*, 141–148. [CrossRef]

2. Jurkschat, K.; Mehring, M. Organometallic polymers of germanium, tin and lead. In *The Chemistry of Organic Germanium, Tin and Lead Compounds*; Rappoport, Z., Ed.; Wiley: Chichester, UK; New York, NY, USA, 2002; Volume 2, pp. 1–130.

3. Amadoruge, M.L.; Weinert, C.S. Singly bonded catenated germanes: Eighty years of progress. *Chem. Rev.* **2008**, *108*, 4253–4294. [CrossRef] [PubMed]

4. Weinert, C.S. Syntheses, structures and properties of linear and branched oligogermanes. *Dalton Trans.* **2009**, 1691–1699. [CrossRef] [PubMed]

5. Marschner, C.; Hlina, J. Catenated Compounds–Group 14 (Ge, Sn, Pb). In *Comprehensive Inorganic Chemistry II*; Reedijk, J., Poeppelmeier, K., Eds.; Elsevier B.V.: Amsterdam, the Netherlands, 2013; Volume 1, pp. 83–117.

6. Gillman, H.; Cartledge, F.K. Tetrakis(triphenylstannyl)tin. *J. Organomet. Chem.* **1966**, *5*, 48–56. [CrossRef]

7. Jurkschat, K.; Klaus, C.; Dargatz, M.; Tzschach, A.; Meunier-Piret, J.; Mahieu, B. Reaction of 3-dimethylamino-(1,1-dimethyl)propyl magnesium chloride with tin(IV) and tin(II) chlorides. Stabilization of a SnCl$^+$ cation in the new tin cluster [Me$_2$NCH$_2$CH$_2$C(Me$_2$)SnCl]$_3$· SnCl$_2$. *Z. Anorg. Allg. Chem.* **1989**, *577*, 122–134. [CrossRef]

8. Sita, L.R. Structure/property relationships of polystannanes. *Adv. Organomet. Chem.* **1995**, *38*, 189–243.

9. Willemsens, L.C.; van der Kerk, G.J.M. Studies in group IV organometallic chemistry: XIII. Organometallic compounds with five metal atoms in neopentane configuration. *J. Organomet. Chem.* **1964**, *2*, 260–264. [CrossRef]

10. Willemsens, L.C.; van der Kerk, G.J.M. Investigations on organolead compounds: I. A novel red organolead compound a reinvestigation of krause's red diphenyllead. *J. Organomet. Chem.* **1964**, *2*, 271–276. [CrossRef]

11. Stabenow, F.; Saak, W.; Weidenbruch, M. Tris(triphenylplumbyl)plumbate: An anion with three stretched lead–lead bonds. *Chem. Commun.* **2003**, *18*, 2342–2343. [CrossRef]

12. Wang, Y.; Quillian, B.; Wei, P.; Yang, X.-J.; Robinson, G.H. New Pb–Pb bonds: Syntheses and molecular structures of hexabiphenyldiplumbane and tri(trisbiphenylplumbyl)plumbate. *Chem. Commun.* **2004**, *19*, 2224–2225. [CrossRef] [PubMed]

13. Wells, W.L.; Brown, T.L. An investigation of methyltin–lithium compounds preparation and properties of tris(trimethylstannyl)stannyllithium tris(tetrahydrofuran). *J. Organomet. Chem.* **1968**, *11*, 271–280. [CrossRef]

14. Kobayashi, K.; Kawanisi, M.; Hitomi, T.; Kozima, S. Mechanistic studies of decomposition of trialkylstannyllithiums. *J. Organomet. Chem.* **1982**, *233*, 299–311. [CrossRef]

15. Westerhausen, M. Synthesis and crystal structure of calcium bis(trimethylstannanide)·4 THF. *Angew. Chem. Int. Ed. Engl.* **1994**, *33*, 1493–1495. [CrossRef]

16. Englich, U.; Ruhlandt-Senge, K.; Uhlig, F. Novel triphenyltin substituted derivatives of heavier alkaline earth metals. *J. Organomet. Chem.* **2000**, *613*, 139–147. [CrossRef]

17. Flacke, F.; Jacobs, H. [Li(NH$_3$)$_4$][Sn(SnPh$_3$)$_3$]·C$_6$H$_6$, crystal structure of a stannide with trigonal pyramidal tin skeleton. *Eur. J. Solid State Inorg. Chem.* **1997**, *34*, 495–501.

18. Bochkarev, L.N.; Grachev, O.V.; Ziltsov, S.F.; Zakharov, L.N.; Struchkov, Y.T. Synthesis and structure of organotin complexes of ytterbium. *J. Organomet. Chem.* **1992**, *436*, 299–311. [CrossRef]

19. Bochkarev, L.N.; Grachev, O.V.; Molosnova, N.E.; Ziltsov, S.F.; Zakharov, L.N.; Fukin, G.K. Novel polynuclear organotin complexes of samarium and ytterbium. *J. Organomet. Chem.* **1993**, *443*, C26–C28. [CrossRef]

20. Drost, C.; Hildebrand, M.; Lönnecke, P. Synthesis and crystal structure of a novel distannylstannanediyl and a rare pentastannapropellane. *Main Group Met. Chem.* **2002**, *25*, 93–98. [CrossRef]

21. Eichler, B.E.; Phillips, A.D.; Power, P.P. Reactions of phenyllithium with the stannylene Ar*SnPh (Ar* = C$_6$H$_3$-2,6-Trip$_2$; Trip = C$_6$H$_2$-2,4,6-^{i}Pr$_3$) and the synthesis of the distannylstannylene Sn(SnPh$_2$Ar*)$_2$: Contrasting behavior in methyl and phenyl derivatives. *Organometallics* **2003**, *22*, 5423–5426. [CrossRef]

22. Zeckert, K.; Zahn, S.; Kirchner, B. Tin–lanthanoid donor–acceptor bonds. *Chem. Commun.* **2010**, *46*, 2638–2640. [CrossRef]

23. Zeckert, K. Syntheses and structures of lanthanoid(II) complexes featuring Sn–M (M = Al, Ga, In) bonds. *Dalton Trans.* **2012**, *41*, 14101–14106. [CrossRef] [PubMed]

24. Reichart, F.; Kischel, M.; Zeckert, K. Lanthanide(II) complexes of a dual functional tris(2-pyridyl)stannate derivative. *Chem. Eur. J.* **2009**, *15*, 10018–10020. [CrossRef] [PubMed]

25. Leung, W.-P.; Weng, L.-H.; Kwok, W.-H.; Zhou, Z.-Y.; Zhang, Z.-Y.; Mak, T.C.W. Synthesis of a novel binuclear chlorotin(II) alkyl and a lithium trialkylstannate zwitterionic cage molecule: Crystal structures of [Sn(Cl)R^N]$_2$ and [(SnR$^N_3$)Li(µ^3-Cl)(Li(tmeda)]$_2$ [R^N = CH(SiBu^{t}Me$_2$)C$_5$H$_4$N-2]. *Organometallics* **1999**, *18*, 1482–1485. [CrossRef]

26. Drost, C.; Lönnecke, P.; Sieler, J. Stannylplumbylenes: Bonding between tetravalent tin and divalent lead. *Chem. Commun.* **2012**, *48*, 3778–3780. [CrossRef] [PubMed]

27. Cardin, C.J.; Cardin, D.J.; Constantine, S.P.; Todd, A.K.; Teat, S.J.; Coles, S. The first structurally authenticated compound containing a bond between divalent tin and tetravalent tin. *Organometallics* **1998**, *17*, 2144–2146. [CrossRef]

28. Phillips, A.D.; Hino, S.; Power, P.P. A reversible valence equilibrium in a heavier main group compound. *J. Am. Chem. Soc.* **2003**, *125*, 7520–7521. [CrossRef] [PubMed]

29. Setaka, W.; Hirai, K.; Tomioka, H.; Sakamoto, K.; Kira, M. Formation of a stannylstannylene via intramolecular carbene addition of a transient stannaacetylene (RSn≡CR′). *Chem. Commun.* **2008**, *28*, 6558–6560. [CrossRef] [PubMed]

30. Henning, J.; Eichele, K.; Fink, R.F.; Wesemann, L. Structural and spectroscopic characterization of tin–tin double bonds in cyclic distannenes. *Organometallics* **2014**, *33*, 3904–3918. [CrossRef]

31. Jurkschat, K.; Abicht, H.-P.; Tzschach, A.; Mahieu, B. Zur Umsetzung von o-Brommagnesiumbenzyldiphenylphosphin mit SnCl$_2$; Isolierung einer Spezies mit direkter Sn^{II}–Sn^{IV}-Bindung. *J. Organomet. Chem.* **1986**, *309*, C47–C50. [CrossRef]

32. Kennedy, J.D.; McFarlane, W. Nuclear spin–spin coupling between directly bound tin atoms studied by a novel form of INDOR spectroscopy. *Dalton Trans.* **1976**, 1219–1223. [CrossRef]

33. Schneider, C.; Dräger, M. Über gemischte Gruppe 14-Gruppe 14-Bindungen IV. Hexa-p-tolylethan-Analoga p-ToI$_6$Sn$_2$, p-ToI$_6$PbSn und p-ToI$_6$Pb$_2$: Darstellung, Homöotype Strukturen, NMR-Kopplungen und Valenzschwingungen. *J. Organomet. Chem.* **1991**, *415*, 349–362. [CrossRef]

34. Schneider, C.; Behrends, K.; Dräger, M. Über gemischte Gruppe 14–Gruppe 14-Bindungen: VI. Hexa-o-tolylethan-Analoga o-ToI$_6$Sn$_2$, o-ToI$_6$PbSn und o-ToI$_6$Pb$_2$: Ein Vergleich von Bindungsstärke und Polarität in der Reihung Sn–Sn, Pb–Sn, Pb–Pb. *J. Organomet. Chem.* **1993**, *448*, 29–38. [CrossRef]

35. Riviere, P.; Castel, A.; Riviere-Baudet, M. Alkaline and alkaline earth metal-14 compounds: Preparation, spectroscopy, structure and reactivity. In *The Chemistry of Organic Germanium, Tin and Lead Compounds*; Rappoport, Z., Ed.; Wiley: Chichester, UK, New York, NY, USA, 2002; Volume 2, pp. 653–748.

36. Zeckert, K.; Griebel, J.; Kirmse, R.; Weiß, M.; Denecke, R. Versatile reactivity of a lithium tris(aryl)plumbate(II) towards organolanthanoid compounds: Stable lead–lanthanoid-metal bonds or redox processes. *Chem. Eur. J.* **2013**, *19*, 7718–7722. [CrossRef] [PubMed]

37. García-Rodríguez, R.; Wright, D.S. Direct synthesis of the Janus-head ligand (^{Me}Py)$_3$Sn–Sn(^{Me}Py)$_3$ using an unusual pyridyl-transfer reaction (^{Me}Py = 6-methyl-2-pyridyl). *Dalton Trans.* **2014**, *43*, 14529–14532. [CrossRef] [PubMed]

38. Gynane, M.J.S.; Harris, D.H.; Lappert, M.F.; Power, P.P.; Rivière, P.; Rivière-Baudet, M. Subvalent group 4B metal alkyls and amides. Part 5. The synthesis and physical properties of thermally stable amides of germanium(II), tin(II), and lead(II). *Dalton Trans.* **1977**, 2004–2009. [CrossRef]

39. Sheldrick, G.M. *SHELXS-97 and SHELXL-97, Programs for Crystal Structure Analysis*; University of Göttingen: Göttingen, Germany, 1997.

40. Sheldrick, G.M. A short history of SHELX. *Acta Cryst.* **2008**, *A64*, 112–122. [CrossRef] [PubMed]

3

Mesoporous WN/WO$_3$-Composite Nanosheets for the Chemiresistive Detection of NO$_2$ at Room Temperature

Fengdong Qu [1], Bo He [1], Rohiverth Guarecuco [2] and Minghui Yang [1,*]

[1]   Dalian Institute of Chemical Physics, Chinese Academy of Sciences, Dalian 116023, China; qufd@dicp.ac.cn (F.Q.); hebomk@dicp.ac.cn (B.H.)

[2]   Department of Chemical Engineering, Massachusetts Institute of Technology, Cambridge, MA 02139-4307, USA; rog2029@med.cornell.edu

*   Correspondence: myang@dicp.ac.cn

Academic Editor: Rainer Niewa

Abstract: Composite materials, which can optimally use the advantages of different materials, have been studied extensively. Herein, hybrid tungsten nitride and oxide (WN/WO$_3$) composites were prepared through a simple aqueous solution route followed by nitriding in NH$_3$, for application as novel sensing materials. We found that the introduction of WN can improve the electrical properties of the composites, thus improving the gas sensing properties of the composites when compared with bare WO$_3$. The highest sensing response was up to 21.3 for 100 ppb NO$_2$ with a fast response time of ~50 s at room temperature, and the low detection limit was 1.28 ppb, which is far below the level that is immediately dangerous to life or health (IDLH) values (NO$_2$: 20 ppm) defined by the U.S. National Institute for Occupational Safety and Health (NIOSH). In addition, the composites successfully lower the optimum temperature of WO$_3$ from 300 °C to room temperature, and the composites-based sensor presents good long-term stability for NO$_2$ of 100 ppb. Furthermore, a possible sensing mechanism is proposed.

Keywords: WN/WO$_3$ composite; nanosheets; gas sensor; NO$_2$; room temperature

---

1. Introduction

Gas sensors have been in increasing demand in recent years because of their wide applications, not only in environmental monitoring, but also in human health [1–4]. Metal oxide semiconductors (MOSs) based gas sensors have been intensely investigated for their high sensitivity, excellent selectivity, good portability, and low cost [5–7]. However, for detecting lower concentrations of toxic and disease signal gases, composite materials have shown greater potential than bare materials which exhibit low or no sensitivity in low gas concentrations (sub-ppm), as well as poor selectivity [8,9]. These composite materials include heterostructural materials, doped materials, and noble metal decorated materials, among which the heterostructural materials have been widely investigated because of super injection of carriers [10,11]. When different materials hybridize to form a heterojunction, an electron depletion layer (EDL) or an electron accumulation layer (EAL) will form at the interfaces. It is well acknowledged that the heterojunctions can serve as a lever in electron transfer, which can facilitate or restrain the electron transfer, resulting in enhanced sensing performance of a gas sensor [12]. Thus, designing composite sensing materials with heterostructures is regarded as one of the best strategies to achieve excellent gas sensing characteristics.

In the past few years, various sensing materials have been prepared that have exhibited excellent gas sensing characteristics [13,14]. However, to date and to the best of our knowledge, there have been very scarce reports about heterostructural metal nitride and oxide composites for sensing applications [15]. Metal nitrides can have varied properties: from metal-like, to semiconducting, to that of insulators. Thus, introducing metal nitrides into metal oxides could tune electrical properties that have great influence on gas sensing characteristics, such as carrier mobility and electrical resistivity [16]. In addition, other reports have demonstrated that establishing heterojunction could largely decrease the working temperature of sensing materials [17–19]. Thus, designing composites consisting of metal nitrides and oxides, and innovatively applying the composites as gas sensor materials, is of vital significance.

Tungsten trioxide (WO_3) is a multifunctional material that has been widely investigated as a photocatalyst [20,21] and a gas sensor [22,23]. When applied as a gas sensing material, WO_3 exhibits sensitivity to several gases, and has especially high-sensitivity to NO_2, making it an ideal material for detecting NO_2 [23–25]. However, because of its wide band gap, (2.5–3.5 eV) WO_3-based sensors have to work at high working temperature or with the assistance of ultraviolet light [19,26,27]. Therefore, efforts should be focused on lowering working temperature while achieving optimal sensitivity. Tungsten nitride (WN) shows metallic properties with a resistivity of 10.9×10^{-5} $\Omega \cdot m$ at room temperature [28]. Therefore, synthesis of WN-composite WO_3 materials for detecting NO_2 gas at low working temperature may be feasible and is highly desirable.

Herein, for the first time, we present an applicable strategy for the efficient synthesis of new gas sensing materials made of hybrid WN/WO_3 composites. The strategy includes the synthesis of a porous WO_3 nanosheets precursor and subsequent transformation to WN/WO_3 composites by thermal annealing of the as-prepared precursor in NH_3. The WN/WO_3 composites sensing materials exhibit sensitive, selective and reliable detection of NO_2 at room temperature.

2. Results and Discussion

2.1. Structure, Composition and Morphology

Powder X-ray diffraction (XRD) was used to characterize the phase composition and purity of the as-synthesized products. The yellow powder (precursors) obtained from the oil bath reaction agreed well with reported orthorhombic tungsten oxide hydrate ($WO_3 \cdot H_2O$, JCPDS No. 84-0886), as shown in Figure 1a (Pattern A) [29]. The $WO_3 \cdot H_2O$ precursors were converted into WN and WN/WO_3 composites by nitridation in NH_3 atmosphere and oxidation in air, under heat treatment at 700 °C and 200 °C for 3 h, respectively. Figure 1a (Patterns B and C) shows the XRD of WN and WN/WO_3 composites. As shown in pattern b, all the diffraction peaks matched well with those of hexagonal WN (JCPDS No. 65-2898). For the composite (Pattern C), the crystal phase was a mixture of hexagonal WN (JCPDS No. 65-2898), a big convex appearing in the position of 2θ 20°–30°, and a small convex in the position of 2θ 50°–60°, which can be attributed to amorphous WO_3. Thus, the final product is a composite material composed of hexagonal WN and amorphous WO_3 (WN/WO_3). The composites underwent a mass change during the transformation from WN to WO_3 by calcination in air. By using this mass change, we can calculate the ratio of WN to WO_3. Thermogravimetric analyses of the products at temperature range of 0–1000 °C are shown in Figure 1b. The DSC-TGA results of the WN/WO_3 composite nanosheets are displayed in Figure 1b. It was observed from the TGA curve that the specimen shows a weight loss of about 1%, which relates to adsorbed water and other trace substances. When the temperature reached about 211.5 °C, the weight of the specimen began to increase. When the temperature reaches about 457 °C, the weight nearly reaches stability. Moreover, the DSC curve exhibited an obvious exothermal peak at 426 °C, which corresponded to the obvious weight gain of approximately 8.86% between 211 and 457 °C in the TGA curves. The significant weight gain was attributed to the reaction of the WN with oxygen to generate WO_3. According to the weight gain of approximately 8.86%, the WN content of the composites is calculated to be 55.9%.

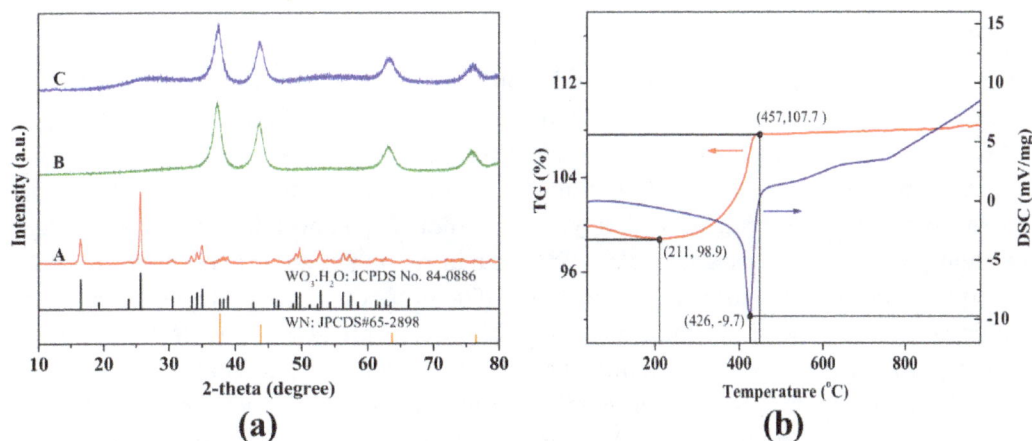

Figure 1. (**a**) Powder X-ray diffraction (XRD) patterns of tungsten oxide hydrate (WO$_3$·H$_2$O) A, tungsten nitride (WN) B and hybrid tungsten nitride and oxide (WN/WO$_3$) composites C; (**b**) differential scanning calorimetry-thermal gravimetric analysis (DSC-TGA) curves of WN/WO$_3$ composites.

The morphology of WN/WO$_3$ composites was characterized through SEM and TEM analysis, and the results are shown in Figure 2. According to Figure 2a–c, the WN/WO$_3$ composites exhibited square shaped nanosheets with uniform lateral dimensions mainly ranging from 80 to 150 nm (Figure 2a,b) and thickness of 10 to 30 nm (Figure 2c). In addition, relatively homogeneous pores with average diameter of 10 nm were dispersed uniformly at the surfaces of WN/WO$_3$ composites. The porous feature may have contributed to the volume shrinkage caused by transformation of orthorhombic WO$_3$·H$_2$O to hexagonal WN. Figure 2d shows the HRTEM image of the WN/WO$_3$ composites, where the lattice fringes d-spacings of 0.239 and 0.204 nm are identified and correspond to the (111) and (200) planes of hexagonal WN (JCPDS No. 65-2898), respectively. The inset of Figure 2d presents the selected area electron diffraction (SEAD) of WN/WO$_3$ composites. The distinctive diffraction rings and spots on the SEAD pattern match the crystal structure identification obtained from the HRTEM and XRD. The diffraction rings are indexed to cubic WN (JCPDS No. 65-2898) (111), (200), (220), (311), (400), and (331) planes, respectively.

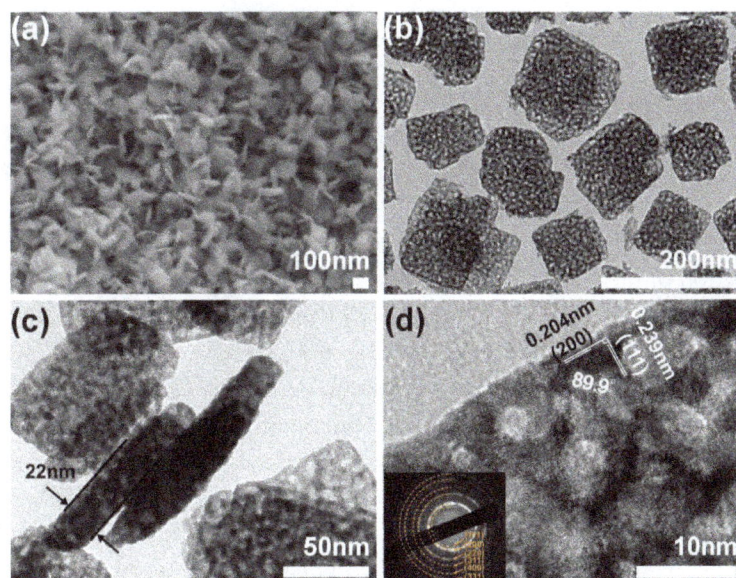

Figure 2. SEM images (**a**); top/side-view transmission electron microscopy (TEM) images (**b–c**); and high-resolution transmission electron microscopy (HRTEM) images of WN/WO$_3$ composites (**d**). Inset: selected area electron diffraction (SEAD) pattern of WN/WO$_3$ composites.

2.2. Gas Sensing Characteristics

Dynamic sensing performance of WN/WO$_3$ composites-based sensors to various concentrations of NO$_2$ at room temperature is shown in Figure 3a. NO$_2$ is an oxidizing gas, so when an NO$_2$ molecule adsorbs on the surface of WN/WO$_3$ composites, it captures electrons from the composites, resulting in the increase of resistance of the composites-based sensors upon exposure to NO$_2$. Figure 3b plots the sensor response as a function of NO$_2$ concentration, and shows that with an increase in NO$_2$ concentration, the response of the sensor also exhibits an increasing trend. Interestingly, the sensitivities to NO$_2$ concentrations from 5 to 100 ppb and 100 to 1000 ppb exhibit linear dependence, fitting well into two linear curves which have different slopes. It is well understood that electrical conductance shows strong dependence on carrier and mobility ($\sigma = ne\mu_n$, where n is the concentration of electrons, e is electronic charge, and μ_n is electron mobility), and that electrical resistivity is the reciprocal of electrical conductance. When the WN/WO$_3$ composites sensor is exposed to low concentration of NO$_2$, the amount of adsorbed NO$_2$ is so sparse that the degradation of carrier could be almost negligible, and the response of the sensor exhibits a "high increase model". However, when exposed to high concentration of NO$_2$, the effect of NO$_2$ adsorption will be enhanced, and the degradation of carrier mobility becomes relatively remarkable, resulting in a decrease in the rate of sensitivity increase under abundant NO$_2$ adsorption [30]. The detection limit of the WN/WO$_3$ composites sensor for NO$_2$ is calculated to be approximately 1.28 ppb based on signal-to-noise ratio of 3 (see the Section "Calculation of Theoretical Limit of Detection Using Signal/Noise Ratio" in Supplementary Materials), which is far below the immediately dangerous to life or health (IDLH) values (NO$_2$: 20 ppm) defined by the U.S. National Institute for Occupational Safety and Health (NIOSH). The recommended NO$_2$ exposure limit defined by the NIOSH is 1 ppm [31].

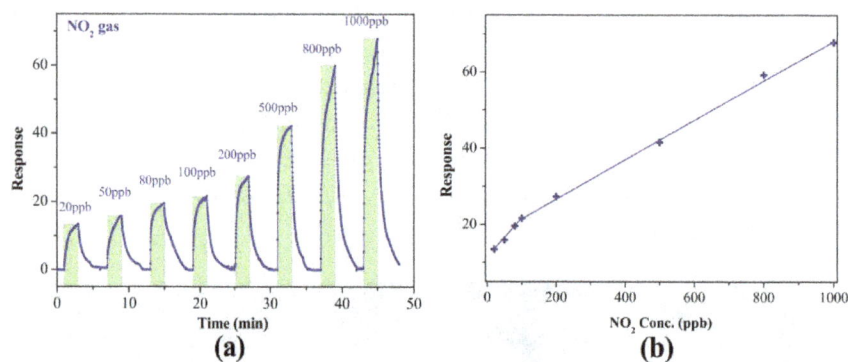

Figure 3. (a) Dynamic sensing performance of WN/WO$_3$ composites sensor to NO$_2$ with concentration from 20 to 1000 ppb at room temperature (RT); (b) calibration curves of WN/WO$_3$ composites sensor towards various concentrations of NO$_2$ at RT.

Selective detection of target gas from various interference gases is still an unsolved drawback for chemiresistive gas sensors [32]. The responses of the sensor based on WN/WO$_3$ composites to common vehicle exhaust, including SO$_2$, CO$_2$, moisture, CO, NH$_3$, ethanol and NO$_2$, were measured to quantify the selectivity, which is shown in Figure 4. As clearly illustrated in Figure 4, the sensor shows a high response to NO$_2$, which is more than 2 times higher than that for other interference gases, indicating the excellent selectivity to NO$_2$ as opposed to other selected interference gases. The selective detection to NO$_2$ gas might be due to the high reactivity and large electron affinity (2.28 eV) of NO$_2$, in comparison with pre-adsorbed oxygen (0.43 eV) and other test gases [33]. In addition, the responses of the sensor based on WN/WO$_3$ composites to 200 ppb NO$_2$ at RT under different humidities were tested and the result is shown in Figure S1. It can be observed that, with increasing relative humidity, the response of the sensor decreases because of the competition between water molecules and NO$_2$ molecules for the reacting sites [34]. However, the sensor shows a high response of 17.8 even at high humidity of 90 RH%.

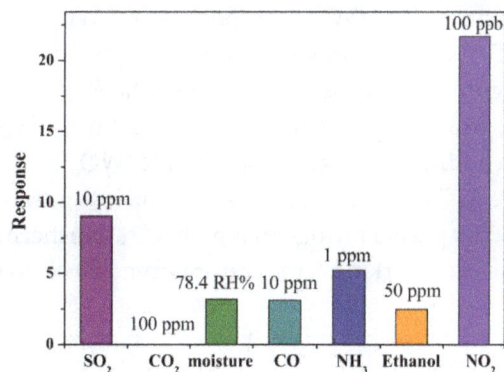

Figure 4. Cross-response of WN/WO$_3$ composites sensor to various gases.

Figure 5a shows the response and recovery times of the WN/WO$_3$ composites-based sensor. With increasing NO$_2$ concentration, the response time and recovery time both experience a decrease. The reason why the response and recovery times are substantially longer at low concentration is unclear and needs further study. The following can be considered as a plausible explanation. At low concentration of ambient NO$_2$, the NO$_2$ molecules occupy a lower percentage of air relative to oxygen molecules, so the probability of NO$_2$ molecules arriving and being captured by the surface of the sensing materials (WN/WO$_3$) is relatively smaller, resulting in a long response. When the concentration of NO$_2$ increases, the probability of NO$_2$ molecules arriving and being captured by the surface of sensing materials becomes higher, decreasing the response time. As for the recovery time: at low concentration of NO$_2$, the amount of adsorbed NO$_2$ is too small, making it difficult to desorb completely, resulting in a relatively longer recovery time. When the concentration is higher, the concentration difference between the surface of WN/WO$_3$ and ambient atmosphere is large enough to desorb NO$_2$ molecules quickly, reducing the recovery time. Similar observations have also been reported by other investigators [35,36]. In addition, the sensor shows excellent reversibility properties with a response time of less than 80 s and a recovery time of less than 180 s, regardless of the NO$_2$ concentration. Stability is another key quality indicator in the development of gas sensors for real markets. Thus, we measured the response of the WN/WO$_3$ composites-based sensor for 7 weeks, once a day (every 24 h) at room temperature for a week and then once a week for 6 weeks, shown in Figure 5b. We can observe that the sensor experiences a loss in response of less than 12.27% after 7 weeks of aging. The response of the sensor comes into saturation after 2 weeks. The results indicate that the WN/WO$_3$ composites sensor exhibits good long-term stability.

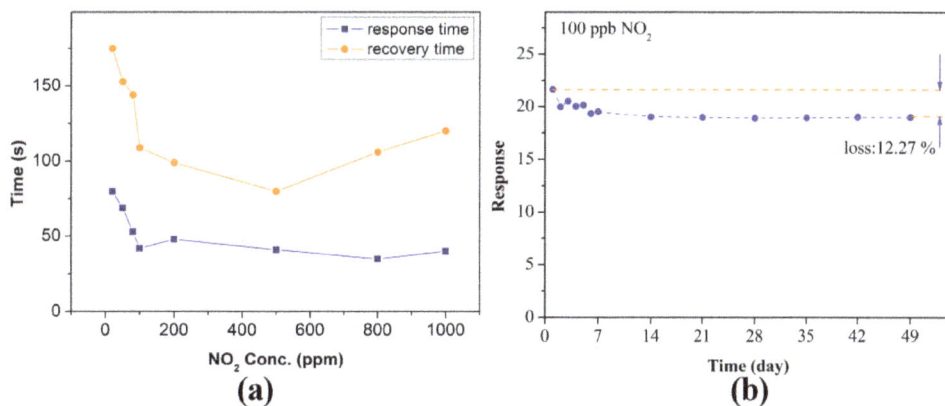

Figure 5. (**a**) Response and recovery time of WN/WO$_3$ composites sensor to various concentrations of NO$_2$ at RT. (**b**) Stability measurement of the WN/WO$_3$ composites sensor to 100 ppb NO$_2$ at RT for 7 weeks.

Dynamic sensing transients of WN/WO₃ composites and WO₃ nanosheets to 100 ppb NO₂, at room temperature and 300 °C (the optimum working temperature of WO₃, data shown in Figure S2) respectively, are shown in Figure 6. It can be observed that the R_a of WN/WO₃ composites is lower than that of WO₃ nanosheets, which confirms that the introduction of WN into WO₃ can decrease the resistance of the materials. In addition, the responses of WN/WO₃ composites and WO₃ nanosheets sensors to 100 ppb NO₂ were 23.7 and 6.2, respectively, which suggests that gas sensing responses towards NO₂ can be enhanced by partial nitrogenizing of WO₃. Furthermore, the introduction of WN into WO₃ apparently decreases the working temperature from 300 °C to room temperature.

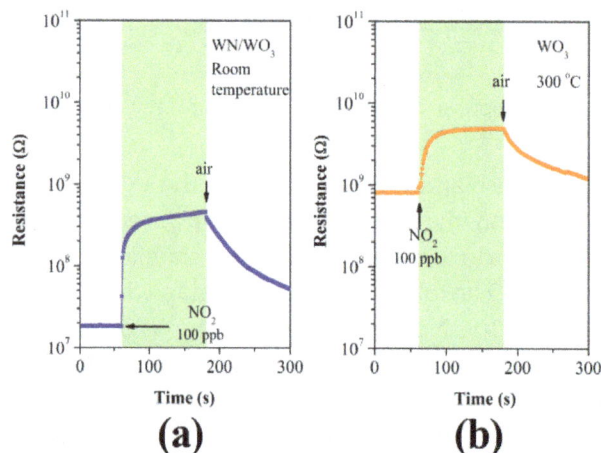

Figure 6. Dynamic sensing transients to 100 ppb NO₂ of (**a**) WN/WO₃ composite nanosheets sensors at room temperature; and (**b**) WO₃ nanosheets sensors at 300 °C.

2.3. Gas Sensing Mechanism

Typically, the gas sensing mechanism of WO₃ belongs to surface-controlled type, where the adsorption and desorption of NO₂ molecules on the surface of WO₃ play a vital role [24]. In ambient air, the oxygen molecules will adsorb on the surface of WO₃ and trap electrons from the conduction band of WO₃ to form ionic oxygen species (O^{2-}, O^- and O^{2-}), decreasing the electron concentration of the surface of WO₃, resulting in a high-resistance depletion layer at the surface [37]. When the NO₂ gases come in, the NO₂ molecules, which are highly electrophilic, will capture electrons from the conduction band of WO₃, and will also react with the adsorbed ionic oxygen species. This will lead to a further decrease in electron concentration, increasing the resistance of WO₃. The reactions taking place on the surface of WO₃ are as follows.

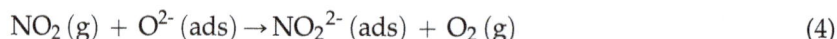

$$O_2 \, (g) \rightarrow O_2 \, (ads) \tag{1}$$

$$O_2 \, (ads) + 2e^- \rightarrow O^{2-} \, (ads) \tag{2}$$

$$NO_2 \, (g) + e \rightarrow NO_2 \, (ads)^- \tag{3}$$

$$NO_2 \, (g) + O^{2-} \, (ads) \rightarrow NO_2{}^{2-} \, (ads) + O_2 \, (g) \tag{4}$$

The as-fabricated WN/WO₃ composites-based sensor, when compared with WO₃ sensor, exhibited enhanced sensing performances. The reasons why WN/WO₃ composites-based sensor shows enhanced gas sensing characteristics are still unclear and worthy of further investigation. The following can be considered as a plausible explanation based on our experimental results. Firstly, WN exhibits metallic properties with a high electron concentration of about $(5–6) \times 10^{20} \, cm^{-3}$ and a resistivity of $(1–2) \times 10^{-1} \, \Omega$ at room temperature [38]. However, WO₃ is an n-type semiconductor, whose resistivity is far larger than that of WN. Thus, the electrons will transfer from the high concentration side to the low concentration side because of diffusion effect, resulting in a lower resistivity of the composites when compared with bare WO_3. The lower baseline of resistivity of WN/WO_3 composites will eventually increase the response ($S = (R_g/R_a - 1) \times 100\%$) of the sensor. In addition, WN and WO_3 themselves did not show any sensing properties for NO_2 at room temperature (Figure S3). However, the WN/WO_3 composites exhibit excellent sensing characteristics to NO_2 at room temperature. This may largely prove that there might be some synergetic effect in our WN/WO_3 composites, leading to room temperature sensing for NO_2.

3. Materials and Methods

3.1. Materials

Sodium tungstate dihydrate ($Na_2WO_4 \cdot 2H_2O$, AR) of $\geqslant 99.5\%$ purity was purchased from Aladdin Chemical Reagent Co., Ltd. (Shanghai, China). Hydrochloric acid (HCl) and oxalic dihydrate ($H_2C_2O_4 \cdot 2H_2O$) of analytical reagent grade were purchased from Sinopharm Chemical Reagent Co., Ltd. (Shanghai, China).

3.2. Synthesis

$Na_2WO_4 \cdot 2H_2O$ (3.2985 g, 0.01 mol) was dissolved in 50 mL deionized water with continuous stirring for 30 min at room temperature. 50 mL of HCl aqueous solution (2 M) was added dropwise into the above solution under continuous stirring. A certain amount of $H_2C_2O_4 \cdot 2H_2O$, with a mole ratio of 0.25 $Na_2WO_4 \cdot 2H_2O$ to $H_2C_2O_4 \cdot 2H_2O$, was then added into the above system. The reaction vessel was then transferred to a 90 °C oil bath for 3 h. The yellow precipitate obtained was filtered and washed with deionized water and absolute ethanol several times, and subsequently dried at 60 °C in a vacuum oven for 24 h. Then, the yellow powder was nitrided at 700 °C at a ramp rate of 10 °C· min^{-1} under NH_3 gas flow rate of 300 sccm (standard-state cubic centimeter per minute) for 3 h. Finally, the nitride was calcined at 420 °C in air for 3 h at a ramp rate of 10 °C· min^{-1}.

3.3. Materials Characterizations

X-ray diffraction (XRD) analysis was conducted on a Rigaku MiniFlex 600 powder X-ray diffractometer (Rigaku Corporation, Tokyo, Japan) with Cu Kα radiation (λ = 1.5418 Å, accelerating voltage 40 kV, applied current 15 mA) at scanning rate of 1 °/min. Scanning electron microscopy (SEM) images were performed on a JSM-7800F (Japan Electron Optics Laboratory Co., Ltd., Tokyo, Japan) instrument (accelerating voltage 3.0 kV). Transmission electron microscopy (TEM) and high-resolution transmission electron microscopy (HRTEM) observations were conducted on an FEI TecnaiG2 F30 (Japan Electron Optics Laboratory Co., Ltd.). Differential Scanning Calorimetry (DSC) and thermal gravimetric analysis (TGA) were done on a NETZSCH/STA 449 F1 (Netzsch, Selb, Freistaat Bayern, Germany).

3.4. Fabrication and Measurement of Gas Sensor

Typically, 50 mg· mL^{-1} WN/WO_3 composites-ethanol solution was deposited on the surface of the device and calcined at 100 °C for 3 h. Gas sensing performances were measured by a homemade sensor testing system (a cylindrical glass chamber with a volume of 100 mL). A gas mixing line equipped with mass flow controllers was designed to prepare target gases at specific concentrations in the testing chamber as shown in Figure S4. The resistance changes of sensor in air or tested gas were monitored by a high-resistance meter (Victor, 86E, Shenzhen, China). The response value (S) was defined as $S = (|R_g - R_a|/R_a) \times 100\%$, where R_a and R_g denoted the resistance of the sensors in the absence and presence of the target gases (reducing gases), respectively. The time taken by the sensor to achieve 90% of the total resistance change was defined as the response time in the case of response (target gas adsorption) or the recovery time in the case of recovery (target gas desorption).

4. Conclusions

In conclusion, we successfully synthesized WN/WO$_3$ composites through a simple strategy, and innovatively utilized them as gas sensing materials. The WN/WO$_3$ composites-based sensor exhibited sensitivity and high selectivity to NO$_2$ at room temperature. Overall, the excellent NO$_2$ sensing performance of the WN/WO$_3$ composites-based sensor supplies exciting opportunities for environmental monitoring and disease diagnosing. Furthermore, we expect our findings to bring up new promising gas sensing materials, and to inspire rational synthesis of other transition metal nitride and oxide hybrids for high-performance gas sensing.

Supplementary Materials:
Figure S1: Response of WN/WO$_3$ composites sensor at RT upon exposure to 200 ppb NO$_2$ concentration at various relative humidities (RH), Figure S2: Response of the sensor based on WO$_3$ nanosheets to 100 ppb NO$_2$ as a function of the operating temperature, Figure S3: Resistances of the sensors based on WN and WO$_3$ to 100 ppb NO$_2$ at room temperature, Calculation of theoretical limit of detection using signal/noise ratio, Figure S4: The schematic illustration of (**a**) gas sensing analysis system and (**b**) gas mixing line equipment.

Acknowledgments: This work is supported by National Natural Science Foundation of China through grant 21471147 and Liaoning Provincial Natural Science Foundation through grant 2014020087. Minghui Yang would like to thank the National "Thousand Youth Talents" program of China.

Author Contributions: The preparation of the manuscript was made by all authors. Fengdong Qu and Minghui Yang conceived and designed the experiments; Fengdong Qu and Bo He performed the experiments; Fengdong Qu, Bo He and Rohiverth Guarecuco analyzed the data.

References

1. Haick, H.; Broza, Y.Y.; Mochalski, P.; Ruzsanyi, V.; Amann, A. Assessment, origin, and implementation of breath volatile cancer markers. *Chem. Soc. Rev.* **2014**, *43*, 1423–1449. [CrossRef]

2. Hagleitner, C.; Hierlemann, A.; Lange, D.; Kummer, A.; Kerness, N.; Brand, O.; Baltes, H. Smart single-chip gas sensor microsystem. *Nature* **2001**, *414*, 293–296. [CrossRef] [PubMed]

3. Peng, G.; Tisch, U.; Adams, O.; Hakim, M.; Shehada, N.; Broza, Y.Y.; Billan, S.; Abdah-Bortnyak, R.; Kuten, A.; Haick, H. Diagnosing lung cancer in exhaled breath using gold nanoparticles. *Nat. Nanotechnol.* **2009**, *4*, 669–673. [CrossRef] [PubMed]

4. Kolmakov, A.; Klenov, D.; Lilach, Y.; Stemmer, S.; Moskovits, M. Enhanced gas sensing by individual SnO$_2$ nanowires and nanobelts functionalized with Pd catalyst particles. *Nano Lett.* **2005**, *5*, 667–673. [CrossRef] [PubMed]

5. Deng, S.; Tjoa, V.; Fan, H.M.; Tan, H.R.; Sayle, D.C.; Olivo, M.; Mhaisalkar, S.; Wei, J.; Sow, C.H. Reduced graphene oxide conjugated Cu$_2$O nanowire mesocrystals for high-performance NO$_2$ gas sensor. *J. Am. Chem. Soci.* **2012**, *134*, 4905–4917. [CrossRef] [PubMed]

6. Rai, P.; Khan, R.; Raj, S.; Majhi, S.M.; Park, K.-K.; Yu, Y.-T.; Lee, I.-H.; Sekhar, P.K. Au@Cu$_2$O core–shell nanoparticles as chemiresistors for gas sensor applications: Effect of potential barrier modulation on the sensing performance. *Nanoscale* **2014**, *6*, 581–588. [CrossRef] [PubMed]

7. Jing, Z.; Zhan, J. Fabrication and gas-sensing properties of porous ZnO nanoplates. *Adv. Mater.* **2008**, *20*, 4547–4551. [CrossRef]

8. Cheng, W.; Ju, Y.; Payamyar, P.; Primc, D.; Rao, J.; Willa, C.; Koziej, D.; Niederberger, M. Large-area alignment of tungsten oxide nanowires over flat and patterned substrates for room-temperature gas sensing. *Angew. Chem. Int. Ed.* **2015**, *54*, 340–344. [CrossRef] [PubMed]

9. Kim, H.W.; Na, H.G.; Kwon, Y.J.; Cho, H.Y.; Lee, C. Decoration of Co nanoparticles on ZnO-branched SnO$_2$ nanowires to enhance gas sensing. *Sens. Actuators B Chem.* **2015**, *219*, 22–29. [CrossRef]

10. Liang, Y.; Cui, Z.; Zhu, S.; Li, Z.; Yang, X.; Chen, Y.; Ma, J. Design of a highly sensitive ethanol sensor using a nano-coaxial p-Co$_3$O$_4$/n-TiO$_2$ heterojunction synthesized at low temperature. *Nanoscale* **2013**, *5*, 10916–10926. [CrossRef] [PubMed]

11. Xu, S.; Gao, J.; Wang, L.; Kan, K.; Xie, Y.; Shen, P.; Li, L.; Shi, K. Role of the heterojunctions in In$_2$O$_3$-composite SnO$_2$ nanorod sensors and their remarkable gas-sensing performance for NO$_x$ at room temperature. *Nanoscale* **2015**, *7*, 14643–14651. [CrossRef] [PubMed]

12. Zhou, X.; Feng, W.; Wang, C.; Hu, X.; Li, X.; Sun, P.; Shimanoe, K.; Yamazoe, N.; Lu, G. Porous $ZnO/ZnCo_2O_4$ hollow spheres: Synthesis, characterization, and applications in gas sensing. *J. Mater. Chem. A* **2014**, *2*, 17683–17690. [CrossRef]

13. Sun, P.; Zhou, X.; Wang, C.; Shimanoe, K.; Lu, G.; Yamazoe, N. Hollow $SnO_2/\alpha\text{-}Fe_2O_3$ spheres with a double-shell structure for gas sensors. *J. Mater. Chem. A* **2014**, *2*, 1302–1308. [CrossRef]

14. Chan, N.Y.; Zhao, M.; Wang, N.; Au, K.; Wang, J.; Chan, L.W.H.; Dai, J. Palladium nanoparticle enhanced giant photoresponse at $LaAlO_3/SrTiO_3$ two-dimensional electron gas heterostructures. *ACS Nano* **2013**, *7*, 8673–8679. [CrossRef] [PubMed]

15. Pearton, S.; Kang, B.; Kim, S.; Ren, F.; Gila, B.; Abernathy, C.; Lin, J.; Chu, S. Gan-based diodes and transistors for chemical, gas, biological and pressure sensing. *J. Phys. Condens. Matter* **2004**, *16*, R961. [CrossRef]

16. Aliano, A.; Cicero, G.; Catellani, A. Origin of the accumulation layer at the $InN/a\text{-}In_2O_3$ interface. *ACS Appl. Mater. Interfaces* **2015**, *7*, 5415–5419. [CrossRef] [PubMed]

17. Liu, J.; Dai, M.; Wang, T.; Sun, P.; Liang, X.; Lu, G.; Shimanoe, K.; Yamazoe, N. Enhanced gas sensing properties of SnO_2 hollow spheres decorated with CeO_2 nanoparticles heterostructure composite materials. *ACS Appl. Mater. Interfaces* **2016**, *8*, 6669–6677. [CrossRef] [PubMed]

18. Sun, P.; Cai, Y.; Du, S.; Xu, X.; You, L.; Ma, J.; Liu, F.; Liang, X.; Sun, Y.; Lu, G. Hierarchical $\alpha\text{-}Fe_2O_3/SnO_2$ semiconductor composites: Hydrothermal synthesis and gas sensing properties. *Sens. Actuators B Chem.* **2013**, *182*, 336–343. [CrossRef]

19. An, X.; Jimmy, C.Y.; Wang, Y.; Hu, Y.; Yu, X.; Zhang, G. WO_3 nanorods/graphene nanocomposites for high-efficiency visible-light-driven photocatalysis and NO_2 gas sensing. *J. Mater. Chem.* **2012**, *22*, 8525–8531. [CrossRef]

20. Tanaka, A.; Hashimoto, K.; Kominami, H. Visible-light-induced hydrogen and oxygen formation over $Pt/Au/WO_3$ photocatalyst utilizing two types of photoabsorption due to surface plasmon resonance and band-gap excitation. *J. Am. Chem. Soc.* **2014**, *136*, 586–589. [CrossRef] [PubMed]

21. Chen, D.; Ye, J. Hierarchical WO_3 hollow shells: Dendrite, sphere, dumbbell, and their photocatalytic properties. *Adv. Funct. Mater.* **2008**, *18*, 1922–1928. [CrossRef]

22. Li, X.-L.; Lou, T.-J.; Sun, X.-M.; Li, Y.-D. Highly sensitive WO_3 hollow-sphere gas sensors. *Inorg. Chem.* **2004**, *43*, 5442–5449. [CrossRef] [PubMed]

23. Penza, M.; Tagliente, M.; Mirenghi, L.; Gerardi, C.; Martucci, C.; Cassano, G. Tungsten trioxide (WO_3) sputtered thin films for a NO_x gas sensor. *Sens. Actuators B Chem.* **1998**, *50*, 9–18. [CrossRef]

24. Lee, D.-S.; Han, S.-D.; Huh, J.-S.; Lee, D.-D. Nitrogen oxides-sensing characteristics of WO_3-based nanocrystalline thick film gas sensor. *Sens. Actuators B Chem.* **1999**, *60*, 57–63. [CrossRef]

25. Tao, W.-H.; Tsai, C.-H. H_2S sensing properties of noble metal doped WO_3 thin film sensor fabricated by micromachining. *Sens. Actuators B Chem.* **2002**, *81*, 237–247. [CrossRef]

26. Deng, L.; Ding, X.; Zeng, D.; Tian, S.; Li, H.; Xie, C. Visible-light activate mesoporous WO_3 sensors with enhanced formaldehyde-sensing property at room temperature. *Sens. Actuators B Chem.* **2012**, *163*, 260–266. [CrossRef]

27. Wang, G.; Ji, Y.; Huang, X.; Yang, X.; Gouma, P.-I.; Dudley, M. Fabrication and characterization of polycrystalline WO_3 nanofibers and their application for ammonia sensing. *J. Phys. Chem. B* **2006**, *110*, 23777–23782. [CrossRef] [PubMed]

28. Guruvenket, S.; Rao, G.M. Bias induced structural changes in tungsten nitride films deposited by unbalanced magnetron sputtering. *Mater. Sci. Eng. B* **2004**, *106*, 172–176. [CrossRef]

29. Li, G.; Song, J.; Pan, G.; Gao, X. Highly Pt-like electrocatalytic activity of transition metal nitrides for dye-sensitized solar cells. *Energy Environ. Sci.* **2011**, *4*, 1680–1683. [CrossRef]

30. Cui, S.; Pu, H.; Wells, S.A.; Wen, Z.; Mao, S.; Chang, J.; Hersam, M.C.; Chen, J. Ultrahigh sensitivity and layer-dependent sensing performance of phosphorene-based gas sensors. *Nat. Commun.* **2015**, *6*. [CrossRef] [PubMed]

31. Kulkarni, G.S.; Reddy, K.; Zhong, Z.; Fan, X. Graphene nanoelectronic heterodyne sensor for rapid and sensitive vapour detection. *Nat. Commun.* **2014**, *5*, 4376. [CrossRef] [PubMed]

32. Sun, Y.-F.; Liu, S.-B.; Meng, F.-L.; Liu, J.-Y.; Jin, Z.; Kong, L.-T.; Liu, J.-H. Metal oxide nanostructures and their gas sensing properties: A review. *Sensors* **2012**, *12*, 2610–2631. [CrossRef] [PubMed]

33. Ganbavle, V.; Inamdar, S.; Agawane, G.; Kim, J.; Rajpure, K. Synthesis of fast response, highly sensitive and selective Ni:ZnO based NO_2 sensor. *Chem. Eng. J.* **2016**, *286*, 36–47. [CrossRef]

34. Koziej, D.; Bârsan, N.; Weimar, U.; Szuber, J.; Shimanoe, K.; Yamazoe, N. Water–oxygen interplay on tin dioxide surface: Implication on gas sensing. *Chem. Phys. Lett.* **2005**, *410*, 321–323. [CrossRef]

35. Wang, C.; Li, X.; Feng, C.; Sun, Y.; Lu, G. Nanosheets assembled hierarchical flower-like WO_3 nanostructures: Synthesis, characterization, and their gas sensing properties. *Sens. Actuators B Chem.* **2015**, *210*, 75–81. [CrossRef]

36. Wang, L.; Dou, H.; Lou, Z.; Zhang, T. Encapsuled nanoreactors (Au@SnO_2): A new sensing material for chemical sensors. *Nanoscale* **2013**, *5*, 2686–2691. [CrossRef] [PubMed]

37. Bao, M.; Chen, Y.; Li, F.; Ma, J.; Lv, T.; Tang, Y.; Chen, L.; Xu, Z.; Wang, T. Plate-like p–n heterogeneous NiO/WO_3 nanocomposites for high performance room temperature NO_2 sensors. *Nanoscale* **2014**, *6*, 4063–4066. [CrossRef] [PubMed]

38. Nandi, D.K.; Sen, U.K.; Sinha, S.; Dhara, A.; Mitra, S.; Sarkar, S.K. Atomic layer deposited tungsten nitride thin films as a new lithium-ion battery anode. *Phys. Chem. Chem. Phys.* **2015**, *17*, 17445–17453. [CrossRef] [PubMed]

The *Trans* Influence in Unsymmetrical Pincer Palladacycles: An Experimental and Computational Study

Sarote Boonseng [1], Gavin W. Roffe [1], Rhiannon N. Jones [1], Graham J. Tizzard [2], Simon J. Coles [2], John Spencer [1,*] and Hazel Cox [1,*]

[1] Department of Chemistry, School of Life Sciences, University of Sussex, Falmer, Brighton, East Sussex, BN1 9QJ, UK; s.boonseng@sussex.ac.uk (S.B.); gwr20@sussex.ac.uk (G.W.R.); R.N.Jones@sussex.ac.uk (R.N.J.)

[2] UK National Crystallography Service, School of Chemistry, University of Southampton, Highfield, Southampton SO17 1BJ, UK; Graham.Tizzard@soton.ac.uk (G.J.T.); S.J.Coles@soton.ac.uk (S.J.C.)

* Correspondence: j.spencer@sussex.ac.uk (J.S.); h.cox@sussex.ac.uk (H.C.)

Academic Editor: Duncan H. Gregory

Abstract: A library of unsymmetrical SCN pincer palladacycles, [ClPd{2-pyr-6-(RSCH$_2$)C$_6$H$_3$}], R = Et, Pr, Ph, *p*-MePh, and *p*-MeOPh, pyr = pyridine, has been synthesized via C–H bond activation, and used, along with PCN and N'CN unsymmetrical pincer palladacycles previously synthesized by the authors, to determine the extent to which the *trans* influence is exhibited in unsymmetrical pincer palladacycles. The *trans* influence is quantified by analysis of structural changes in the X-ray crystal and density functional theory (DFT) optimized structures and a topological analysis of the electron density using quantum theory of atoms in molecules (QTAIM) to determine the strength of the Pd-donor atom interaction. It is found that the *trans* influence is controlled by the nature of the donor atom and although the substituents on the donor-ligand affect the Pd-donor atom interaction through the varied electronic and steric constraints, they do not influence the bonding of the ligand *trans* to it. The data indicate that the strength of the *trans* influence is P > S > N. Furthermore, the synthetic route to the family of SCN pincer palladacycles presented demonstrates the potential of late stage derivitization for the effective synthesis of ligands towards unsymmetrical pincer palladacycles.

Keywords: pincer palladacycles; density functional theory; atoms in molecules; *trans* influence

1. Introduction

Palladacycles have been extensively studied since their discovery in 1965 by Cope and Siekman [1]. They have been widely used as catalysts or pre-catalysts in organic reactions, such as in Heck and Suzuki–Miyaura cross-couplings [2–5]. Pincer palladacycles are an interesting type of palladacycle, of which there are two different types. The majority of pincer palladacycles studied have been of the symmetrical YCY type, such as NCN [6], SCS [7], PCP [8], and SeCSe [9,10]. There are limited numbers of reported unsymmetrical pincer palladacycles owing to their more difficult synthesis. For example, Szabó et al. synthesized unsymmetrical PCS pincer palladacycles from 1,3-bis(bromomethyl)benzene, albeit in low overall yield (38%) [11]. However, the unsymmetrical PCS pincer palladacycle reported showed enhanced catalytic activity when compared to related symmetrical pincer palladacycles [11,12]. Recently, we reported the synthesis of an unsymmetrical SCN pincer palladacycle by C–H bond activation [13], and novel unsymmetrical PCN and N'CN analogues [14] (Figure 1).

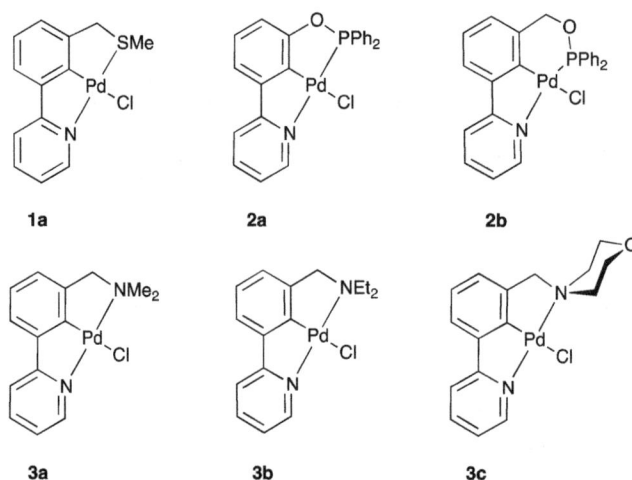

Figure 1. Unsymmetrical pincer palladacylces, SCN (**1a**) [13], PCN (**2a–2b**), and N′CN (**3a–3c**) [14].

The pincer palladacycle structures are stabilized by an intramolecular coordination to the metal of the two donor atoms in the side arms. Their reactivity and other properties are influenced by the donor group around the metal [3]. The attractive feature of pincer palladacycles is the possibility for fine-tuning the catalytic activity by varying the two side arms to modify the palladium environment, by changing the donor atoms and their substituents, providing the opportunity to alter hard/soft acid base properties, or by changing the ring size, giving rise to varying steric hindrance [12]. These factors provide the potential for hemilabile coordination of the donor ligand arms with the metal center, an important consideration in the design of pincer palladacycles [15–17]. This can lead to different physical and chemical properties of the donor ligand arms, resulting in preferential decoordination of one of the ligand arms, providing the opportunity to fine tune the catalytic activity of unsymmetrical pincer palladacycles [18–21]. An excellent example by Wendt and co-workers reported the hemilabile nature of nitrogen and phosphorus donor atoms by reacting with a strong nucleophile (MeLi) [22]. The results showed that the nitrogen donor atom arm decoordinated from the Pd center, while the phosphorus donor atom arm remained coordinated to the Pd center (Scheme 1). It is clear that the different properties of the side arms result in hemilability due to the changing strength and/or nature of interaction between donor atoms and the Pd center.

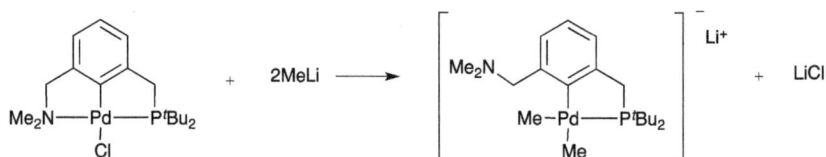

Scheme 1. Reaction of unsymmetrical pincer palladacycles with MeLi showing the hemilability of the nitrogen donor atom arm by Wendt et al. [22].

Pincer palladacycles exist in a square planar configuration at the Pd(II) center, and a key factor determining the strength of the interaction between Pd and the donor atoms is the *trans* influence, potentially affecting hemilability of donor atom arms. The *trans* influence is defined by Pidcock [23] as "the tendency of a ligand to weaken the bond *trans* to itself". The "*trans* influence" affects the structure in the ground, or thermodynamic state. Therefore, sometimes, it is called the thermodynamic *trans* effect, while the "*trans* effect" is related to the kinetic rate of reaction, depending on substitution of the bond *trans* to itself. The *trans* influence has been used to explain the stability of square planar complexes [24], while the *trans* effect has been used to study reaction pathways [25]. There are many experimental studies into the *trans* influence, generally using spectroscopic or X-ray crystallographic methods [26–28]. Additionally, density functional theory (DFT) structure optimization and molecular orbital analysis have been employed in the study of the *trans* influence and give a good explanation of the *trans* influence in organometallic complexes [29–33].

In this study, we have investigated the *trans* influence in both model and experimentally-characterized unsymmetrical pincer palladacycles, using DFT calculations and quantum theory of atoms in molecules (QTAIM) analysis. Additionally, in order to determine the effect of varying the substituents on the donor atom, we have synthesized a library of unsymmetrical SCN pincer palladacycles, providing the opportunity to vary the steric and electronic properties on the sulfur atom. We have then used these palladacycles to further investigate the *trans* influence using DFT.

2. Results and Discussion

2.1. SCN Pincer Palladacycle Synthesis

Our previous work has demonstrated a novel synthetic route to an unsymmetrical SCN pincer palladacycle, via a key biaryl benzyl bromide intermediate **5** (Scheme 2) [13]. By changing the sulfur nucleophile in step c (Scheme 2), the ability to synthesize a library of SCN pincer ligands is possible. This provides the opportunity to vary the thioether substituent to tune the steric and electronic properties of the sulfur atom, which will be bound to the palladium atom in the resulting palladacycle. The SCN ligands then undergo C–H bond activation with in situ-generated $Pd(MeCN)_4(BF_4)$ [13,34] synthesizing a library of SCN pincer palladacycles **1b–1f** (Scheme 2, Table 1).

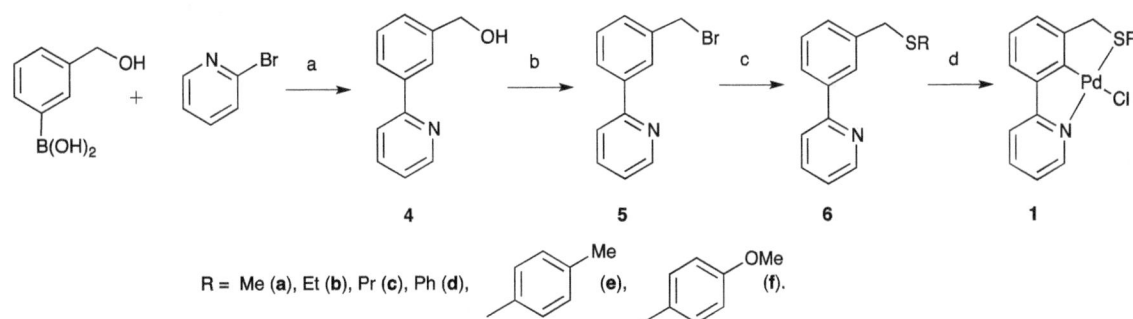

Scheme 2. A synthesis of SCN pincer palladacycles **1b–1f** via a key biaryl benzyl bromide intermediate **5**, based on the previous synthesis of **1a**. (Step a = $Pd(PPh_3)_4$, K_3PO_4, Tol/EtOH/H_2O; b = 48% HBr in H_2O; c = NaSMe, EtOH; d = (i) $PdCl_2$, $AgBF_4$, MeCN; and (ii) NaCl, H_2O/MeCN).

Table 1. SCN pincer palladacycle synthesis yields.

Entry	Palladacycle	Ligand Synthesis Conditions	Ligand Synthesis Isolated Yield/%	Palladacycle Synthesis Yield/%
1	**1b**	A	72	83
2	**1c**	B	>99	85
3	**1d**	B	99	71
4	**1e**	C	51	89
5	**1f**	B	60	54

A = NaH, DMF, MW 150 °C 15 min, B = NaH, DMF, MW 150 °C 20 min, C = NEt_3, EtOH, MW 150 °C 20 min.

The SCN ligand syntheses presented in Table 1 reveal excellent to moderate yields, followed by C–H bond activation, also in moderate to excellent yield. Single crystals of palladacycles **1b–d** and **1f** were obtained by slow evaporation of dichloromethane (DCM) from a saturated solution, and were characterized by X-ray crystallography (Figure 2). The CCDC numbers for the structures are 1486787 for **1b**, 1486788 for **1c**, 1486789 for **1d** and 1486790 for **1f**. The data can be obtained free of charge from The Cambridge Crystallographic Data Centre via www.ccdc.cam.ac.uk/structures.

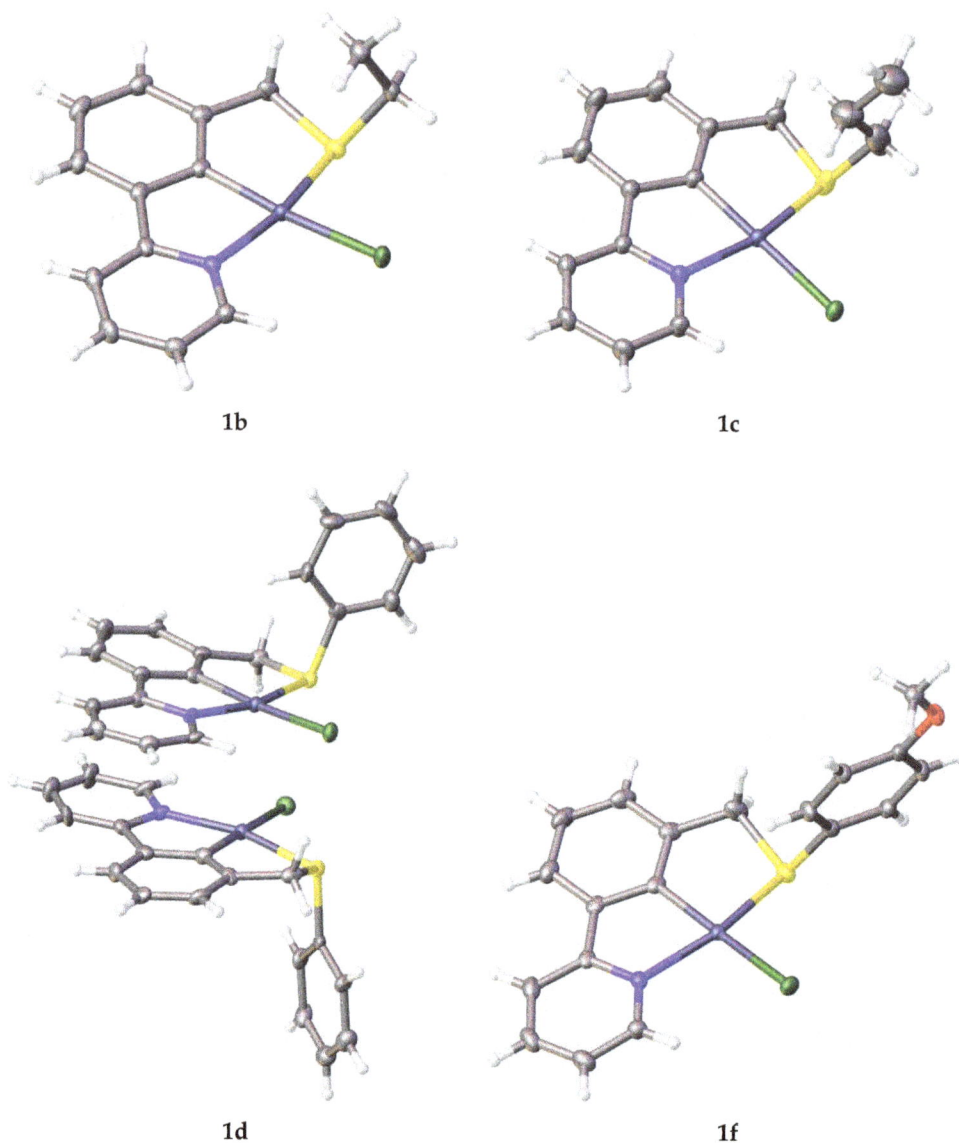

1b   1c

1d   1f

Figure 2. X-ray crystallographic structures of **1b–d**, **1f**.

2.2. Investigaing the Trans Influence

To determine the accuracy of the Perdew–Burke–Ernzerhof exchange–correlation functional (PBE) for optimizing the YCY′ pincer complexes, we have analyzed the mean signed errors (MSE), which is the average of the deviation between calculation and experiment, and the mean unsigned errors (MUE), which is the average of the absolute deviation between calculated and experimental Pd–L bond lengths (L = Y, Y′, C and Cl). The MSE for **1a–1d**, **1f** is 0.001 Å, for **2a–2b** is 0.012 Å and for **3a–3c** is 0.001 Å. The MUE for **1a–1d**, **1f** is 0.017 Å, for **2a–2b** is 0.012 Å and for **3a–3c** is 0.001 Å.

A topological analysis of the electron density was performed using QTAIM. The bond path is a single line of locally-maximum density linking the nuclei of any two bonded atoms in a molecule [35]. The minimum along this path is defined as the bond critical point (BCP) and the magnitude of the electron density $\rho(\mathbf{r})$ at this point can be used to determine the strength of the interaction.

Several AIM parameters determined at the Pd–Y and Pd–Y′ BCPs are provided in the Supplementary Information (electron density $\rho(\mathbf{r})$, Laplacian of the density $\nabla^2\rho(\mathbf{r})$, delocalization index, ellipticity, bond degree parameter, etc.) which can be used to determine the nature and strength of the interaction; the conclusions regarding the nature of the bonding are in complete agreement with our previous work on the nature of the bonding in symmetrical pincer palladacycles and, so, are not presented again here [36]. In the present work, the focus is on the *trans* influence and, so, the magnitude of the electron density $\rho(\mathbf{r})$ at the BCP is used to determine the increase or decrease in the strength of the Pd–Y interaction when the ligand *trans* to it is varied.

2.2.1. *Trans* Influence in Model Palladacycles I–III

Normally, the *trans* influence has been studied in systems with four monodentate ligands coordinated to the metal center to form a square planar complex. Furthermore, a *trans* Pt–Cl bond length, situated *trans* to the donor atom arm, of a square planar complex is normally used to consider the *trans* influence [37–40]. From the unsymmetrical SCN pincer palladacycle structures, we do not have a Cl atom for monitoring the strength of the *trans* influence in this way. Therefore, first, simple model palladacycles I–III (Figure 3) have been studied using DFT to evaluate the *trans* influence in CY palladacycles before studying the unsymmetrical YCY' pincer palladacycles. The models I–III contain a Cl ligand *trans* to a donor atom group (NMe_2, SMe_2, and PMe_2, respectively) to monitor the strength of the *trans* influence. A topological analysis of the electron density was performed using QTAIM and the magnitude of the electron density $\rho(\mathbf{r})$ at the bond critical point (BCP), the minimum along the bond path between interacting atoms, was used to determine the strength of the Pd–Cl interaction. A larger $\rho(\mathbf{r})$ value corresponds to a stronger interaction between atoms [41] and, therefore, can be used to study the *trans* influence in palladacycles I–III. When $\rho(\mathbf{r})$ at the BCP of Pd–Cl bond has a high value (strong interaction), it indicates that the donor atom *trans* to Cl has a weak *trans* influence, whereas a low $\rho(\mathbf{r})$ value (weak interaction) indicates that the donor atom *trans* to the Pd–Cl bond has a strong *trans* influence. A relative change in bond length is a physical manifestation that indicates the strength of the *trans* influence. When the Pd–Cl bond is situated *trans* to a donor atom that exhibits a strong *trans* influence, the Pd–Cl bond lengthens compared to when the Pd–Cl bond is situated *trans* to a weak *trans* influence donor atom. The data provided in Table 2 show that the $\rho(\mathbf{r})$ value of the Pd–Cl bond of III is smaller than that in II, which is smaller than in I, indicating that the *trans* influence of PMe_2 is greater than that of SMe which is greater than NMe_2. The $\rho(\mathbf{r})$ data is supported by the bond lengths, with I having the shortest Pd–Cl bond length and III a significantly longer Pd–Cl bond length than in I and II, again demonstrating the stronger PMe_2 *trans* influence. Based on this analysis the ordering of the *trans* influence series is $PMe_2 > SMe > NMe_2$.

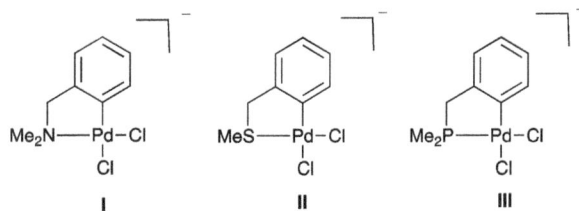

Figure 3. Model palladacycles I–III studied to investigate the *trans* influence.

Table 2. The electron density $\rho(\mathbf{r})$ and Pd–Cl bond lengths.

Compound	$\rho(\mathbf{r})$ of Pd–Cl (a.u.)	Pd–Cl Bond Length (Å)
I	0.080	2.334
II	0.077	2.352
III	0.070	2.395

2.2.2. *Trans* Influence in Model Unsymmetrical YCY' Pincer Palladacycles

In order to extend our investigation of the *trans* influence to unsymmetrical pincer palladacycles, the palladacycles IV–VI have been studied using DFT and QTAIM, and their bond strengths and bond lengths compared to previous results found for symmetrical pincer palladacycles **PdNCN**, **PdSCS** and **PdPCP** [36] (Figure 4). Considering the $\rho(\mathbf{r})$ value at the BCP of the Pd–Y bond in IV–VI, the $\rho(\mathbf{r})$

value of the Pd–P bond of **V** (0.110 a.u) and **VI** (0.114 a.u.) are greater compared to the $\rho(\mathbf{r})$ values for the Pd–P bond in the **PdPCP** (0.101 a.u.) [36]. This is due to the weaker *trans* influence of N and S, compared to P, leading to stronger Pd–P bonds in **V** and **VI** (Table 3). The $\rho(\mathbf{r})$ value of the Pd–S bond of **IV** (0.097 a.u.) increases, whereas the $\rho(\mathbf{r})$ value of the Pd–S bond of **V** (0.082 a.u.) decreases, compared to the $\rho(\mathbf{r})$ value of the Pd–S bond in the **PdSCS** (0.091 a.u.), therefore showing that S has a moderate *trans* influence. Furthermore, the $\rho(\mathbf{r})$ values of the Pd–N bond of **IV** (0.083 a.u.) and **VI** (0.075 a.u.) decrease compared to the $\rho(\mathbf{r})$ value of the Pd–N bond in the **PdNCN** (0.086 a.u.) [36] indicating that P and S exhibit a stronger *trans* influence than N.

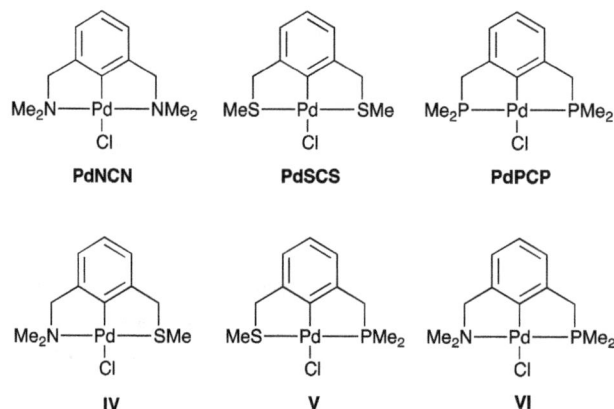

Figure 4. Symmetrical NCN, SCS, and PCP pincer palladacycles (**PdNCN**, **PdSCS** and **PdPCP**) [36] and model unsymmetrical SCN (**IV**), PCS (**V**), and PCN (**VI**) pincer palladacycles.

Table 3. Electron density $\rho(\mathbf{r})$ in symmetrical and unsymmetrical pincer palladacycles (values are in atomic units). The donor atom is shown in bold for each side arm, Y and Y'.

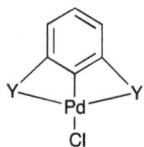

PdYCY'	Y	Y'	$\rho(\mathbf{r})$ of Pd–Y	$\rho(\mathbf{r})$ of Pd–Y'
PdNCN	Me$_2$**N**CH$_2$	Me$_2$**N**CH$_2$	0.086	0.086
PdSCS	Me**S**CH$_2$	Me**S**CH$_2$	0.091	0.091
PdPCP	Me$_2$**P**CH$_2$	Me$_2$**P**CH$_2$	0.101	0.101
IV	Me**S**CH$_2$	Me$_2$**N**CH$_2$	0.097	0.083
V	Me$_2$**P**CH$_2$	Me**S**CH$_2$	0.110	0.082
VI	Me$_2$**P**CH$_2$	Me$_2$**N**CH$_2$	0.114	0.075
1a	Me**S**CH$_2$	2-**N**C$_5$H$_4$	0.096	0.098
1b	Et**S**CH$_2$	2-**N**C$_5$H$_4$	0.095	0.098
1c	Pr**S**CH$_2$	2-**N**C$_5$H$_4$	0.095	0.098
1d	Ph**S**CH$_2$	2-**N**C$_5$H$_4$	0.092	0.098
1e	(p-MeC$_6$H$_4$)**S**CH$_2$	2-**N**C$_5$H$_4$	0.092	0.098
1f	(p-MeOC$_6$H$_4$)**S**CH$_2$	2-**N**C$_5$H$_4$	0.092	0.098
2a	Ph$_2$**P**O	2-**N**C$_5$H$_4$	0.114	0.087
2b	Ph$_2$**P**OCH$_2$	2-**N**C$_5$H$_4$	0.113	0.089
3a	Me$_2$**N**CH$_2$	2-**N**C$_5$H$_4$	0.086	0.102
3b	Et$_2$**N**CH$_2$	2-**N**C$_5$H$_4$	0.085	0.102
3c	(C$_4$H$_8$O)**N**CH$_2$	2-**N**C$_5$H$_4$	0.084	0.102

Supporting the $\rho(\mathbf{r})$ value results, the bond lengths of Pd–Y and Pd–Y' are reported in Table 4. When the donor ligand Y has a *trans* influence the Pd–Y' bond distance increases (and the $\rho(\mathbf{r})$ value decreases) indicating a weakened interaction. By comparing with the symmetrical YCY pincer palladacycles it can be seen that the P donor ligand has a *trans* influence on the S donor ligand

and the N donor ligand, and that the S donor ligand has a *trans* influence on the N donor ligand. For example, in **VI**, the PCN palladacycle, the P donor ligand has a strong influence on the N donor ligand *trans* to it, which manifests as an increased Pd–N (2.203 Å) bond distance compared to the Pd–N bond in **PdNCN** (2.140 Å), and a commensurate decrease in the Pd–P bond distance (2.222 Å) compared to the Pd–P bond length in **PdPCP** (2.287 Å) (Table 4). The results confirm the conclusion from the model systems with Cl as a reference, that P exhibits the greatest *trans* influence and N the least.

Table 4. Calculated and experimental Pd–Y and Pd–Y' bond distances in symmetrical and unsymmetrical pincer palladacycles (bond distances are in Å). The donor atom is shown in bold for each side arm, Y and Y'.

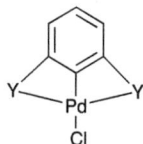

PdYCY'	Y	Y'	Calculation		X-ray		Ref. X-ray
			Pd–Y	Pd–Y'	Pd–Y	Pd–Y'	
PdNCN	Me$_2$**N**CH$_2$	Me$_2$**N**CH$_2$	2.140	2.140	2.103(3)	2.102(3)	[6]
PdSCS	Me**S**CH$_2$	Me**S**CH$_2$	2.313	2.313	2.2831(11)	2.2911(11)	[7]
PdPCP	Me$_2$**P**CH$_2$	Me$_2$**P**CH$_2$	2.287	2.287	n/a	n/a	n/a
IV	Me**S**CH$_2$	Me$_2$**N**CH$_2$	2.285	2.156	n/a	n/a	n/a
V	Me$_2$**P**CH$_2$	Me**S**CH$_2$	2.240	2.364	n/a	n/a	n/a
VI	Me$_2$**P**CH$_2$	Me$_2$**N**CH$_2$	2.222	2.203	n/a	n/a	n/a
1a	Me**S**CH$_2$	2-**N**C$_5$H$_4$	2.288	2.074	2.291(8)	2.09(3)	[13]
1b	Et**S**CH$_2$	2-**N**C$_5$H$_4$	2.290	2.076	2.2638(4)	2.0672(13)	*
1c	Pr**S**CH$_2$	2-**N**C$_5$H$_4$	2.291	2.076	2.2705(7)	2.066(2)	*
1d	Ph**S**CH$_2$	2-**N**C$_5$H$_4$	2.303	2.078	2.2846(17)	2.069(5)	*
1e	(*p*-MeC$_6$H$_4$)**S**CH$_2$	2-**N**C$_5$H$_4$	2.302	2.078	n/a	n/a	n/a
1f	(*p*-MeOC$_6$H$_4$)**S**CH$_2$	2-**N**C$_5$H$_4$	2.303	2.078	2.2674(5)	2.0708(15)	*
2a	Ph$_2$**P**O	2-**N**C$_5$H$_4$	2.219	2.129	2.2028(6)	2.1216(18)	[14]
2b	Ph$_2$**P**OCH$_2$	2-**N**C$_5$H$_4$	2.232	2.114	2.2159(7)	2.103(2)	[14]
3a	Me$_2$**N**CH$_2$	2-**N**C$_5$H$_4$	2.145	2.060	2.105(6)	2.062(5)	[14]
3b	Et$_2$**N**CH$_2$	2-**N**C$_5$H$_4$	2.149	2.063	2.1145(16)	2.0639(16)	[14]
3c	(C$_4$H$_8$O)**N**CH$_2$	2-**N**C$_5$H$_4$	2.154	2.060	2.1239(19)	2.0521(19)	[14]

* The result from this work. n/a: not available.

Based on the $\rho(\mathbf{r})$ values and Pd–Y bond lengths, the ordering of the *trans* influence series is PMe$_2$ > SMe > NMe$_2$. This is in good agreement with that of Kapoor and Kakkar's study [40] into the square planar Pt complexes using DFT calculations. Their results showed a *trans* influence series in order of P > S > N. Moreover, Sajith and Suresh [42] studied the correlation between $\rho(\mathbf{r})$ and *trans* influence in a square planar Pd complex, showing good linear relation between $\rho(\mathbf{r})$ and *trans* influence, with a *trans* influence series of PMe$_3$ > SMe$_2$ > NH$_3$.

2.2.3. *Trans* Influence in Experimentally-Characterized Unsymmetrical YCY' Pincer Palladacycles

In this section DFT and the QTAIM method is used to study the *trans* influence in **1a** (PdSCN), **2a** (PdPCN), and **3a** (PdN'CN). By comparing the Pd–N bond length in the structures **1a**, **2a**, and **3a** (optimized and experimental), the Pd–N bond is longest in **2a** and shortest in **3a** (Table 4). In addition, the smallest $\rho(\mathbf{r})$ values for the Pd–N bond is in **2a** (0.087 a.u.), while the largest is found in **3a** (0.102 a.u.) with **1a** (0.098 a.u.) intermediate (Table 3). The different Pd–N bond lengths and strengths demonstrate the difference in *trans* influence due to the nature of the donor atom of the Pd–Y bond. These results further confirm that the P donor ligand exhibits the strongest *trans* influence, while the N donor ligand has the weakest *trans* influence and that the *trans* influence series for the unsymmetrical pincer palladacycles considered is P > S > N.

The N donor ligand in the experimentally-characterized unsymmetrical SCN pincer palladacycle (Figure 1) is a pyridine rather than the amine considered in the previous section (**IV**). The change in electronic and steric effects when replacing NMe_2 (**IV**) with pyridine (**1a**) in a SCN pincer palladacyle is reflected in the bond strength: $\rho(\mathbf{r})$ value of the Pd–NMe_2 bond is 0.083 a.u. in **IV** whereas the Pd–pyr is 0.098 a.u. in **1a**, and the Pd–NMe_2 bond length is 2.156 Å in **IV** and the Pd–pyr bond length is 2.074 Å in **1a**, demonstrating the stronger Pd–pyridine bond (Tables 3 and 4). However, this does not appear to effect the *trans* influence exerted on the SMe ligand when *trans* to these N donor ligands. The $\rho(\mathbf{r})$ value of the Pd–S bond is 0.091 a.u. in **PdSCS** and increases to 0.097 a.u. in **IV** and 0.096 a.u. in **1a**, and the bond length in **PdSCS** is 2.313 Å and shortens to 2.285 Å in **IV** and 2.288 Å in **1a**. Thus, in both **IV** and **1a** the Pd–S bond is strengthened relative to the symmetric **PdSCS** analog and, thus, can only be attributed to the effect of the N-donor ligand *trans* to it.

Furthermore, by comparing **PdNCN** where N = NMe_2, to PdNCN' (**3a**), where one of the amine ligands has been replaced by pyridine, we can assess the *trans* influence in an unsymmetrical pincer palladacycle where the donor atom is the same (N) for distinctly different donor ligands (NMe_2 and pyr). In **3a** the $\rho(\mathbf{r})$ value of the Pd–NMe_2 bond has not changed and the bond length has increased insignificantly (0.005 Å) from that in the symmetric **PdNCN** palladacycle. Therefore, we can conclude that, although the electronic and steric effects of the pyridine result in a considerably stronger bond to the Pd center, this stronger bond does not exert a *trans* influence on the amine donor ligand. Thus, it would appear the nature of the donor atom is the sole driver for the *trans* influence.

2.2.4. *Trans* Influence on Unsymmetrical Pincer Palladacycles: Donor Atom Substituent Effects

To determine whether the *trans* influence is induced when the substituents on the donor atom are varied, thereby introducing subtle electronic effects, the library of SCN pincer palladacycles synthesized in the present work (**1b–1f**), along with **1a**, have been investigated computationally to determine the influence of the thioether group on the coordinated pyridine *trans* to it. The Pd–N bond distances (experimental and calculated) show very little change when the substituent on the S atom is changed (bond distance differences <0.005 Å, with the exception of the experimental Pd–N bond length for **1a**) (Table 4). Similarly, the $\rho(\mathbf{r})$ values at the Pd–N BCP in the SCN pincer palladacycles are unaffected by changing substitution on donor atom.

Furthermore, when the substituent is changed on the P (**2a** and **2b**) (which both incorporate the phosphinite donor group) or the N (**3a–3c**), Figure 1, it does not alter the *trans* influence on the Pd–pyr interaction within the PCN or N'CN pincer palladacycles. The $\rho(\mathbf{r})$ values for the Pd–pyr bond *trans* to the Pd–N' bond is independent of the nature of the N' ligand, and although the interactions ($\rho(\mathbf{r})$ and bond length) due to the Pd–P ligands exhibit a slight difference (0.002 a.u. and 0.015 Å) they are extremely small.

3. Experimental Section

3.1. General Details

Solvents and chemicals were purchased from Sigma-Aldrich (Merck KGaA, Damstadt, Germany), VWR International (VWR, Radnor, PA, USA), Fisher Scientific (Fisher Scientific UK Ltd., Loughborough, UK) and Fluorochem (Fluorochem Ltd., Hadfield, UK) and used without further purification, with reactions taking place open to atmosphere and moisture.

3.2. Instrumentation

[1]H and ^{13}C spectra were recorded on either a Varian 400 or 500 MHz spectrometer (Agilent Technologies, Yarnton, UK). High resolution mass spectrometry (HRMS) data were obtained on an electrospray ionization (ESI) mass spectrometer using a Bruker Daltonics Apex III (Brucker, Billerica, MA, USA), with source Apollo ESI, using methanol as the spray. Flash chromatography was performed on an automated ISCO RF75 (Teledyne ISCO Inc., Licoln, NE, USA).

Gas chromatography (GC) measurements were obtained using a Perkin Elmer Autosystem XL Gas Chromatograph (PerkinElmer Inc., Waltham, MA, USA), utilizing a flame ionization detector, and a Supelco MDN-5S 30 m × 0.25 mm × 0.25 μm column, with a He mobile phase. Elemental analyses were run by the London Metropolitan University Elemental Analysis Service (ThermoFisher Scientific, Waltham, MA, USA). Crystal structures were obtained by the UK National Crystallography Service at the University of Southampton [43].

3.3. Procedure

2-3-[(Ethylsulfanyl)methyl]phenylpyridine, **6b**.   Under an argon atmosphere, ethanethiol (2.42 mmol, 0.179 mL) and sodium hydride (2.41 mmol, 58 mg) were dissolved in dry DMF (dimethylformamide, 3 mL) and stirred at room temperature in a sealed microwave vial for 15 min. 2-[3-(Bromomethyl)phenyl]pyridine, **5** (1.61 mmol, 400 mg) in dry DMF (3 mL) was then added, and stirred under microwave irradiation (maximum power 300 W, dynamic heating) at 150 °C for 15 min. After cooling, the solvent was removed in vacuo and the crude mixture was diluted in H_2O (25 mL) and DCM (25 mL). The product was extracted with DCM (2 × 25 mL), washed with H_2O (5 × 25 mL) and brine (25 mL). The organic layers were dried over anhydrous $MgSO_4$, filtered, and concentration in vacuo. The crude product was purified using flash column chromatography (7:3 DCM:EtOAc) yielding 263 mg of the expected product, **6c**, as a yellow oil in 71% yield. ^{1}H NMR (500 MHz), Chloroform-d δ (ppm): 8.70 (d, J = 4.8 Hz, 1H), 7.96 (s, 1H), 7.86 (d, J = 7.5 Hz, 1H), 7.77–7.73 (m, 2H), 7.43 (dd, J = 7.5, 7.5 Hz, 1H), 7.39 (d, J = 7.5 Hz, 1H), 7.24 (ddd, 6.3, 4.8, 2.3 Hz, 1H), 3.81 (s, 2H), 2.48 (q, J = 7.5 Hz, 2H), 1.25 (t, J = 7.5 Hz, 3H). ^{13}C NMR (126 MHz), Chloroform-d δ (ppm): 157.3, 149.7, 139.6, 139.2, 136.7, 129.4, 128.9, 127.4, 125.5, 122.1, 120.6, 36.0, 25.4, 14.4. HRMS (m/z). Calc. for $[C_{14}H_{15}NS + H]^+$ 230.0998. Found 230.0998.

2-3-[(Propylsulfanyl)methyl]phenylpyridine, **6c**. Same methodology as **6b**, using propane-1-thiol (1.97 mmol, 0.178 mL), and reacting for 20 min in the microwave. After workup, 300 mg of the expected product, **6c** was found, without purification in >99% yield as a yellow oil. ^{1}H NMR (500 MHz), Chloroform-d δ (ppm): 8.70 (d, J = 4.8 Hz, 1H), 7.96 (s, 1H), 7.86 (d, J = 7.5 Hz, 1H), 7.78–7.73 (m, 2H), 7.43 (dd, J = 7.5, 7.5 Hz, 1H), 7.39 (d, J = 7.5 Hz, 1H), 7.24 (ddd, J = 6.5, 4.8, 2.1 Hz, 1H), 3.79 (s, 2H), 2.44 (t, J = 7.2 Hz, 2H), 1.64–1.57 (m, 2H), 0.96 (t, J = 7.3 Hz, 3H).). ^{13}C NMR (126 MHz), Chloroform-d δ (ppm): 157.2, 149.6, 139.6, 139.3, 136.7, 129.4, 128.8, 127.4, 125.5, 122.1, 120.6, 36.3, 33.6, 22.6, 13.5. HRMS (m/z). Calc. for $[C_{15}H_{17}NS + H]^+$ 244.1154. Found 244.1155.

2-3-[(Phenylsulfanyl)methyl]phenylpyridine, **6d**. Same methodology as **6b**, using benzenethiol (1.86 mmol, 0.190 mL). After workup, 418 mg of the expected product, **6d** as a yellow oil in 99% yield. ^{1}H NMR (500 MHz), Chloroform-d δ (ppm): 8.70 (d, J = 4.8 Hz, 1H), 7.93 (s, 1H), 7.87 (d, J = 7.6 Hz, 1H), 7.76–7.72 (m, 2H), 7.67 (d, J = 8.0 Hz, 1H), 7.39 (dd, J = 7.6, 7.6 Hz, 1H), 7.36–7.33 (m, 2H), 7.27–7.21 (m, 3H), 7.20–7.16 (m, 1H), 4.20 (s, 2H). ^{13}C NMR (126 MHz), Chloroform-d δ (ppm): 157.2, 149.6, 139.6, 138.0, 136.7, 130.0 (2C), 129.3 (2C), 128.9, 128.8 (2C), 127.4, 126.4, 125.8, 122.1, 39.2.

2-(3-[(4-Methylphenyl)sulfanyl]methylphenyl)pyridine, **6e**.   Under an argon atmosphere, 4-methylbenzenethiol (0.70 mmol, 87 mg) and trimethylamine (0.70 mmol, 0.099 mL) were dissolved in dry EtOH (2 mL) and stirred at room temperature in a sealed microwave vial for 15 min. 2-[3-(Bromomethyl)phenyl]pyridine, **5** (0.44 mmol, 110 mg) in dry EtOH (2 mL) was then added and the mixture was stirred under microwave irradiation (maximum power, 300 W, dynamic heating) at 150 °C for 20 min. After cooling, the solvent was removed in vacuo and the crude mixture diluted with H_2O (25 mL) and EtOAc (25 mL). The product was extracted with EtOAc (2 × 25 mL), washed with H_2O (2 × 25 mL) and brine (25 mL). The organic layers were dried over anhydrous $MgSO_4$, filtered and concentrated in vacuo. The crude product was purified by flash chromatography (8:2 hexane:Et_2O) yielding 65 mg of the expected product, **6e** as a yellow oil in 51% yield. ^{1}H NMR (500 MHz), Chloroform-d δ (ppm): 8.70 (d, J = 4.8 Hz, 1H), 7.90 (s, 1H), 7.87 (d, J = 7.6 Hz, 1H), 7.72 (m, 1H), 7.65 (d, J = 8.0 Hz, 1H), 7.39 (dd, J = 7.6, 7.6 Hz, 1H), 7.32 (d, J = 7.6 Hz, 1H), 7.25

(d, J = 8.1 Hz, 2H), 7.21 (ddd, J = 7.4, 4.8, 1.2 Hz), 7.06 (d, J = 8.1 Hz, 2H), 4.15 (s, 2H), 2.30 (s, 3H). ^{13}C NMR (126 MHz), Chloroform-d δ (ppm): 157.2, 149.6, 139.6, 138.3, 136.6, 132.4, 131.5, 130.9 (2C), 129.6 (2C), 129.3, 128.8, 127.4, 125.7, 122.1, 120.5, 39.9, 21.0. HRMS (m/z). Calc. for [C$_{19}$H$_{17}$NS + H]$^+$ 292.1154. Found 292.1151.

2-(3-[(4-Methoxyphenyl)sulfanyl]methylphenyl)pyridine, **6f**.   Same method as **6b**, using 4-methoxybenzenethiol (1.10 mmol, 0.136 mL). The crude product was purified using flash column chromatography (9:1 DCM:hexane) yielding 203 mg of the expected product, **6f** as a yellow oil in 60% yield. ^{1}H NMR (500 MHz), Chloroform-d δ (ppm): 8.69 (ddd, J = 4.8, 1.8, 0.9 Hz, 1H), 7.86 (d, J = 7.7 Hz, 1H), 7.81 (s, 1H), 7.73 (ddd, J = 9.7, 7.9, 1.8 Hz, 1H), 7.65 (d, J = 7.7 Hz, 1H), 7.37 (dd, J = 7.7, 7.7 Hz, 1H), 7.28 (d, J = 8.8 Hz, 2H), 7.25–7.21 (m, 2H), 6.79 (d, J = 8.8 Hz, 2H), 4.07 (s, 2H), 3.76 (s, 3H). ^{13}C NMR (126 MHz), Chloroform-d δ (ppm): 159.3, 157.2, 149.6, 139.5, 138.6, 136.6, 134.2 (2C), 129.4, 128.8, 127.5, 126.0 (2C), 125.6, 122.1, 120.6, 114.5, 55.3, 41.3. HRMS (m/z). Calc. for [C$_{19}$H$_{17}$NOS + H]$^+$ 308.1104. Found 308.1109.

2-3-[(Ethylsulfanyl)methyl]phenylpyridine chloro-palladacycle, **1b**. Under an argon atmosphere, PdCl$_2$ (1.17 mmol, 208 mg) was dissolved in dry MeCN (10 mL) and heated under reflux until a red solution had formed. AgBF$_4$ (2.36 mmol, 460 mg) in dry MeCN (5 mL) was added to the PdCl$_2$ solution and heated under reflux for 2 h, forming a white precipitate. The precipitate was filtered off, and **6b** (1.13 mmol, 260 mg) dissolved in dry MeCN (10 mL), was added to the filtrate and heated under reflux for 4 h. The solution was cooled to room temperature, filtered over celite, and the solvent removed in vacuo. The crude solid was dissolved in MeCN (5 mL), and NaCl (26.0 mmol, 1.52 g) dissolved in H$_2$O (5 mL) was added, and stirred at room temperature for 3 h. The solvent was removed in vacuo, and the crude mixture dissolved in DCM (25 mL) and H$_2$O (25 mL). The crude product was extracted with DCM (2 × 25 mL), washed with H$_2$O (2 × 25 mL) and brine (25 mL), and dried over anhydrous Na$_2$SO$_4$. The mixture was filtered over celite, and the solvent removed in vacuo, yielding 347 mg of the expected product, **1b** as a yellow solid in 83% yield. ^{1}H NMR (500 MHz), Chloroform-d δ (ppm): 9.15 (d, J = 5.5 Hz, 1H), 7.84 (ddd, J = 7.8, 7.8 Hz, 1H), 7.64 (d, J = 7.8 Hz, 1H), 7.33 (d, J = 7.7 Hz, 1H), 7.26–7.23 (m, 1H), 7.08 (dd, J = 7.7, 7.7 Hz, 1H), 7.03 (d, J = 7.7 Hz, 1H), 4.25 (bs, 2H), 3.20 (q, J = 7.4 Hz, 2H), 1.57 (t, J = 7.4 Hz, 3H). ^{13}C NMR (126 MHz), Chloroform-d δ (ppm): 165.5, 165.3, 150.5, 148.1, 144.4, 139.0, 125.0, 124.7, 122.9, 122.2, 118.7, 45.8, 33.8, 14.8. HRMS (m/z). Calc. for [C$_{14}$H$_{14}$NPdS]$^+$ 333.9876. Found 333.9878. Elemental Analysis. Calc. (%) for C$_{14}$H$_{14}$NPdSCl: C 45.42, H 3.81, N 3.78; found C 45.50, H 3.75, N 3.83.

2-3-[(Propylsulfanyl)methyl]phenylpyridine chloro-palladacycle, **1c**. Same method as **1b** using **6c** (0.55 mmol, 113 mg), yielding 179 mg of the expected product, **1c** as a yellow solid in 85% yield. ^{1}H NMR (500 MHz), Chloroform-d δ (ppm): 9.11 (d, J = 5.5 Hz, 1H), 7.82 (ddd, J = 7.8, 7.8, 1.7 Hz, 1H), 7.62 (d, J = 7.7 Hz, 1H), 7.30 (d, J = 7.8 Hz, 1H), 7.22 (ddd, J = 7.5, 5.5, 1.3 Hz, 1H), 7.05 (dd, 7.7, 7.7 Hz, 1H), 7.00 (d, J = 7.7 Hz, 1H), 4.27 (bs, 2H), 3.15 (t, J = 7.8 Hz, 2H), 1.96 (m, 2H), 1.07 (t, J = 7.4 Hz, 3H). ^{13}C NMR (126 MHz), Chloroform-d δ (ppm): 165.5, 165.3, 150.5, 148.1, 144.4, 139.0, 125.0, 124.6, 122.9, 122.1, 118.7, 46.6, 41.4, 23.3, 13.3. HRMS (m/z). Calc. for [C$_{15}$H$_{16}$NPdS]$^+$ 348.0033. Found 348.0032. Elemental Analysis. Calc. (%) for C$_{15}$H$_{16}$NPdSCl: C 46.89, H 4.20, N 3.65; found: C 47.02, H 4.08, N 3.56.

2-3-[(Phenylsulfanyl)methyl]phenylpyridine chloro-palladacycle, **1d**. Same method as **1b** using **6d** (1.51 mmol, 418 mg). The crude product was purified using flash column chromatography (100% consisting of 98:2 DCM:MeOH) yielding 446 mg of the expected product **1d** as a yellow solid in 71% yield. ^{1}H NMR (500 MHz), Chloroform-d δ (ppm): 9.14 (d, J = 5.5 Hz, 1H), 7.91–7.89 (m, 2H), 7.83 (ddd, J = 7.7, 7.7, 1.7 Hz, 1H), 7.62 (d, J = 7.7 Hz, 1H), 7.36–7.33 (m, 3H), 7.29 (d, J = 7.7 Hz, 1H), 7.20 (ddd, J = 7.7, 5.5, 1.2 Hz, 1H), 7.06 (dd, J = 7.7, 7.7 Hz, 1H), 7.00 (d, J = 7.7 Hz, 1H), 4.63 (s, 2H). ^{13}C NMR (126 MHz), Chloroform-d δ (ppm): 166.0, 165.5, 150.8, 147.8, 144.6, 139.1, 132.8, 131.9 (2C), 129.9, 129.6 (2C), 124.9, 124.8, 122.9, 122.3, 118.8, 53.1. HRMS (m/z). Calc. for [C$_{18}$H$_{14}$NPdS]$^+$ 381.9876.

Found 381.9876. Elemental Analysis. Calc. (%) for $C_{18}H_{14}NPdSCl$: C 51.69, H 3.37, N 3.35; found C 51.50, H 3.28, N 3.41.

2-(3-[(4-Methylphenyl)sulfanyl]methylphenyl)pyridine, **1e**.   Same method as **1b**, using **6e** (0.54 mmol, 158 mg). ^{1}H NMR (500 MHz), Chloroform-d δ (ppm): 9.19 (d, J = 5.5 Hz, 1H), 7.84 (dd, J = 7.6, 7.6 Hz, 1H), 7.79 (d, J = 8.1 Hz, 2H), 7.64 (d, J = 7.6 Hz, 1H), 7.33 (d, J = 7.6 Hz, 1H), 7.23–7.21 (m, 1H), 7.16 (d, J = 8.1 Hz, 2H), 7.08 (dd, J = 7.6, 7.6 Hz, 1H), 7.00 (d, J = 7.6 z, 1H), 4.60 (bs, 2H), 2.32 (s, 3H). ^{13}C NMR (126 MHz), Chloroform-d δ (ppm): 166.0, 165.5, 150.7, 147.9, 144.5, 140.4, 139.1, 132.0 (2C), 130.3 (2C), 129.4, 124.8, 122.9, 122.2, 118.8, 53.5, 21.2. HRMS (m/z). Calc. for $[C_{19}H_{16}NPdS]^+$ 396.0033. Found 396.0050. Elemental Analysis. Calc. (%) for $C_{19}H_{16}NPdSCl$: C 52.63, H 3.84, N 3.29; found C 52.79, H 3.73, N 3.24.

2-(3-[(4-Methoxyphenyl)sulfanyl]methylphenyl)pyridine, **1f**. Same method as **1b**, using **1e** (0.62 mmol, 190 mg). ^{1}H NMR (500 MHz), Chloroform-d δ (ppm): 9.18 (d, J = 5.5 Hz, 1H), 7.86–7.83 (m, 3H), 7.65 (d, J = 7.9 Hz, 1H), 7.33 (d, J = 7.6 Hz, 1H), 7.23 (ddd, J = 7.6, 5.5, 1.4 Hz, 1H), 7.08 (dd, J = 7.6, 7.6 Hz, 1H), 6.99 (d, J = 7.6 Hz, 1H), 6.87 (d, J = 8.9 Hz, 1H), 4.58 (s, 2H), 3.77 (s, 3H). ^{13}C NMR (126 MHz), Chloroform-d δ (ppm): 165.9, 165.5, 161.2, 150.8, 147.8, 144.5, 139.1, 134.0 (2C), 124.8, 123.5, 122.9, 122.2, 118.8, 115.1 (2C), 55.5, 54.4. HRMS (m/z). Calc. for $[C_{19}H_{16}NOPdS]^+$ 411.9982. Found 411.9991. Elemental Analysis. Calc. (%) for $C_{19}H_{16}NOPdSCl$: C 50.91, H 3.60, N 3.12; found C 50.80, H 3.47, N 3.19.

4. Computational Section

Geometry optimization calculations were performed using Gaussian09 [44], in the gas-phase. The minimized structures were confirmed by the absence of any imaginary modes of vibration using frequency analysis. All structures were optimized using the generalized gradient approximation (GGA) PBE density functional [45,46]. The SDD ECP basis set was used for Pd, and the 6-31+G(d,p) basis set was used for all other atoms (PBE/6-31+G(d,p)[SDD]). This methodology has been validated in our previous study into the structures of symmetrical pincer palladacycles [36]. The topological analysis using quantum theory of atoms in molecules (QTAIM) was performed using the Multiwfn program [47]. The ωB97XD[48]/6-311+G(2df,2p)[DGDZVP] model chemistry was used for these calculations. The all-electron relativistic DGDZVP basis set was used to treat Pd [49] as the bond path cannot be traced when treated using ECP.

5. Conclusions

It has been shown that the *trans* influence plays a key role in the stability of unsymmetrical pincer palladacycles, with the bond strength, and the bond length of the Pd-donor atom interaction affected significantly when *trans* to a ligand exhibiting a strong *trans* influence. The topological analysis of the electron density at the bond critical point, and the structure determination, show that the strength of the *trans* influence is in the order P > S > N. This is in agreement with previous work [40,42].

A library of SCN pincer palladacycles were synthesized via C–H bond activation and characterized using X-ray crystallography, demonstrating the utility of late stage derivitization. These SCN palladacycles, along with PCN and N'CN previously synthesized by the authors, were used to investigate the driving force for the *trans* influence. It was shown, by investigating the electron density at the bond critical point and changes in the Pd-donor ligand bond length, that it is the donor atom that is responsible for the *trans* influence. The electronic and steric factors of the ligand do not influence significantly the bond strength of the ligand *trans* to it. This demonstrates the important role of unsymmetrical pincer palladacycles, with different donor atoms, in the search for harnessing and exploiting hemilability in the design of effective new palladacycle catalysts.

Acknowledgments: We would like to thank the University of Sussex for a studentship (GWR), Royal Thai Government for scholarship (SB) and Christopher Dadswell for the support and use of the GC equipment, and Johnson Matthey for the loan of palladium salts. We would also like to thank Alaa Abdul-Sada at the University of Sussex and the EPSRC Mass Spectroscopy Service (University of Swansea) for mass spectrometry services.

Author Contributions: Sarote Boonseng performed the calculations, interpreted the data and wrote the theoretical section. Gavin W. Roffe performed the experiments, interpreted their data and wrote the experimental section. Rhiannon N. Jones synthesized **1d**. Simon J. Coles and Graham J. Tizzard performed X-ray crystal structure analysis. Hazel Cox and John Spencer provided the idea, supervised the project and revised the manuscript.

Abbreviations

Pyr	Pyridine
DFT	Density Functional Theory
QTAIM	Quantum Theory of Atoms in Molecules
Tol	Toluene
HRMS	High Resolution Mass Spectrometry
ESI	Electrospray Ionization
GC	Gas Chromatography
DMF	Dimethylformamide
DCM	Dichloromethane
PBE	Perdew Burke Ernzerhof Exchange–Correlation Functional
MSE	Mean Signed Error
MUE	Mean Unsigned Error
BCP	Bond Critical Point
GGA	Generalized Gradient Approximation
SDD	Stuttgart-Dresden
ECP	Effective Core Potentials

References

1. Cope, A.C.; Siekman, R.W. Formation of covalent bonds from platinum or palladium to carbon by direct substitution. *J. Am. Chem. Soc.* **1965**, *87*, 3272–3273. [CrossRef]

2. Dupont, J.; Consorti, C.S.; Spencer, J. The potential of palladacycles: More than just precatalysts. *Chem. Rev.* **2005**, *105*, 2527–2571. [CrossRef] [PubMed]

3. Niu, J.-L.; Hao, X.-Q.; Gong, J.-F.; Song, M.-P. Symmetrical and unsymmetrical pincer complexes with group 10 metals: Synthesis via aryl C–H activation and some catalytic applications. *Dalton Trans.* **2011**, *40*, 5135–5150. [CrossRef] [PubMed]

4. Selander, N.; Szabó, K.J. Catalysis by palladium pincer complexes. *Chem. Rev.* **2011**, *111*, 2048–2076. [CrossRef] [PubMed]

5. Morales-Morales, D. Pincer complexes. Applications in catalysis. *Rev. Soc. Quim. Mex.* **2004**, *48*, 338–346. [CrossRef]

6. Liu, B.-B.; Wang, X.-R.; Guo, Z.-F.; Lu, Z.-L. Mononuclear versus dinuclear palladacycles derived from 1,3-bis(*N,N*-dimethylaminomethyl)benzene: Structures and catalytic activity. *Inorg. Chem. Commun.* **2010**, *13*, 814–817. [CrossRef]

7. Kruithof, C.A.; Dijkstra, H.P.; Lutz, M.; Spek, A.L.; Gebbink, R.J.M.K.; van Koten, G. X-Ray and NMR study of the structural features of SCS-pincer metal complexes of the group 10 triad. *Organometallics* **2008**, *27*, 4928–4937. [CrossRef]

8. Kjellgren, J.; Aydin, J.; Wallner, O.A.; Saltanova, I.V.; Szabó, K.J. Palladium pincer complex catalyzed cross-coupling of vinyl epoxides and aziridines with organoboronic acids. *Chem. Eur. J.* **2005**, *11*, 5260–5268. [CrossRef] [PubMed]

9. Yao, Q.; Sheets, M. A SeCSe−Pd(II) pincer complex as a highly efficient catalyst for allylation of aldehydes with allyltributyltin. *J. Org. Chem.* **2006**, *71*, 5384–5387. [CrossRef] [PubMed]

10. Aydin, J.; Selander, N.; Szabó, K.J. Strategies for fine-tuning the catalytic activity of pincer-complexes. *Tetrahedron Lett.* **2006**, *47*, 8999–9001. [CrossRef]

11. Gagliardo, M.; Selander, N.; Mehendale, N.C.; Van Koten, G.; Klein Gebbink, R.J.M.; Szabó, K.J. Catalytic performance of symmetrical and unsymmetrical sulfur-containing pincer complexes: Synthesis and tandem catalytic activity of the first PCS-pincer palladium complex. *Chem. Eur. J.* **2008**, *14*, 4800–4809. [CrossRef] [PubMed]

12. Moreno, I.; SanMartin, R.; Ines, B.; Herrero, M.T.; Domínguez, E. Recent advances in the use of unsymmetrical palladium pincer complexes. *Curr. Org. Chem.* **2009**, *13*, 878–895.

13. Roffe, G.W.; Boonseng, S.; Baltus, C.B.; Coles, S.J.; Day, I.J.; Jones, R.N.; Press, N.J.; Ruiz, M.; Tizzard, G.J.; Cox, H.; et al. A synthetic, catalytic and theoretical investigation of an unsymmetrical SCN pincer palladacycle. *R. Soc. Open Sci.* **2016**, *3*. [CrossRef] [PubMed]

14. Roffe, G.W.; Tizzard, G.J.; Coles, S.J.; Cox, H.; Spencer, J. Synthesis of unsymmetrical N'CN and PCN pincer palladacycles and their catalytic evaluation compared with a related SCN pincer palladacycle. *Org. Chem. Front.* **2016**, *3*, 957–965. [CrossRef]

15. Braunstein, P.; Naud, F. Hemilability of hybrid ligands and the coordination chemistry of oxazoline-based systems. *Angew. Chem. Int. Ed.* **2001**, *40*, 680–699. [CrossRef]

16. Khusnutdinova, J.R.; Milstein, D. Metal-ligand cooperation. *Angew. Chem. Int. Ed.* **2015**, *54*, 12236–12273. [CrossRef] [PubMed]

17. Zhang, W.H.; Chien, S.W.; Hor, T.S.A. Recent advances in metal catalysts with hybrid ligands. *Coord. Chem. Rev.* **2011**, *255*, 1991–2024. [CrossRef]

18. Ramírez-Rave, S.; Estudiante-Negrete, F.; Toscano, R.A.; Hernández-Ortega, S.; Morales-Morales, D.; Grévy, J.M. Synthesis and characterization of new Pd(II) non-symmetrical Pincer complexes derived from thioether functionalized iminophosphoranes. Evaluation of their catalytic activity in the Suzuki–Miyaura couplings. *J. Organomet. Chem.* **2014**, *749*, 287–295. [CrossRef]

19. Saha, D.; Verma, R.; Kumar, D.; Pathak, S.; Bhunya, S.; Sarkar, A. A "hemilabile" palladium–carbon bond: Characterization and its implication in catalysis. *Organometallics* **2014**, *33*, 3243–3246. [CrossRef]

20. Poverenov, E.; Gandelman, M.; Shimon, L.J.W.; Rozenberg, H.; Ben-David, Y.; Milstein, D. Pincer "hemilabile" effect. PCN platinum(II) complexes with different amine "arm length". *Organometallics* **2005**, *24*, 1082–1090.

21. Gargir, M.; Ben-David, Y.; Leitus, G.; Diskin-Posner, Y.; Shimon, L.J.W.; Milstein, D. PNS-Type ruthenium pincer complexes. *Organometallics* **2012**, *31*, 6207–6214. [CrossRef]

22. Fleckhaus, A.; Mousa, A.H.; Lawal, N.S.; Kazemifar, N.K.; Wendt, O.F. Aromatic PCN palladium pincer complexes. Probing the hemilability through reactions with nucleophiles. *Organometallics* **2015**, *34*, 1627–1634. [CrossRef]

23. Pidcock, A.; Richards, R.E.; Venanzi, L.M. ^{195}Pt–^{31}P nuclear spin coupling constants and the nature of the *trans*-effect in platinum complexes. *J. Chem. Soc. A Inorg. Phys. Theor.* **1966**, 1707–1710. [CrossRef]

24. Appleton, T.G.; Clark, H.C.; Manzer, L.E. The trans-influence: Its measurement and significance. *Coord. Chem. Rev.* **1973**, *10*, 335–422. [CrossRef]

25. Quagliano, J.V.; Schubert, L. The *trans* effect in complex inorganic compounds. *Chem. Rev.* **1952**, *50*, 201–260. [CrossRef]

26. Rigamonti, L.; Rusconi, M.; Manassero, C.; Manassero, M.; Pasini, A. Quantification of *cis* and *trans* influences in [PtX(PPh$_3$)$_3$]$^+$ complexes. A ^{31}P NMR study. *Inorg. Chim. Acta* **2010**, *363*, 3498–3505. [CrossRef]

27. Randaccio, L.; Bresciani-Pahor, N.; Toscano, P.J.; Marzilli, L.G. Bonding mode and *trans* influence of the nitromethyl ligand. Structure of *trans*-bis(dimethylglyoximato)(nitromethyl)(pyridine)cobalt(III). *Inorg. Chem.* **1981**, *20*, 2722–2724. [CrossRef]

28. Otto, S.; Johansson, M.H. Quantifying the *trans* influence of triphenylarsine. Crystal and molecular structures of *cis*-[PtCl$_2$(SMe$_2$)(AsPh$_3$)] and *cis*-[PtCl$_2$(AsPh$_3$)$_2$]·CHCl$_3$. *Inorg. Chim. Acta* **2002**, *329*, 135–140. [CrossRef]

29. Kaltsoyannis, N.; Mountford, P. Theoretical study of the geometric and electronic structures of pseudo-octahedral d^0 imido compounds of titanium: The *trans* influence in *mer*-[Ti(NR)Cl$_2$(NH$_3$)$_3$] (R = Bu^t, C$_6$H$_5$ or C$_6$H$_4$NO$_2$-4). *Dalt. Trans.* **1999**, 781–790. [CrossRef]

30. Lyne, P.D.; P. Mingos, D.M. The effects of back-bonding to phosphines on the *trans* influence in [Mo(NH)Cl$_3$(PR$_3$)$_2$]$^{0,\pm1}$ (R = H, Me and F). *J. Organomet. Chem.* **1994**, *478*, 141–151. [CrossRef]

31. Jacobsen, H.; Berke, H. Tuning of the transition metal hydrogen bond: How do trans ligands influence bond strength and hydridicity? *Dalt. Trans.* **2002**, 3117–3122. [CrossRef]

32. Deeth, R.J. The trans influence in [RH(Ph$_3$)$_3$Cl]: A density functional theory study. *Dalt. Trans.* **1993**, 3711–3713. [CrossRef]

33. Burdett, J.K.; Albright, T.A. *Trans* influence and mutual influence of ligands coordinated to a central atom. *Inorg. Chem.* **1979**, *18*, 2112–2120. [CrossRef]

34. Loeb, S.J.; Shimizu, G.K.H.; Wisner, J.A. Mono- versus dipalladation of the durene-based tetrathioether Ligand 1,2,4,5-(tBuSCH$_2$)$_4$C$_6$H$_2$. Structures of [PdCl((tBuSCH$_2$)$_4$C$_6$H)] and [Pd$_2$((tBuSCH$_2$)$_4$C$_6$)(MeCN)$_2$][BF$_4$]$_2$. *Organometallics* **1998**, *17*, 2324–2327. [CrossRef]

35. Bader, R.F.W. A bond path: A universal indicator of bonded interactions. *J. Phys. Chem. A* **1998**, *5639*, 7314–7323. [CrossRef]

36. Boonseng, S.; Roffe, G.W.; Spencer, J.; Cox, H. The nature of the bonding in symmetrical pincer palladacycles. *Dalt. Trans.* **2015**, *44*, 7570–7577. [CrossRef] [PubMed]

37. Hartley, F.R. The *cis*- and *trans*-effects of ligands. *Chem. Soc. Rev.* **1973**, *2*. [CrossRef]

38. Sajith, P.K.; Suresh, C.H. Bond dissociation energies of ligands in square planar Pd(II) and Pt(II) complexes: An assessment using *trans* influence. *J. Organomet. Chem.* **2011**, *696*, 2086–2092. [CrossRef]

39. Manojlovic-Muir, L.J.; Muir, K.W. The *trans*-influence of ligands in platinum(II) complexes. The significance of the bond length data. *Inorg. Chim. Acta* **1974**, *10*, 47–49. [CrossRef]

40. Kapoor, P.N.; Kakkar, R. *Trans* and *cis* influence in square planar Pt(II) complexes: A density functional study of [PtClX(dms)$_2$] and related complexes. *J. Mol. Struct. Theochem.* **2004**, *679*, 149–156. [CrossRef]

41. Bader, R.F.W. *Atoms in Molecules: A Quantum Theory*; Oxford University Press: Oxford, UK, 1990.

42. Sajith, P.K.; Suresh, C.H. Quantification of mutual trans influence of ligands in Pd(II) complexes: A combined approach using isodesmic reactions and AIM analysis. *Dalton Trans.* **2010**, *39*, 815–822. [CrossRef] [PubMed]

43. Coles, S.J.; Gale, P.A. Changing and challenging times for service crystallography. *Chem. Sci.* **2012**, *3*, 683–689. [CrossRef]

44. Frisch, M.J.; Trucks, G.W.; Schlegel, H.B.; Scuseria, G.E.; Robb, M.A.; Cheeseman, J.R.; Scalmani, G.; Barone, V.; Mennucci, B.; Petersson, G.A.; et al. Gaussian 09, Revision B.01. Gaussian, Inc.: Wallingford, CT, USA, 2009.

45. Perdew, J.P.; Burke, K.; Ernzerhof, M. Generalized gradient approximation made simple . *Phys. Rev. Lett.* **1996**, *77*, 3865–3868. [CrossRef] [PubMed]

46. Perdew, J.P.; Burke, K.; Ernzerhof, M. Generalized gradient approximation made simple [*Phys. Rev. Lett. 77*, 3865 (1996)]. *Phys. Rev. Lett.* **1997**, *78*, 1396. [CrossRef]

47. Lu, T.; Chen, F. Multiwfn: A multifunctional wavefunction analyzer. *J. Comput. Chem.* **2012**, *33*, 580–592. [CrossRef] [PubMed]

48. Chai, J.-D.; Head-Gordon, M. Long-Range corrected hybrid density functionals with damped atom–atom dispersion corrections. *Phys. Chem. Chem. Phys.* **2008**, *10*, 6615–6620. [CrossRef] [PubMed]

49. Sajith, P.K.; Suresh, C.H. Mechanisms of reductive eliminations in square planar Pd(II) complexes: Nature of eliminated bonds and role of *trans* influence. *Inorg. Chem.* **2011**, *50*, 8085–8093. [CrossRef] [PubMed]

5

Single Crystal Growth and Anisotropic Magnetic Properties of $Li_2Sr[Li_{1-x}Fe_xN]_2$

Peter Höhn [1], Tanita J. Ballé [2], Manuel Fix [2], Yurii Prots [1] and Anton Jesche [2,*]

[1] Max-Planck-Institut für Chemische Physik fester Stoffe, Nöthnitzer Str. 40, D-01187 Dresden, Germany; peter.hoehn@cpfs.mpg.de (P.H.); yurii.prots@cpfs.mpg.de (Y.P.)

[2] EP VI, Center for Electronic Correlations and Magnetism, Augsburg University, D-86159 Augsburg, Germany; tanita.balle@physik.uni-augsburg.de (T.J.B.); manuel.fix@physik.uni-augsburg.de (M.F.)

* Correspondence: anton.jesche@physik.uni-augsburg.de

Academic Editor: Rainer Niewa

Abstract: Up to now, investigation of physical properties of ternary and higher nitridometalates has been severely hampered by challenges concerning phase purity and crystal size. Employing a modified lithium flux technique, we are now able to prepare sufficiently large single crystals of the highly air and moisture sensitive nitridoferrate $Li_2Sr[Li_{1-x}Fe_xN]_2$ for anisotropic magnetization measurements. The magnetic properties are most remarkable: large anisotropy and coercivity fields of 7 Tesla at $T = 2$ K indicate a significant orbital contribution to the magnetic moment of iron. Altogether, the novel growth method opens a route towards interesting phases in the comparatively recent research field of nitridometalates and should be applicable to various other materials.

Keywords: solution growth; magnetically hard materials; unquenched orbital moment; single crystal; flux growth

1. Introduction

Binary transition metal nitrides attract considerable interest due to their valuable mechanical, electrical and magnetic properties. In contrast, chemistry and physics of multinary nitrides have been far less thoroughly explored [1]. Nitridometalates of d metals T represent an interesting class of solid state phases, which contain nitrogen as isolated anions N^{3-} or feature complex anions $[T_xN_y]^{z-}$ of different dimensionalities with coordination numbers of T by N typically between two and four and oxidation states of the transition metals being comparatively low. Whereas the bonding within these complex anions and frameworks is essentially covalent, nitridometalates are stabilized by predominantly ionic bonding through counterions of electropositive metals like alkali (A) or alkaline-earth (AE) cations.

Iron in nitridoferrates of alkali and alkaline-earth metals may be coordinated tetrahedrally ($Li_3[Fe^{III}N_{4/2}]$ [2,3]) or trigonal-planar in isolated units (($Ca_3N)_2[Fe^{III}N_3]$ [4], $Sr_3[Fe^{III}N_3]$ [5], $Ba_3[Fe^{III}N_3]$ [6], $(Sr_{1-x}Ba_x)_3[Fe^{III}N_3]$ [7], $Sr_8[Fe^{III}N_3][Fe^{II}N_2]$ [8,9]) as well as in oligomers ($Ca_2[Fe^{II}N_2]$ [10], $Sr_2[Fe^{II}N_2]$ [10]) and 1D chains ($LiSr_2[Fe_2^{II}N_3]$ [11], $LiBa_2[Fe_2^{II}N_3]$ [11]). Linear coordination is observed as linear dumbbells in $Sr_8[Fe^{III}N_3][Fe^{II}N_2]$ [8,9], $Sr_8[Mn^{III}N_3][Fe^{II}N_2]$ [12], $Sr_2[Fe^{II}N_2]$ [10], and $Li_4[Fe^{II}N_2]$ [13], and in linear substituted chains $[(Li_{1-x}Fe_x^I)N]^{2-}$ in $Li_2[(Li_{1-x}Fe_x^I)N]$ [14–16], $Li_2Ca[(Li_{1-x}Fe_x^I)N]_2$ [17], $Li_2Sr[(Li_{1-x}Fe_x^I)N]_2$ [17], $LiCa_2[(Fe_{1-x}^I Li_x)N_2]$ [18] and $LiSr_2[(Fe_{1-x}^I Li_x)N_2]$ [19]. Structural data for the majority of nitridoferrates reported up to now were derived from single crystal data, whereas in most cases single phase powder samples had to be employed for the investigation of physical properties, with the exception of $Li_2[(Li_{1-x}Fe_x^I)N]$ [20] and $LiSr_2[(Fe_{1-x}^I Li_x)N_2]$ [19], where single crystals of sufficient size were available for physical properties investigations.

The single crystal growth of nitrides is often challenging due to the large dissociation energy of the N_2 molecule, the reactivity of the starting materials and enhanced vapor pressures. Only recently, Li-rich flux was successfully used for the growth of large single crystals of LiCaN, Li_3N and $Li_2(Li_{1-x}T_x)N$ with $T = Mn, Fe, Co$ [20] and $T = Ni$ [21]. Further development of the high temperature centrifugation aided filtration technique [22–24] by addition of Na and NaN_3 to increase the basicity of the flux also enabled the growth of large single crystals of nitride metalides like $Li_{16}Sr_6Ge_6N$ [25]. However, the extent of application of this method towards more complex systems, in particular transition metal rich ternary or quaternary nitridometalates that also contain alkaline earth elements, has not been investigated in detail. Magnetic properties were reported only for few nitridometalates containing alkaline-earth metals. Some of these phases show ferromagnetic ($LiSr_2[CoN_2]$ [19]) or antiferromagnetic ($LiSr_2[FeN_2]$ [19]) ordering. Furthermore, for many phases (for example $Sr_8[MnN_3]_2[MnN_2]$ [26]) a large spin-orbit coupling together with low or intermediate spin states is being discussed.

We shall present results on Fe-substituted Li_4SrN_2. The peculiar (almost) linear, two-fold coordination of Fe makes this material particularly promising since unprecedented magnetic coercivity and anisotropy were found in Fe-substituted Li_3N, which shares the same structural feature [16,27]. Synthesis and crystal structure of Fe-substituted Li_4SrN_2 was first reported by Klatyk and Kniep [17]: Small single crystals sufficient for X-Ray diffraction were obtained by reaction of Li, $Li_2(Li_{0.66}Fe_{0.33})N$ and Sr_2N in a molar ratio of 7:6:4. The compound crystallizes in a tetragonal lattice, space group $I4_1/amd$ (No. 141) with $a = 3.7909(2)$ Å, and $c = 27.719(3)$ Å. As indicated by the notation of the chemical formula, $Li_2Sr[(Li_{1-x}Fe_x)N]_2$ with $x = 0.46$, the substituted Fe atoms occupy only one of the two Li-sites (the one in two-fold coordination of N, see Figure 1). The Fe-coordination is not strictly linear. Rather the N–Fe–N angle amounts to $177.4°$ as inferred from the reported crystal structure [17]. The Fe-substitution causes a decrease of a but an increase of c by 0.8% and 2.5%, respectively, compared to the parent compound Li_4SrN_2 [28]. No physical properties have been reported so far.

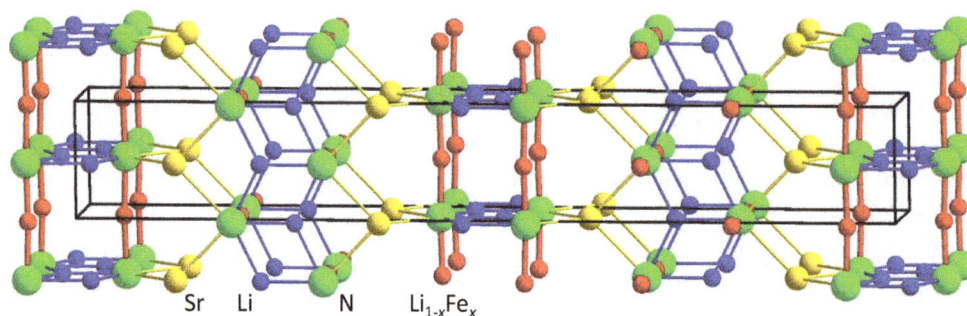

Figure 1. Crystal structure of $Li_2Sr[(Li_{1-x}Fe_x)N]_2$, the unit cell is indicated by the black lines (space group $I4_1/amd$). Fe is substituted in linear, two-fold coordination between N.

Here we show that $Li_2Sr[(Li_{1-x}Fe_x)N]_2$ single crystals of several millimeter along a side can be obtained from Li-rich flux. The large magnetic anisotropy and coercivity revealed a significant orbital contribution to the magnetic moment of Fe. Cu-substituted Li_4SrN_2 was investigated as a non-local-moment-bearing reference compound.

2. Results

2.1. Single Crystal Growth

Due to the air and moisture sensitivity of both the reactants (Li, Sr_2N, NaN_3) and the final product $Li_2Sr[(Li_{1-x}Fe_x)N]_2$ all manipulations including grinding and weighing, as well as complete sample preparations for measurements were carried out in an inert gas glove box (Ar, O_2 and $H_2O \leq 1$ ppm). The title compound was obtained in form of large, black single crystals from the reaction of Sr_2N, Fe, Li and NaN_3 in molar ratio 1:1.8:36:1 with NaN_3 acting as nitrogen source and Li acting as flux and mineralizer. The mixtures with a total mass of roughly 0.6 g were placed in a tantalum ampule [29]. The whole device was sealed by arc welding under inert atmosphere of 700 mbar argon and subsequently encapsulated in a quartz tube with an internal argon pressure of 300 mbar in order to prevent oxidization of the tantalum. The sample was heated from room temperature to $T = 700\,°C$ within 7 h, annealed for 2 h, then cooled to $T = 300\,°C$ within 400 h, and finally centrifuged with 3000 min^{-1} to separate the single crystals of a blackish color from the excess flux. Besides several large single crystals of the title phase, small amounts of single crystalline Li_3N and $LiSr_2[(Fe_{1-x}Li_x)N_2]$ [19] were also obtained. Using a two-probe multimeter, the different phases showed remarkably different resisitivities enabling the discrimination of the various phases.

A representative $Li_2Sr[(Li_{1-x}Fe_x)N]_2$ single crystal is shown in Figure 2. The samples show a plate-like habit with the crystallographic c-axis oriented perpendicular to the large surface as confirmed by Laue-back-reflection (right panel in Figure 2). The spot-size of the X-Ray beam was similar to the sample size.

Figure 2. $Li_2Sr[(Li_{1-x}Fe_x)N]_2$ single crystal ($x = 0.41$) on a millimeter grid (**left**) and corresponding Laue-back-reflection pattern to the **right** showing the four-fold rotational symmetry along the crystallographic c-axis.

An Fe-concentration of $x = 0.42$ was determined by energy-dispersive X-Ray analysis (EDX) based on the Fe:Sr ratio and assuming fully occupied Sr-sites. An almost identical value of $x = 0.41$ was found by chemical analysis by means of inductively coupled plasma optical emission spectroscopy (ICP-OES; accessible is the Li:Sr:Fe ratio) and was confirmed with a second sample taken from the same batch. The observed slight Li-excess of 0.15 per formula unit (0.19 for the other sample), is attributed to small amounts of Li-rich flux remnants.

2.2. Crystal Structure

Figure 3 shows the X-Ray powder diffraction pattern measured on ground $Li_2Sr[(Li_{1-x}Fe_x)N]_2$ single crystals. No foreign phases were detected. The region between $2\theta = 20°$–$25°$ was excluded due to amorphous constituents that are created by degradation during the measurement. Lattice parameters of $a = 3.79536(9)$ Å and $c = 27.6492(13)$ Å and an Fe occupancy of $x = 0.32$ were obtained by Rietveld refinement. The decrease in a and increase in c in comparison to the parent compound Li_4SrN_2 ($a = 3.822(2)$ Å and $c = 27.042(9)$ Å [28]) are slightly smaller than reported in Ref. [17] ($x = 0.46, a = 3.7909(2)$ Å and $c = 27.719(3)$ Å) in accordance with a somewhat lower Fe concentration.

Smaller crystals were obtained from crushed samples and selected for single crystal X-Ray diffraction. The results are summarized in Table 1. Powder as well as single crystal X-Ray diffraction revealed an Fe-concentration of $x = 0.32$. This is somewhat smaller than the values obtained by EDX and chemical analysis and indicates that the Fe concentration may vary between different samples; the slight difference between powder and single crystal data may also stem from slight inhomogeneities within the samples. The single crystal used for the magnetization measurements (see below) was directly analyzed by ICP-OES and showed an iron concentration of $x = 0.41$.

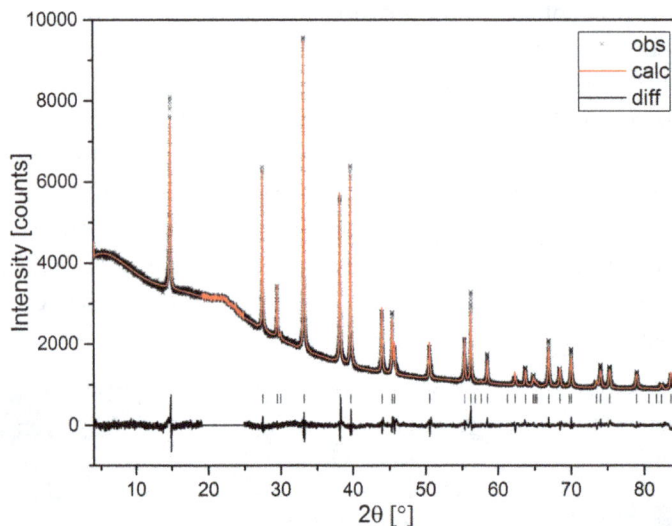

Figure 3. X-Ray diffraction pattern measured on ground single crystals of $Li_2Sr[(Li_{1-x}Fe_x)N]_2$ (Co-$K\alpha$ radiation, $\lambda = 1.78892$ Å). Tick marks correspond to reflection positions of $Li_2Sr[(Li_{1-x}Fe_x)N]_2$.

Table 1. Single crystal structure refinement for $Li_2Sr[(Li_{1-x}Fe_x)N]_2$.

crystal system	tetragonal
space group	$I4_1/amd$ (No. 141)
a (Å)	3.8011(1)
c (Å)	27.586(3)
Fe occupancy (x)	0.32(1)
cell volume (Å^3)	398.57(5)
Z	4
ρ_{calcd} (gcm^{-3})	2.907
crystal color, habit	black, tetragonal column
crystal size (mm)	$0.02 \times 0.02 \times 0.07$
μ(Mo$_{K\alpha}$, mm^{-1})	15.52
2θ range (°)	5.90–59.60
diffractometer	RIGAKU AFC7
wavelength (Å)	0.71069 (Mo-$K\alpha$)
monochromator	graphite
temperature	293 K
scan mode	profile data from ϕ scans
measured reflections	3355
observed reflections [$F_o > 4\sigma(F_o)$]	2919
independent reflections	187
R_{int}	0.041
structure solution method	direct
number of parameters	17
goodness-of-fit on F^2	1.081
$wR2$	0.078
$R1$ [$F_o > 4\sigma(F_o)$]	0.030
$R1$ (all data)	0.035
residual electron density (e$\times 10^{-6}$ pm^{-3})	1.24/-1.70

2.3. Magnetic Properties

The temperature-dependent magnetic susceptibility $\chi(T) = M/H$ measured parallel and perpendicular to the crystallographic c-axis is shown in Figure 4a (field-cooled in $\mu_0 H = 7$ T). A temperature-independent contribution of a ferromagnetic impurity phase (with a Curie temperature significantly above room temperature, presumably elemental Fe or Fe_3O_4) was subtracted by assuming that the intrinsic, local-moment contribution of the title compound is linear in field at $T = 300$ K (analogous to the Honda-Owen method). A pronounced anisotropy is observed over the whole temperature range investigated. The ratio of $\chi_{\perp c}/\chi_{\|c}$ increases upon cooling from a value of 2 at $T = 300$ K to 12 at $T = 2$ K.

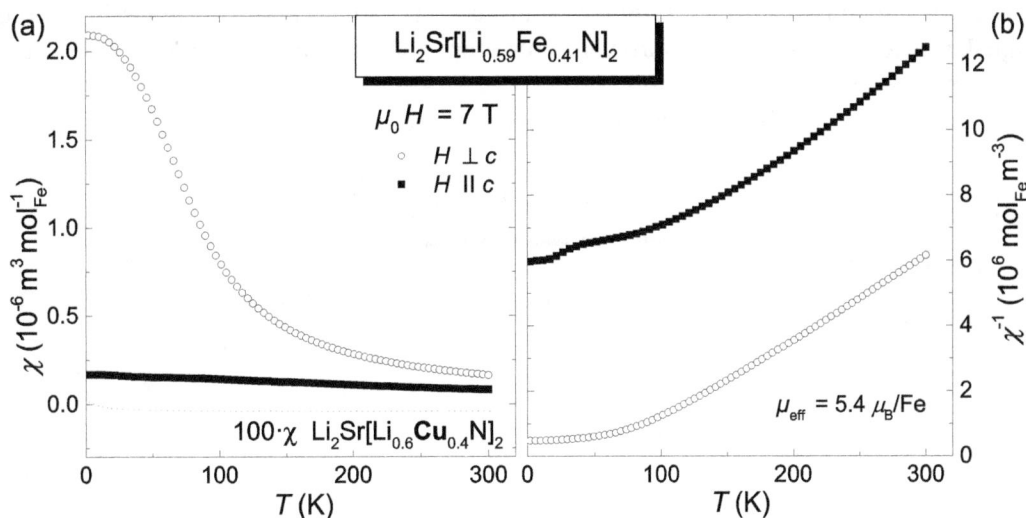

Figure 4. Magnetic susceptibility $\chi = M/H$ per mol Fe as a function of temperature for field applied parallel and perpendicular to the crystallographic c-axis. (**a**) A pronounced anisotropy with larger χ for $H \perp c$ (open, red symbols) is observed up to room temperature. The molar susceptibility of $Li_2Sr[(Li_{0.6}Cu_{0.4})N]_2$ multiplied by a factor of 100 is shown for comparison (blue, dotted line, $H \perp c$); (**b**) The temperature dependence of the inverse susceptibility roughly follows a Curie-Weiss law for $T > 150$ K.

In order to confirm the proposed, unusual valence state of Fe(I) [17], we performed further magnetization measurements on isotypic $Li_2Sr[(Li_{0.6}Cu_{0.4})N]_2$ (the sample was grown similar to the title compound $Li_2Sr[(Li_{1-x}Fe_x)N]_2$). The local moment behavior associated with the spin-1/2 of Cu(II) is supposed to be markedly different from the one of Cu(I). As shown by the blue, dotted line in Figure 4a, the susceptibility of $Li_2Sr[(Li_{0.6}Cu_{0.4})N]_2$ does not show local moment behavior and is indeed negligibly small compared to the one of the Fe-substituted homologue. The largely temperature independent value of $\chi = -5(1) \times 10^{-10}$ m$^3 \cdot$ mol^{-1} is in reasonable agreement with the ionic diamagnetic contribution of the Li_4SrN_2 host material of $\chi = -6 \times 10^{-10}$ m$^3 \cdot$ mol^{-1} {assuming $\chi(Li^{1+}) = -0.09 \times 10^{-10}$ m$^3 \cdot$ mol^{-1} [30], $\chi(Sr^{2+}) = -2 \times 10^{-10}$ m$^3 \cdot$ mol^{-1} [31] and $\chi(N^{3-}) = -1.6 \times 10^{-10}$ m$^3 \cdot$ mol^{-1} [32]}. Accordingly, the observed non-local moment behavior of $Li_2Sr[(Li_{1-x}Cu_x)N]_2$ implies the presence of Cu(I) and supports a valence state of Fe(I).

The inverse susceptibility roughly follows a Curie-Weiss law for $T = 150$–300 K (Figure 4b). For $H \perp c$ an effective moment of $\mu_{eff} = 5.4\,\mu_B$ per Fe and a ferromagnetic Weiss temperature of $\Theta_W = 49$ K were obtained. The fit to the data considered a minor diamagnetic contribution of $\chi_0 = -1.5 \times 10^{-8}$ m$^3 \cdot$ mol^{-1} (10% of the absolute value at room temperature). The slope of χ^{-1} suggests a similar value of the effective moment for $H \| c$, however, the small absolute value of χ in combination with a large antiferromagnetic Weiss temperature prohibits an accurate estimate.

　　　The isothermal magnetization in μ_B per Fe as a function of an applied magnetic field at $T = 10\,\mathrm{K}$ is shown in Figure 5a. For $H \parallel c$ the magnetization increases slowly with the applied field in a linear fashion without any appreciable hysteresis. The magnetization is significantly larger for $H \perp c$ and exceeds values of $2\,\mu_B$ per Fe. However, no saturation is observed even at the largest available field of $\mu_0 H = 7\,\mathrm{T}$. In accordance with the large anisotropy, a pronounced hysteresis loop with a coercive field of $\mu_0 H_c = 1.2\,\mathrm{T}$ forms for $H \perp c$. The coercivity field increases rapidly upon cooling (Figure 5b). At the lowest accessible temperature $T = 2\,\mathrm{K}$ the coercivity reaches almost $7\,\mathrm{T}$. The hysteresis vanishes for temperatures higher than $T \approx 16\,\mathrm{K}$. Furthermore, the $M - H$ loops are asymmetric for $T < 14\,\mathrm{K}$: The (field cooled) value in $\mu_0 H = +7\,\mathrm{T}$ is larger than the corresponding value found at $\mu_0 H = -7\,\mathrm{T}$.

　　　The small anomalies at $H \sim 0$ and at $M \sim 1\,\mu_B$ result from the subtraction of the ferromagnetic impurity contribution (which is not fully temperature independent) and a zero-crossing of the raw data signal, respectively. A significant in-plane anisotropy for different orientations perpendicular to the c-axis could be present, in particular $\langle 100 \rangle$ vs. $\langle 110 \rangle$ (see discussion). The closer investigation of this anisotropy, however, is beyond the scope of this publication.

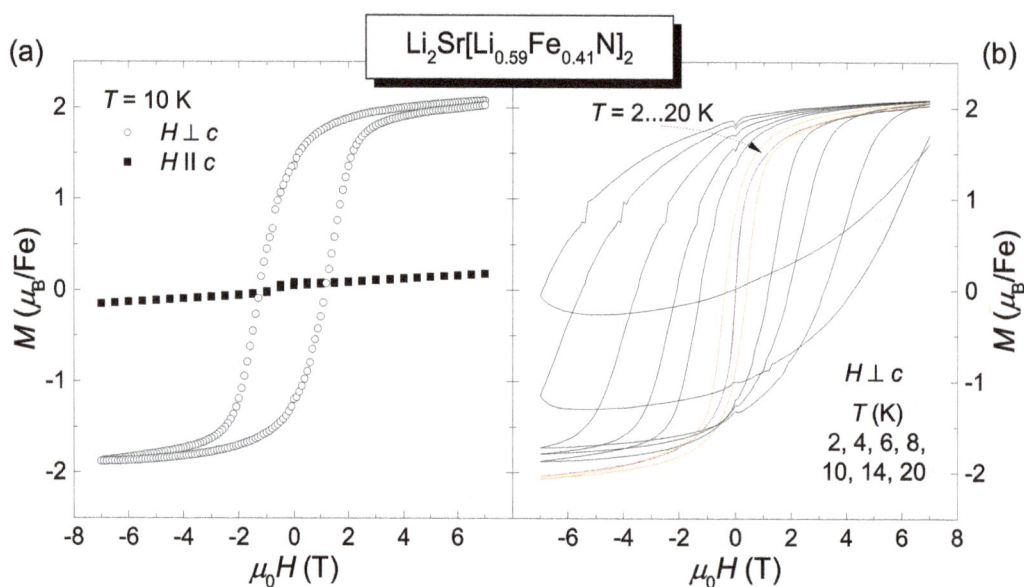

Figure 5. Isothermal magnetization in μ_B per Fe at $T = 10\,\mathrm{K}$; **(a)** Large hysteresis emerges for field applied perpendicular to the c-axis (which is perpendicular to the N–Fe–N 'molecular axis'); **(b)** The hysteresis for $H \perp c$ increases rapidly with decreasing temperature. Spontaneous magnetization disappears for temperatures larger than $T{\sim}16\,\mathrm{K}$.

3. Discussion

　　　The successful use of a Li-rich flux for the growth of large single crystals of Fe-substituted Li_4SrN_2 has been anticipated since there are only a few binaries known that may compete with the formation of this compound. Among those, Li–Sr binaries are not considered exceedingly stable as indicated by their low peritectic decomposition temperatures ($<200\,°\mathrm{C}$) [33]. Nevertheless, a good solubility of Sr in Li is inferred. Only a few Sr–N binary compounds are known and non of those seem to be stable in the presence of Li. No binary compounds of Sr–Fe or Sr–Ta (which may prevent the use of Ta crucibles) are known. Furthermore, our work shows that Li_3N is not so stable that it prevents the formation of other nitrides. Whether the related compounds $LiSr_2Fe_2N_3$ [11] or $LiSr_2FeN_2$ [19] (isotypic to $LiSr_2CoN_2$ [34]) can be grown as large single crystals by adjusting temperature profile and/or ratio of the starting materials is subject of ongoing research. The situation is similar for ternary and multinary nitrides that contain other alkaline earth and/or other transition metals and it seems not unlikely to find a wide range of applications for the Li-flux method.

The magnetic properties of $Li_2Sr[(Li_{1-x}Fe_x)N]_2$, in particular the large anisotropy and coercivity, are most remarkably. There are only a few materials known that show coercivity fields in the range of $\mu_0 H_c \sim 7$ T or above. Those are melt-spun ribbons of rare-earth-based Dy–Fe–B and Tb–Fe–B alloys (reported are $\mu_0 H_c = 6.4$ T for both materials [35] and $\mu_0 H_c = 7.7$ T for the latter one [36]). For transition-metal-based compounds, the only examples we are aware of are $LuFe_2O_4$ ($\mu_0 H_c = 9$ T [37] and $\mu_0 H_c = 11$ T [38]) and Fe-substituted Li_3N ($\mu_0 H_c = 11.6$ T [27]). The large magnetic anisotropy and coercivity of both materials is caused by a significant orbital contribution to the magnetic moment of Fe (see [39] for the former and [21,40–42] for the latter one). Whereas the emergence of the orbital moment in $LuFe_2O_4$ is a result of a complex interplay between charge ordering, ferroelectricity and ferrimagnetism [43], the orbital moment in Fe-substituted Li_3N seems to be directly linked to the linear, two-fold coordination of Fe [27]. Such a linear molecule or linear chain is not subject to a Jahn-Teller distortion [44,45] which is the driving force for the quenching of the orbital magnetic moment that is usually observed in transition metals.

As found for Fe-substituted Li_3N, the magnetic hard axis of $Li_2Sr[(Li_{1-x}Fe_x)N]_2$ is oriented perpendicular to the N–Fe–N "molecular axis" (see Figures 1 and 5). Accordingly, there is no unique easy-axis present in $Li_2Sr[(Li_{1-x}Fe_x)N]_2$ since the N–Fe–N molecules run along both the a- and the b-axes (in contrast to Fe-substituted Li_3N where the N–Fe–N molecules do define the easy axis and are oriented along the unique, hexagonal c-axis). The magnetization in the a–b plane is therefore expected to reach between $1/2$ and $1/\sqrt{2}$ of the saturation magnetization of Fe corresponding to field along $\langle 100 \rangle$ and $\langle 110 \rangle$, respectively. The increase of the magnetization for $H \perp c$ is only slightly larger than for $H \parallel c$ (see Figure 5a). This implies similar large magnetic anisotropy energies for $\langle 110 \rangle$ and $\langle 001 \rangle$ (with respect to $\langle 100 \rangle$).

To summarize, the large magnetic anisotropy and coercivity observed in $Li_2Sr[(Li_{1-x}Fe_x)N]_2$ give strong evidence for a significant orbital contribution to the magnetic moment of Fe. The similarities to $Li_2(Li_{1-x}Fe_x)N$ [21,27] are apparent. The origin of this behavior is attributed to the (almost) linear, two-fold coordination Fe. Our work shows that small deviations from linearity, that is a N–Fe–N angle of $\sim 177°$ instead of $180°$, do not prevent the formation of unquenched orbital moments. This finding significantly increases the number of materials that are promising for the investigation of orbital moment formation in transition metal compounds.

4. Materials and Methods

Starting materials were lithium rod (Evochem, 99%), iron powder (Alfa Aesar 99.998%), strontium nitride (Sr_2N) powder (prepared from strontium metal (Alfa Aesar, distilled dentritic pieces, 99.8%) and nitrogen (Praxair, 99.999%, additionally purified by molecular sieves)), and sodium azide (NaN_3) powder (Roth, 99%) as a further nitrogen source. Tantalum ampules were produced on-site from pre-cleaned tantalum tube and tantalum sheet in an arc furnace located within a glove box.

Laboratory powder X-ray diffraction data of finely ground (gray) powder samples were collected on a Huber G670 imaging plate Guinier camera using a curved germanium (111) monochromator and Cu-$K\alpha_1$ radiation in the range $4° \leq 2\theta \leq 100°$ with an increment of $0.005°$ at 293(1) K. The powder samples were placed between Kapton foils to avoid degradation in air. Preliminary data processing was done using the WinXPow program package [46]. Rietveld refinement of the structure of $Li_2Sr[(Li_{1-x}Fe_x)N]_2$ was performed with the software package Jana2006 [47]. Small intervals of the diffraction pattern between $2\theta = 20°-25°$ corresponding to amorphous degradation products were excluded during the refinement. After background correction, profile and lattice parameters were refined by using Pseudo-Voigt profile functions and Berar-Baldinozzi's asymmetry model before refining atomic positions and isotropic thermal displacement factors.

The crystal structure and the composition of the title compound $Li_2Sr[(Li_{1-x}Fe_x)N]_2$ was refined from single-crystal X-ray diffraction data which were collected at room temperature on a Rigaku AFC7 four circle diffractometer equipped with a Mercury-CCD detector (Mo-$K\alpha$

radiation, graphite monochromator). After data collection the structures were solved by direct methods, using SHELXS-97 [48] and subsequently refined by using the full-matrix least-squares procedure with SHELXL-97 [49]. Further details on the crystal structure investigations may be obtained from the Fachinformationszentrum Karlsruhe, 76344 Eggenstein-Leopoldshafen, Germany (Fax: +49-7247-808-666; e-mail: crysdata@fiz-karlsruhe.de), on quoting the depository number CSD-432385 (powder) or CSD-432386 (single crystal), the names of the authors, and the journal citation. The morphology of the $Li_2Sr[(Li_{1-x}Fe_x)N]_2$ sample and its metal composition were investigated using a scanning electron microscope Philips XL30 equipped with a Bruker Quantax EDX-System (Silicon drift detector, LaB_6 cathode). The EDX data were processed using the Esprit-Software. The composition of the samples was further analyzed by inductively coupled plasma optical emission spectroscopy (ICP-OES) using a Varian Vista-MPX. To this extent the samples were dissolved in dilute hydrochloric acid solution (4 mL of 37% hydrochloric acid added to 46 mL deionized water). Laue back reflection pattern were taken with a digital Dual FDI NTX camera manufactured by Photonic Science (tungsten anode, $U = 20$ kV). The magnetization was measured using a 7 T Magnetic Property Measurement System (MPMS3) manufactured by Quantum Design.

Acknowledgments: The authors thank Ulrich Burkhardt, Petra Scheppan, and Steffen Hückmann for experimental assistance. Andrea Mohs, Alexander Herrnberger and Klaus Wiedenmann are acknowledged for technical support. This work was supported by the Deutsche Forschungsgemeinschaft (DFG, German Research Foundation) - Grant No. JE 748/1.

Author Contributions: Peter Höhn grew the single crystals and analyzed powder and single crystal X-Ray diffraction data that were collected by Yurii Prots; Tanita J. Ballé performed the magnetization measurements and interpreted the data; Manuel Fix collected and analyzed Laue-back-reflection data; Anton Jesche wrote the paper with the help of all authors.

References

1. Kniep, R.; Höhn, P. Low-Valency Nitridometalates. In *Comprehensive Inorganic Chemistry {II}*, 2nd ed.; Reedijk, J., Poeppelmeier, K., Eds.; Elsevier: Amsterdam, The Netherlands, 2013; pp. 137–160.

2. Gudat, A.; Kniep, R.; Rabenau, A.; Bronger, W.; Ruschewitz, U. Li_3FeN_2, a ternary nitride with $\frac{1}{\infty}[FeN_{4/2}^{3-}]$ chains: Crystal structure and magnetic properties. *J. Less Common Met.* **1990**, *161*, 31–36.

3. Nishijima, M.; Takeda, Y.; Imanishi, N.; Yamamoto, O.; Takano, M. Li Deintercalation and Structural Change in the Lithium Transition Metal Nitride Li_3FeN_2. *J. Solid State Chem.* **1994**, *113*, 205–210.

4. Cordier, G.; Kniep, R.; Höhn, P.; Rabenau, A. Ca_6GaN_5 und Ca_6FeN_5. Verbindungen mit $[CO_3]^{2-}$-isosteren Anionen $[GaN_3]^{6-}$ und $[GaN_3]^{6-}$. *Z. Anorg. Allg. Chem.* **1990**, *591*, 58–66.

5. Bendyna, J.K.; Höhn, P.; Kniep, R. Crystal structure of tristrontium trinitridoferrate(III), $Sr_3[FeN_3]$. *Z. Kristallogr. New Cryst. Struct.* **2008**, *223*, 109–110.

6. Höhn, P.; Kniep, R.; Rabenau, A. $Ba_3[FeN_3]$, new nitridoferrate(III) with $[CO_3]^{2-}$ isosteric anions $[FeN_3]^{6-}$. *Z. Kristallogr.* **1991**, *196*, 153–158.

7. Höhn, P. Ternäre und quaternäre Nitridometallate: Verbindungen in den Systemen Li-Erdalkalimetall-Übergangsmetall-Stickstoff (Übergangsmetall = Ta, Mo, W, Fe, Co). Ph.D. Thesis, TH Darmstadt, Darmstadt, Germany, 1992; pp. 1–284.

8. Bendyna, J.K.; Höhn, P.; Kniep, R. Crystal structure of octastrontium bistrinitridoferrate(III) dinitridoferrate(II), $Sr_8[FeN_3]_2[FeN_2]$. *Z. Kristallogr. New Cryst. Struct.* **2008**, *223*, 181–182.

9. Bendyna, J.K. New Developments in Nitridometalates and Cyanamides: Chemical, Structural and Physical Properties. Ph.D. Thesis, TU Dresden, Dresden, Germany, 2009; pp. 1–227.

10. Höhn, P.; Kniep, R. $Ca_2[FeN_2]$ and $Sr_2[FeN_2]$: Nitridoferrate(II) mit isolierten Anionen $[Fe_2N_4]^{8-}$ und $[FeN_2]^{4-}$. *Z. Naturforsch. B* **1992**, *47*, 477–481.

11. Höhn, P.; Haag, S.; Milius, W.; Kniep, R. $Sr_2Li[Fe_2N_3]$ und $Ba_2Li[Fe_2N_3]$: Nitridoferrate(II) Isotypes with $\frac{1}{\infty}[(FeN_{3/2})_2^{5-}]$ Anions. *Angew. Chem. Int. Ed.* **1991**, *30*, 831–832.

12. Bendyna, J.K.; Höhn, P.; Kniep, R. Crystal structure of octastrontium bistrinitridomanganate(III) dinitridoferrate(II), $Sr_8[MnN_3]_2[FeN_2]$. *Z. Kristallogr. New Cryst. Struct.* **2008**, *223*, 183–184.

13. Gudat, A.; Kniep, R.; Rabenau, A. $Li_4[FeN_2]$: A Nitridoferrate(II) with Anions Isosteric with CO_2. A Defect Variant of the Li_3N Type Structure. *Angew. Chem. Int. Ed.* **1991**, *30*, 199–200.

14. Niewa, R.; Huang, Z.L.; Schnelle, W.; Hu, Z.; Kniep, R. Preparation, Crystallographic, Spectroscopic and Magnetic Characterization of Low-Valency Nitridometalates $Li_2[(Li_{1-x}M_x)N]$ with M = Cu, Ni. *Z. Anorg. Allg. Chem.* **2003**, *629*, 1778–1786.

15. Niewa, R.; Hu, Z.; Kniep, R. Mn and Fe K-edge XAS Spectra of Manganese and Iron Nitrido Compounds. *Eur. J. Inorg. Chem.* **2003**, *2003*, 1632–1634.

16. Klatyk, J.; Schnelle, W.; Wagner, F.R.; Niewa, R.; Novák, P.; Kniep, R.; Waldeck, M.; Ksenofontov, V.; Gütlich, P. Large Orbital Moments and Internal Magnetic Fields in Lithium Nitridoferrate(I). *Phys. Rev. Lett.* **2002**, *88*, 207202.

17. Klatyk, J.; Kniep, R. Crystal structure of alkaline earth dilithium bis(nitridolithiate/ferrates(I)), $Ca\{Li_2[(Li_{1-x}Fe_x)N]_2\}$, $x = 0.30$ and $Sr\{Li_2[(Li_{1-x}Fe_x)N]_2\}$, $x = 0.46$. *Z. Kristallogr. New Cryst. Struct.* **1999**, *214*, 449–450.

18. Klatyk, J.; Kniep, R. Crystal structure of dicalcium (dinitridolithiate/ferrate(I)), $Ca_2\{Li[(Li_{1-x}Fe_x)N_2]\}$, $x = 0.82$. *Z. Kristallogr. New Cryst. Struct.* **1999**, *214*, 451–452.

19. Höhn, P.; Schnelle, W.; Zechel, K. *Book of Abstracts, SCTE-19, Genoa*; SCTE: Genoa, Italy, 2014; p. 71.

20. Jesche, A.; Canfield, P.C. Single crystal growth from light, volatile and reactive materials using lithium and calcium flux. *Philos. Mag.* **2014**, *94*, 2372–2402.

21. Jesche, A.; Ke, L.; Jacobs, J.L.; Harmon, B.; Houk, R.S.; Canfield, P.C. Alternating magnetic anisotropy of $Li_2(Li_{1-x}T_x)N$ with T = Mn, Fe, Co, and Ni. *Phys. Rev. B* **2015**, *91*, 180403.

22. Fisk, Z.; Remeika, J.P. Growth of single crystals from molten metal fluxes. In *Handbook on the Physics and Chemistry of Rare Earths*; Gschneidner, K.A., Eyring, L., Eds.; Elsevier: Amsterdam, The Netherlands, 1989; Volume 12.

23. Canfield, P.C.; Fisk, Z. Growth of single crystals from metallic fluxes. *Philos. Mag. B* **1992**, *65*, 1117–1123.

24. Boström, M.; Hovmöller, S. Preparation and crystal structure of the novel decagonal approximant $Mn_{123}Ga_{137}$. *J. Alloys Compd.* **2001**, *314*, 154–159.

25. Pathak, M.; Bobnar, M.; Schnelle, W.; Höhn, P.; Grin, Y. $Li_{16}Sr_6M_6N$ (M = Ge, Sn, Pb)—Cubic Tetrelide-Nitrides with an Ordered Ir_4Sc_{11} Type Structure. *Z. Anorg. Allg. Chem.* **2016**, *642*, 1075.

26. Ovchinnikov, A.; Schnelle, W.; Bobnar, M.; Borrmann, H.; Sichelschmidt, J.; Grin, Y.; Höhn, P. Extended anionic frameworks in the AE-Mn-N systems. Synthesis, structure and physical properties of new nitridomanganates. In Proceedings of the European Conference on Solid State Chemistry, Vienna, Austria, 23–26 August 2015; p. 288.

27. Jesche, A.; McCallum, R.W.; Thimmaiah, S.; Jacobs, J.L.; Taufour, V.; Kreyssig, A.; Houk, R.S.; Bud'ko, S.L.; Canfield, P.C. Giant magnetic anisotropy and tunnelling of the magnetization in $Li_2(Li_{1-x}Fe_x)N$. *Nat. Commun.* **2014**, *5*, doi:10.1038/ncomms4333.

28. Cordier, G.; Gudat, A.; Kniep, R.; Rabenau, A. LiCaN and Li_4SrN_2, Derivatives of the Fluorite and Lithium Nitride Structures. *Angew. Chem. Int. Ed.* **1989**, *28*, 1702–1703.

29. Canfield, P.C.; Fisher, I.R. High-temperature solution growth of intermetallic single crystals and quasicrystals. *J. Cryst. Growth* **2001**, *225*, 155–161.

30. Banhart, J.; Ebert, H.; Voitländer, J.; Winter, H. Diamagnetic susceptibility of pure metals and binary alloys. *J. Magn. Magn. Mater.* **1986**, *61*, 221–224.

31. Myers, W.R. The Diamagnetism of Ions. *Rev. Mod. Phys.* **1952**, *24*, 15–27.

32. Höhn, P.; Hoffmann, S.; Hunger, J.; Leoni, S.; Nitsche, F.; Schnelle, W.; Kniep, R. β-Ca_3N_2, a Metastable Nitride in the System Ca–N. *Chem. Eur. J.* **2009**, *15*, 3419–3425.

33. Massalski, T.B. (Ed.) *Binary Alloy Phase Diagrams*; ASM International: Metals Park, OH, USA, 1996; Volume 3.

34. Höhn, P.; Kniep, R. $LiSr_2[CoN_2]$: Ein Nitridocobaltat mit gestreckten Anionen $[CoN_2]^{5-}$. *Z. Naturforsch. B* **1992**, *47*, 434–436.

35. Pinkerton, F. High coercivity in melt-spun Dy–Fe–B and Tb–Fe–B alloys. *J. Magn. Magn. Mater.* **1986**, *54*, 579–582.

36. Liu, R.M.; Yue, M.; Na, R.; Deng, Y.W.; Zhang, D.T.; Liu, W.Q.; Zhang, J.X. Ultrahigh coercivity in ternary Tb–Fe–B melt-spun ribbons. *J. Appl. Phys.* **2011**, *109*, 07A760, doi:10.1063/1.3567144.

37. Wu, W.; Kiryukhin, V.; Noh, H.J.; Ko, K.T.; Park, J.H.; Ratcliff, W.; Sharma, P.A.; Harrison, N.; Choi, Y.J.; Horibe, Y.; et al. Formation of Pancakelike Ising Domains and Giant Magnetic Coercivity in Ferrimagnetic LuFe$_2$O$_4$. *Phys. Rev. Lett.* **2008**, *101*, 137203.

38. Iida, J.; Tanaka, M.; Nakagawa, Y.; Funahashi, S.; Kimizuka, N.; Takekawa, S. Magnetization and Spin Correlation of Two-Dimensional Triangular Antiferromagnet LuFe$_2$O$_4$. *J. Phys. Soc. Jpn.* **1993**, *62*, 1723–1735.

39. Ko, K.T.; Noh, H.J.; Kim, J.Y.; Park, B.G.; Park, J.H.; Tanaka, A.; Kim, S.B.; Zhang, C.L.; Cheong, S.W. Electronic Origin of Giant Magnetic Anisotropy in Multiferroic LuFe$_2$O$_4$. *Phys. Rev. Lett.* **2009**, *103*, 207202.

40. Novák, P.; Wagner, F.R. Electronic structure of lithium nitridoferrate: Effects of correlation and spin-orbit coupling. *Phys. Rev. B* **2002**, *66*, 184434.

41. Antropov, V.P.; Antonov, V.N. Colossal anisotropy of the magnetic properties of doped lithium nitrodometalates. *Phys. Rev. B* **2014**, *90*, 094406.

42. Ke, L.; van Schilfgaarde, M. Band-filling effect on magnetic anisotropy using a Green's function method. *Phys. Rev. B* **2015**, *92*, 014423.

43. Yang, I.K.; Kim, J.; Lee, S.H.; Cheong, S.W.; Jeong, Y.H. Charge ordering, ferroelectric, and magnetic domains in LuFe$_2$O$_4$ observed by scanning probe microscopy. *Appl. Phys. Lett.* **2015**, *106*, 152902.

44. Jahn, H.A.; Teller, E. Stability of Polyatomic Molecules in Degenerate Electronic States. I. Orbital Degeneracy. *Proc. R. Soc. London, Ser. A* **1937**, *161*, 220–235.

45. Kugel', K.I.; Khomskiĭ, D.I. The Jahn-Teller effect and magnetism: Transition metal compounds. *Sov. Phys. Uspekhi* **1982**, *25*, 231.

46. *WinXPow*; STOE & Cie GmbH: Darmstadt, Germany, 2003.

47. Petříček, V.; Dušek, M.; Palatinus, L. Crystallographic Computing System JANA2006: General features. *Z. Kristallogr.* **2014**, *229*, 345–352.

48. Sheldrick, G.M. *SHELXS-97-2*; Program for Solution of Crystal Structures; University of Göttingen: Göttingen, Germany, 1997.

49. Sheldrick, G.M. *SHELXL-97-2*; Program for Refinement of Crystal Structures; University of Göttingen: Göttingen, Germany, 1997.

6

Computing Free Energies of Hydroxylated Silica Nanoclusters: Forcefield versus Density Functional Calculations

Antoni Macià Escatllar [1] (ID)**, Piero Ugliengo** [2] (ID) **and Stefan T. Bromley** [1,3,*] (ID)

[1] Departament de Ciència de Materials i Química Física and Institut de Química Teòrica i Computacional (IQTCUB), Universitat de Barcelona, E-08028 Barcelona, Spain; tonimacia@gmail.com

[2] Dipartimento di Chimica and NIS Centre, Università degli Studi di Torino, Via P. Giuria 7, I-10125 Torino, Italy; piero.ugliengo@unito.it

[3] Institució Catalana de Recerca i Estudis Avançats (ICREA), E-08010 Barcelona, Spain

* Correspondence: s.bromley@ub.edu

Academic Editor: Duncan H. Gregory

Abstract: We assess the feasibility of efficiently calculating accurate thermodynamic properties of $(SiO_2)_n \cdot (H_2O)_m$ nanoclusters, using classical interatomic forcefields (FFs). Specifically, we use a recently parameterized FF for hydroxylated bulk silica systems (FFSiOH) to calculate zero-point energies and thermal contributions to vibrational internal energy and entropy, in order to estimate the free energy correction to the internal electronic energy of these nanoclusters. The performance of FFSiOH is then benchmarked against the results of corresponding calculations using density functional theory (DFT) calculations employing the B3LYP functional. Results are reported first for a set of $(SiO_2)_n \cdot (H_2O)_m$ clusters with n = 4, 8 and 16, each possessing three different degrees of hydroxylation ($R = m/n$): 0.0, 0.25 and 0.5. Secondly, we consider five distinct hydroxylated nanocluster isomers with the same $(SiO_2)_{16} \cdot (H_2O)_4$ composition. Finally, the free energies for the progressive hydroxylation of three nanoclusters with R = 0–0.5 are also calculated. Our results demonstrate that, in all cases, the use of FFSiOH can provide estimates of thermodynamic properties with an accuracy close to that of DFT calculations, and at a fraction of the computational cost.

Keywords: hydroxylated silica; nanoclusters; free energy; density functional theory; forcefields

1. Introduction

The energy of a chemical system standardly calculated via ab initio methods of quantum chemistry, such as density functional theory (DFT), only provides the internal energy U_{elec} (which includes the potential and kinetic energy contributions from electrons and their interaction with nuclei). U_{elec}, however, does not take into account the contribution arising from the quantum nature of the nuclear motion, i.e., the zero point energy (ZPE) for nuclei in their lowest vibrational energy level. More importantly, at finite temperatures U_{elec} also omits contributions from all the different microstates occupied by the system at a temperature, T, and volume, V. Using statistical mechanics, these microstates can be appropriately counted and their weighted contributions evaluated and summed to form the partition function, Q. In turn, from Q we can calculate the thermodynamic properties of the system, namely the entropy, S, and ultimately, the free energy, G. Differences in G (i.e., ΔG), often stemming from entropic factors rather than internal energy differences, determine the most likely pathways followed by (bio)chemical processes (e.g., reactions, solvation, and protein docking). Sophisticated methods, which efficiently and accurately estimate ΔG values, are thus understandably essential for the realistic computational modelling of such systems [1].

In the simplest case of a molecule in its electronic ground state residing in a local energy minimum, the estimation of free energy can be greatly simplified. Assuming that the behavior of a molecule can be represented by that of an ideal gas and that the different occupiable microstates (i.e., degrees of freedom) are independent from each other, one can obtain thermodynamic quantities by dividing Q into individual contributions from translations (q_{trans}), rotations (q_{rot}), and vibrations (q_{vib}). Computing the q_{trans} and q_{rot} under these assumptions is straightforward (see below), but evaluation of q_{vib} requires computation of the different vibrational modes. Assuming small amplitude vibrations around a minimum energy configuration, a harmonic oscillator model is a good description of the nuclear vibrations, so that one can obtain the vibrational levels provided that the Hessian matrix of the force constants (second derivative of the total energy with respect to atomic displacements) computed, either analytically or numerically, is available. In this most simple of cases, and for relatively small molecules, the estimate of free energy can be computed by efficient ab initio quantum chemical methods, such as those based on density functional theory (DFT) (thanks to efficient implementation of the analytical second derivatives of the total energy). However, the computational cost of performing these calculations quickly rises with increasing system size, becoming prohibitive for routine calculations for systems possessing more than approximately 100 atoms. This is particularly so for a system in which the Hessian matrix should be computed numerically, either by using analytical gradient or, even worse, when the gradient energy also has to be evaluated numerically. Thus, in order to be able to perform free energy calculations at a reasonable computational cost, classical interatomic forcefields (FFs) are usually employed.

Sampling the energy landscape of configurations of a (bio)molecular system using FFs is typically at least 10,000 times quicker than when done using DFT. Although allowing for faster calculations, it is essential that the FFs employed should be as accurate as possible, as not to negate the increased computational efficiency. FFs developed for modelling (bio)molecular systems have historically been parameterized using experimentally known structural properties and results from *ab initio* quantum chemical calculations [2]. More recently, with the increasingly widespread use of (bio)chemical free energy calculations, such FFs have used experimental free energy data in their parameterization [3,4]. In materials science, especially for inorganic materials and metals/alloys, FFs are also widely employed to study structural and mechanical properties [5]. For such materials, unlike in the biochemical world, entropic contributions to the free energy of the system are not generally thought to be very important for bulk properties. However, for small system sizes (e.g., inorganic nanoclusters) explicit consideration of all degrees of freedom can become significant for describing the structure and evolution of such species under a given set of physical conditions. Studies of inorganic nanocluster systems in a variety of contexts (e.g., nucleation [6–9], presence of a reactive atmosphere [10–12], solvation [13–16]) have confirmed the importance of explicitly considering the entropic contributions to the free energies. These cases have tended to look at rather small species (≤ 40 atoms) and, when taken into account, calculate the entropic contribution to the free energy via DFT using the harmonic approximation for vibrations. As noted above, such calculations quickly become prohibitive with increasing system size and, following the progress in the field of biochemical modelling, accurately parameterizing FFs for inorganic nanosystems would allow for huge increases in computational efficiency. FFs for inorganic materials are usually parameterized for bulk structural and mechanical properties, and thus tend not to be ideally suited for treating nanoclusters (which tend to have non-bulk-like structures and properties). Although some efforts have been made to parameterize FFs (specifically for inorganic nanoclusters) [17], or to make size-transferable FFs for inorganic materials [18], only very few FFs have explicitly included thermodynamic data in their parameterization [19,20] (in order to treat these systems more accurately) [21].

The objective of this work is to assess whether an existing FF, that was not parameterized using thermodynamic data, can be used to provide free energy estimates to a similar accuracy as that provided by DFT. Specifically, we examine the performance of FFSiOH [22], which is a polarisable ionic FF, that was parametrized with respect to the results of DFT calculations of silica bulk phases and hydroxylated silica surfaces. We have recently shown that FFSiOH can provide results in good agreement with DFT calculations (with respect to structure, relative energies, and vibrational frequencies) for a range of hydroxylated silica nanoclusters of different sizes and degrees of hydroxylation [23]. Following this work, we have selected hydroxylated silica nanoclusters as a suitable test case for the present study. Free energies in solution of these species are of particular interest for gaining insight into nucleation, growth, and dissolution of silicate materials [24,25]. The free energies of small silica nanoclusters (or oligomers) have been calculated using DFT a number of times in the literature [13–16], employing the harmonic approximation for vibrations and various solvation models. Considering that both: (i) our objective is to evaluate the performance of the FFSiOH potential (i.e., rather than different solvation models); and (ii) the most computationally expensive part of evaluating free energies in these cases is the evaluation of the vibrational entropy; we focus on the ability of FFSiOH to provide free energies in vacuum.

We calculate ZPE, vibrational entropies, and free energies (both total and of hydroxylation) of a range of hydroxylated silica clusters with a range of sizes and degrees of hydroxylation, using FFSiOH. Overall, we find good agreement with corresponding DFT calculations, indicating that FFSiOH could provide an extremely computationally efficient means through which to calculate accurate free energies of nano-silicate species.

2. Methodology

We first review how to obtain thermodynamic properties from the partition function for a silica nanocluster, residing in a local energy minimum in its electronic ground state. We will only focus on main assumptions and provide the necessary equations by which the results can be achieved (full details can be found in Reference [26]). Assuming that: (1) the nanocluster behaves as an ideal gas; and (2) that all energy levels can be grouped into either translations, rotations, vibrations, or electronic levels, and that they are uncoupled from each other, Q can be defined as:

$$Q = q_{trans} \cdot q_{rot} \cdot q_{vib} \cdot q_{elec} \qquad (1)$$

where each q is the partition function associated with the translational, rotational, vibrational, and electronic energy, respectively. Each q can be calculated with the following expressions derived for a particle in a box of volume V, at temperature T:

$$q_{trans} = \left(\frac{2\pi m k_B T}{h^2} \right)^{3/2} V$$

$$q_{rot} = \frac{\pi^{\frac{1}{2}}}{\sigma_r} \left(\frac{T^{\frac{3}{2}}}{(\Theta_{r,x}\Theta_{r,y}\Theta_{r,z})^{1/2}} \right) \qquad (2)$$

$$q_{vib} = \prod_K \frac{e^{-\Theta_{v,K}/2T}}{1 - e^{-\Theta_{v,K}/T}}$$

where m is the mass, k_B is the Boltzmann constant, h is Planck's constant, σ_r is the symmetry number for rotation, and the Θ correspond to characteristic temperatures for rotations Θ_r and vibrations Θ_v. The product q_{vib} runs over each vibrational mode K. Since the energy to reach electronically excited states is much higher than the thermal energy in our system, and we deal with closed shell molecules, the term q_{elec} becomes 1, and does not contribute to the entropy. The rotational and vibrational characteristic temperatures are defined by:

$$\Theta_r = \frac{h^2}{8\pi^2 I k_B}$$

$$\Theta_{v,K} = \frac{h\nu_K}{k_B} \qquad (3)$$

where I is the moment of inertia and K represents each individual vibration. Assuming ideal gas behavior, we can derive the expressions of internal energy (E_{trans}, E_{rot}, E_{vib}, and E_{elec}), and entropy (S_{trans}, S_{rot}, S_{vib}, and S_{elec}), using the corresponding partition functions, where we assume the zero of energy to be the electronic energy:

$$E_{trans} = 3/2 \cdot R$$

$$E_{rot} = R \cdot T$$

$$E_{vib} = R \sum_K \Theta_{v,K} \left(1/2 + \frac{1}{e^{\Theta_{v,K}/T} - 1} \right)$$

$$S_{trans} = R(lnq_{trans} + 1 + 3/2)$$

$$S_{rot} = R(lnq + 3/2)$$

$$S_{vib} = R \cdot \sum_K \left(\frac{\Theta_{v,K}/T}{e^{\Theta_{v,K}/T} - 1} - ln\left(1 - e^{-\Theta_{v,K}/T}\right) \right)$$

(4)

where R is the ideal gas constant ($R = N \cdot k_B$). The term $R \cdot \sum_K \Theta_{v,K}/2$ in E_{vib} corresponds to the energy of vibrations at 0 K (i.e., the ZPE). In the Results section, we show the ZPE and the $R \cdot \sum_K \frac{1}{e^{\Theta_{v,K}/T-1}}$ factor separately, with the latter defined as the thermal contribution to the total energy (U_{vib}). After obtaining the thermal and entropic contribution of each term, we add them to the electronic energy in the following way to obtain the zero Kelvin energy ($U(T_{0K})$), the internal energy at a given temperature ($U(T)$), the enthalpy ($H(T)$), and the free energy ($G(T)$):

$$U(0K) = U_{elec} + ZPE$$

$$U(T) = U_{elec} + E_{rot} + E_{trans} + E_{vib}$$

$$H(T) = U(T) + k_B T$$

$$G(T) = H(T) - T(S_{trans} + S_{rot} + S_{vib})$$

(5)

In order to give a reasonable estimate of the broad performance of FFSiOH, we perform our calculations on two sets of clusters (A and B) that provide complementary information. Set A contains nine globally minimized clusters from previous work [27–30], with stoichiometries $(SiO_2)_n \cdot (H_2O)_m$ with n = 4, 8, and 16 and m corresponding to different values which represent three degrees of hydroxylation (i.e., $R = m/n$) of: 0.0, 0.25 and 0.5. This entails that for n = 4 we have m = 0, 1, 2, for n = 8 we have m = 0, 2, 4 and for n = 16 we have m = 0, 4, 8. These ratios correspond to anhydrous, low, and high degrees of hydroxylation. The structures in set A can be seen in Figure 1. Set A allows us to assess the dependency of the FFSiOH versus DFT error in the calculated thermodynamic properties (on size and degree of hydroxylation). The contributions from S_{rot} and S_{trans} depend on the geometry and mass of the system, which are essentially identical for FFSiOH and DFT calculations (except for very small displacements of the atoms). The differences between the calculated values from DFT and FFSiOH should thus be very small in these cases. We will therefore focus on the global values of the thermodynamic properties and the vibrational contributions.

The second set (B) of structures consist of five isomers with $(SiO_2)_{16} \cdot (H_2O)_4$ stoichiometry (see Figure 2). These structures are selected from the dataset in Reference [23] as representative low energy minima, as found by classical IP-based global optimization searches, followed by DFT refinement (see details in References [27–30]). The relative energies of the structures calculated at a DFT level using the B3LYP functional are noted in Figure 2. Moreover, this set of structures displays a wide spread of differences between the relative energies provided by FFSiOH and DFT. Set B will thus help establish whether the performance of FFSiOH with respect to reproducing DFT-calculated energy differences also propagates to the free energy corrections or not.

Finally, we compare the results for the free energy of hydroxylation (i.e., reaction with water) of clusters for each size from their anhydrous state to that of low and subsequently high hydroxylation (see Figure 1). We compare the free energy of hydroxylation using: (a) the results directly obtained from DFT calculations; and (b) the electronic energy from DFT results, and the correction to the free energy from FFSiOH. Since FFSiOH is not parameterized to describe the water molecule, we use the Gibbs free energy obtained from DFT for this quantity throughout.

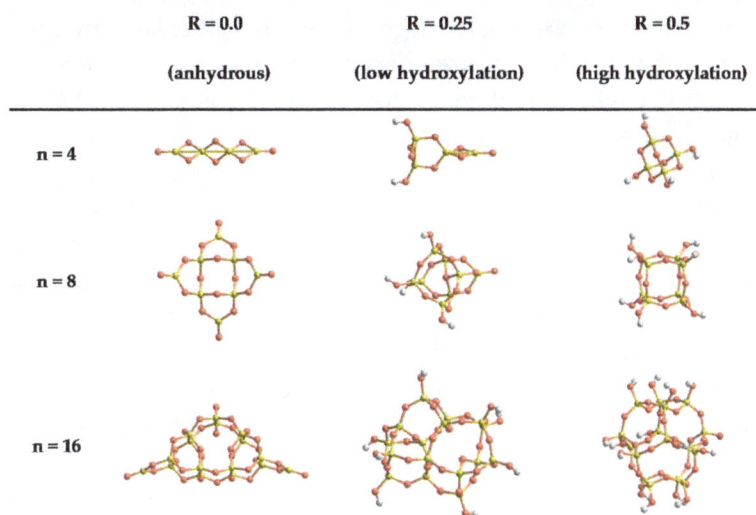

Figure 1. Set A: structures of global minima $(SiO_2)_n \cdot (H_2O)_m$ nanoclusters with sizes ($n = 4, 8,$ and 16) and degrees of hydroxylation ($R = 0.0, 0.25$ and 0.5)—taken from References [27–30].

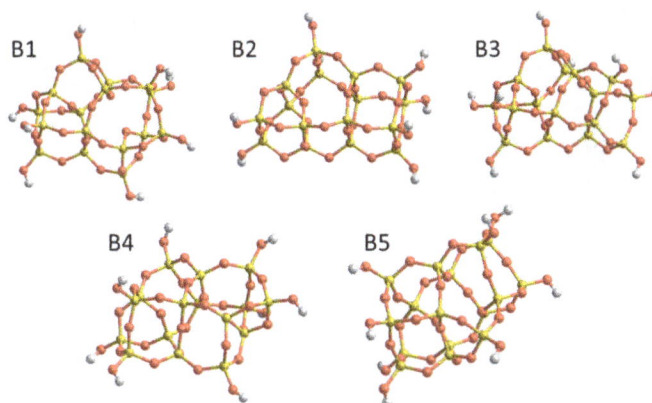

Figure 2. Set B: low energy $(SiO_2)_{16} \cdot (H_2O)_4$ nanoclusters. Relative U_{elec} values calculated using DFT in kJ/mol: B1 (global minimum)—0.0 kJ/mol, B2—13.5 kJ/mol, B3—38.6 kJ/mol, B4—48.2 kJ/mol, B5—95.5 kJ/mol. Structures were taken from the Hydroxylated Nanosilica Dataset in Ref. [23].

All DFT calculations were performed using the Gaussian09 code [31] using the B3LYP [32] functional and 6–31g(d,p) basis set. This combination of functional and basis set has proven to provide a good compromise between computational efficiency and accuracy for hydroxylated silica systems in previous work [14–16,27,33]. Since our systems contain low frequency vibrational modes, we use tight cut-offs and an ultrafine integration grid for the geometry optimizations, using the Berny algorithm [34]. The FFSiOH calculations were performed using the GULP code [35], employing the rational function optimization method for geometry optimizations. All thermodynamic data were calculated at standard conditions (i.e., $T = 298.15$ K and 1 atm pressure). We note that all reported free energy terms are calculated using the structures optimized, using the respective method (i.e.,

FFSiOH-calculated free energies use FFSiOH-optimised nanocluster structures and DFT-calculated free energies use DFT-optimized nanocluster structures).

3. Results and Discussion

3.1. Thermodynamic Data with Respect to Size and Hydroxylation

In Figure 3 we plot the FFSiOH-DFT differences in ZPE, U_{vib}, U correction, $T{\cdot}S_{vib}$, and the G correction of the nanoclusters, with respect to degree of hydroxylation and grouped by size. Energies are given reported in kJ/mol per SiO_2 in order to compare quantities calculated for nanoclusters of different sizes. In the ZPE, the largest difference between the FFSiOH and DFT data is found for the anhydrous $(SiO_2)_4$ nanocluster, with a difference of 2.61 kJ/mol per SiO_2 unit. The smallest difference observed is for $(SiO_2)_8{\cdot}(H_2O)_8$, with a difference of 0.04 kJ/mol per SiO_2 unit. Both increases in size and degree of hydroxylation lower the respective ZPE differences. The higher error in the anhydrous cases can be explained with the fact that at these very small sizes, the proportion of strained rings and terminal oxygens is large. These defects are not present in periodic bulk-calculations, from which the FFSiOH was parametrized, and therefore the higher errors in the anhydrous systems are not unexpected. More precisely, we identified a red-shift in the frequencies of the anhydrous systems when calculated using FFSiOH with respect to DFT (an average difference of 58.1 cm^{-1} for $(SiO_2)_4$, 34.5 cm^{-1} for $(SiO_2)_8$ and 28.9 cm^{-1} for $(SiO_2)_{16}$). When added together, these frequency differences cause an increase in the respective ZPE differences. In comparison, the nanoclusters with 25% hydroxylation have an average ZPE difference of 0.84 kJ/mol per SiO_2 unit, while for 50% hydroxylation the average ZPE difference is 0.4 kJ/mol per SiO_2 unit.

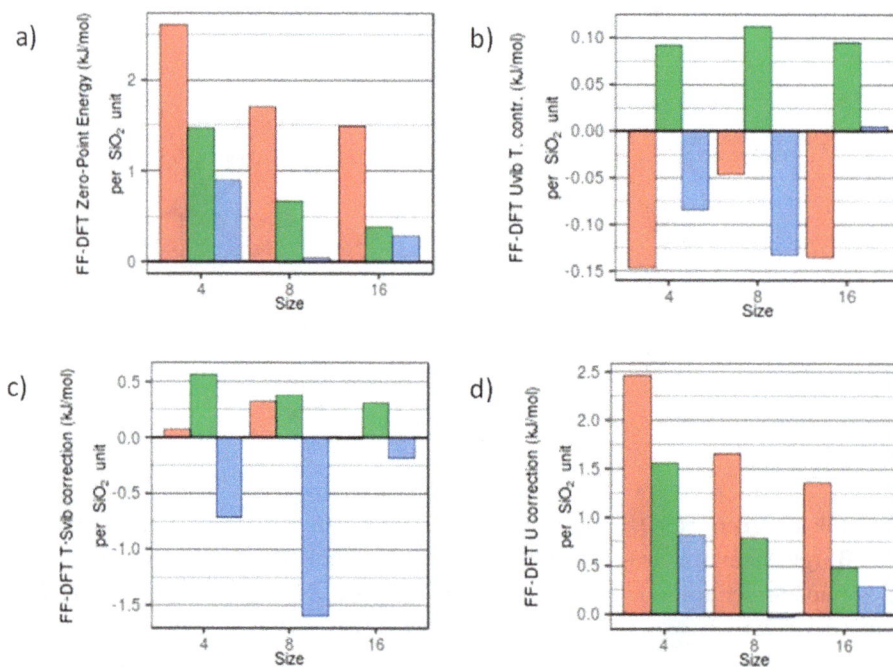

Figure 3. FFSiOH-DFT differences in: (a) ZPE; (b) U_{vib}; (c) S_{vib}; and (d) U at $T = 298.15$K and $P = 1$ atm. Red, green, and blue bars refer to the degree of hydroxylation, $R = 0.0$, 0.25 and 0.5, respectively.

The thermal contribution to U_{vib} from the FFSiOH calculations is in very good agreement with DFT values (see Figure 3b). The FFSiOH-DFT difference averaged over all structures is 0.09 kJ/mol per SiO_2 unit. The U_{vib} is dominated by the low frequencies, due to the negative exponential. The U_{vib} contribution is also smaller (from 13% to 24% the value of ZPE). The differences here do not seem to obey any relationship with either size or degree of hydroxylation. The contributions to the internal energy from translation and rotation are independent of the properties of the molecule, and therefore have the same value. Since the ZPE is five times bigger than the U_{vib} thermal correction, and U_{vib} is very close to DFT values, the total correction to U follows the same trends as the ZPE, and the differences among DFT and FFSiOH values are almost the same as for ZPE.

The $T \cdot S_{vib}$ term calculated using FFSiOH, for most of the systems is, again, in very good agreement with DFT-calculated values (see Figure 3c), with an average absolute FFSiOH-DFT difference of 0.46 kJ/mol per SiO_2 unit. As for the thermal contribution to U_{vib}, the S_{vib} term is dominated by low frequency modes. Here, however, the largest discrepancy between DFT and FFSiOH data is for the $(SiO_2)_8 \cdot (H_2O)_4$ structure, where the DFT-calculated entropy is 1.60 kJ/mol SiO_2 per unit higher than that calculated using FFSiOH. The average absolute FFSiOH-DFT difference of the $T \cdot S_{vib}$ term (without taking into account the $(SiO_2)_8 \cdot (H_2O)_4$ species) is 0.32 kJ/mol per SiO_2 unit.

The average absolute differences between FFSiOH results and DFT for the rotational and translational entropies are 0.05 kJ/mol per SiO_2 unit and 0.03 kJ/mol prer SiO_2 unit respectively. These results are not surprising, since S_{rot} and S_{trans} depend only on the optimized geometric structure of the respective nanocluster, which is very similar when calculated with DFT and FFSiOH (see Reference [23]).

The differences between free energies calculated using DFT and FFSiOH fall in the range 0.25 kJ/mol to 2.68 kJ/mol per SiO_2 unit (see Figure 4). Generally, as size increases, all the FFSiOH-DFT differences in the predicted G correction become smaller, but to a different extent with respect to the degree of hydroxylation. The percentage difference between FFSiOH and DFT results for the G correction are generally found to be <15%. We note that for the smallest cluster considered, the percentage difference is found to be relatively large due to very small DFT-calculated values of G (with respect to the fairly constant FFSiOH-DFT difference). The G correction of the highly hydroxylated $(SiO_2)_8 \cdot (H_2O)_4$ particle is found to be relatively large, due to the overestimation of the S_{vib} contribution (see above). Apart from in this case, the G correction is more accurately calculated for hydroxylated systems than for anhydrous systems. However, we find that nanoclusters with lower (25%) hydroxylation are those for which FFSiOH provides the best match with the DFT-calculated G corrections. This result is in line with the capability of FFSiOH to better estimate relative U_{elec} values (with respect to DFT values) for this degree of hydroxylation [23].

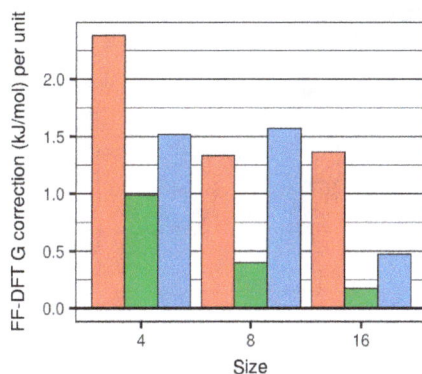

Figure 4. FFSiOH-DFT differences for the free energy correction, G. Red, green and blue bars refers to the degree of hydroxylation $R = 0.0$, 0.25 and 0.50, respectively.

3.2. Free Energy in a Set of Isomers

The FFSiOH-DFT differences in calculated thermodynamic properties among the $(SiO_2)_{16} \cdot (H_2O)_4$ nanocluster isomers in set B are very low, as can be seen in Figure 5 (note that energies are given in kJ/mol). The largest difference is found for the $T \cdot S_{vib}$ correction, with a value of 3.5 kJ/mol. This is due to the fact that, although different in structure, throughout this set the number of SiO_4 tetrahedra is invariant and the hydroxyl groups are in very similar environments for each isomer. The vibration frequencies are therefore quite similar (standard deviation of 49 cm^{-1}), and the vibrational partition functions are roughly the same, implying very similar thermodynamic properties. The differences between FFSiOH and DFT for the ZPE are between 3.7–6.2 kJ/mol, with an average of 4.9 kJ/mol (see Figure 5a). For the U_{vib} thermal contribution, the differences generally are of the order 1 kJ/mol (see Figure 5b). The total U thermal correction, as it is the sum of the ZPE and the U_{vib} thermal contribution, results in an increased difference between DFT and FFSiOH (see Figure 5c). The $T \cdot S_{vib}$ values show the largest variation in the differences between DFT and FFSiOH over the set of isomers, which can be as high as 3.3 kJ/mol (see Figure 5d). Although all FFSiOH-DFT differences for all these individual thermodynamic components are fairly constant and positive, the size of the FFSIOH-DFT differences of the total G corrections is largely offset by the $T \cdot S_{vib}$ term in the equation for G (see above and Figure 5e). Overall, the FFSiOH-DFT differences in free energies for this set are very small (<4 kJ/mol), and vary in a manner that is independent of the pattern of FFSiOH-DFT electronic energy differences (U_{elec}—see Figure 2). In terms of stability of nanoclusters, this means that only if the energy difference among isomers of this size would be of the order 4 kJ/mol in U_{elec}, the free energy ordering of structures obtained using FFSiOH could be different from the one obtained from DFT calculations. Considering that the U_{elec} energy differences between the low energy isomers in set B are all >13 kJ/mol, the error in the FFSiOH-predicted free energies will not alter the stability ordering.

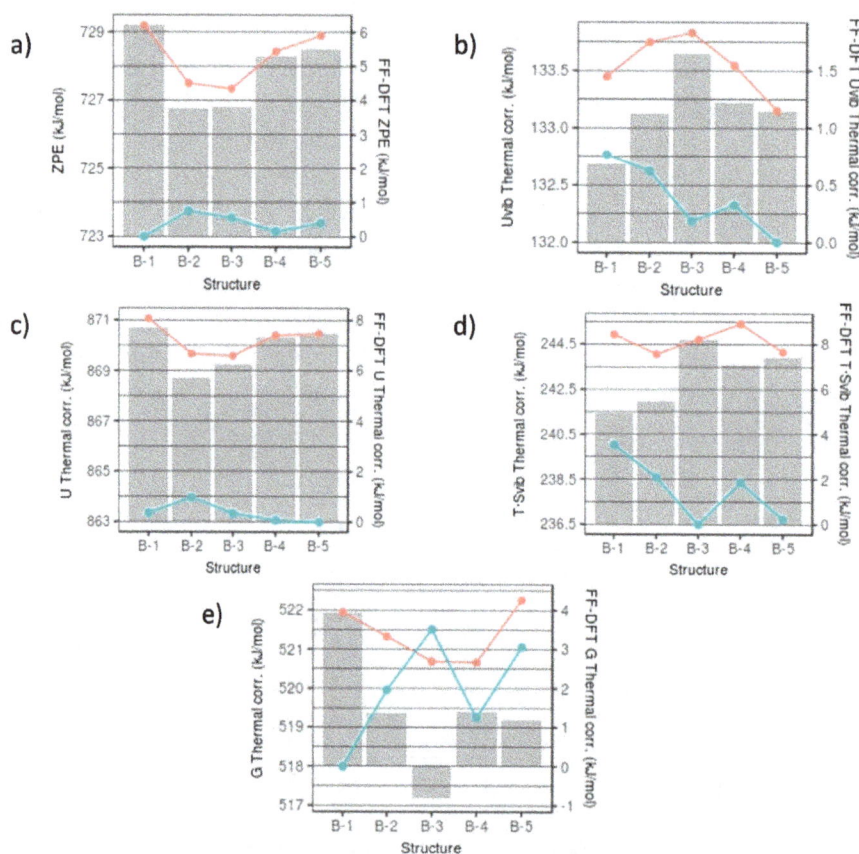

Figure 5. FFSiOH (red line) and DFT (blue line) thermodynamic quantities and their differences (FF-DFT) for nanocluster set B for: (**a**) ZPE; (**b**) U_{vib}; (**c**) S_{vib}; (**d**) U; and (**e**) G. Lines map the absolute values of each thermodynamic quantity. Bars refer to the FFSiOH-DFT differences.

3.3. Free Energy of Hydroxylation

The progressive free energies of hydroxylation of all nanoclusters (considered as calculated using both FFSiOH and DFT) are shown in Table 1. Specifically, for each cluster size we consider two hydroxylation processes: firstly, from the respective anhydrous $(SiO_2)_n$ nanocluster to a $(SiO_2)_n \cdot (H_2O)_m$

nanocluster with low hydroxylation (i.e., $R = 0.25$); and secondly, we then consider the hydroxylation of this latter nanocluster to a $(SiO_2)_n \cdot (H_2O)_m$ nanocluster with a high degree of hydroxylation (i.e., $R = 0.5$). These processes correspond to the evolution of the clusters going from left to right in rows 2–4 in Figure 1. Note that for nanoclusters sizes $n = 8, 16$ these hydroxylation energies relate to dissociative chemisorption of more than one water molecule, and hydroxylation free energies are given per water molecule to allow comparison. For hydroxylation of the smallest $n = 4$ anhydrous nanocluster, the free energies relate to the dissociative chemisorption of a single water molecule at each step. The DFT-calculated free energy differences are -323.0 kJ/mol for the initial hydroxylation step and a further -461.8 kJ/mol for the second hydroxylation reaction. The FFSiOH-calculated values differ by 5.5 kJ/mol and -2.1 kJ/mol, correspondingly, with respect to these DFT values. For the $n = 8$ anhydrous cluster the initial reaction with water to a low hydroxylation state has free energy per water molecule between that of the two reactions considered for the $n = 4$ case (i.e., -392.7 kJ/mol from the DFT calculations). For further hydroxylation, the free energy per water molecule decreases to -244.3 kJ/mol, according to the DFT calculations. For these reactions, the FFSiOH calculations reproduce these values to within 6 kJ/mol. For the largest $n = 16$ cluster considered, the initial anhydrous-low hydroxylation free energy per water molecule has a magnitude well in line with those found for smaller cluster sizes (i.e., -343.2 kJ/mol from the DFT calculations). However, for the low-to-high hydroxylation process, the free energy of hydroxylation drops to only -47.9 kJ/mol per water molecule. From the study in Reference [27] this sharp reduction in hydroxylation energy per molecule is related to the reduced energetic advantage of hydroxylation of finite clusters with increasing hydroxylation above a certain threshold. The respect free energies of these processes as calculated by FFSiOH is found to be within 5 kJ/mol of the DFT-calculated free energies. In all cases the hydroxylation free energy values calculated using FFSiOH are well within the range of typical expected accuracy of DFT calculations with respect to experiment (i.e., 5–15 kJ/mol). This confirms that hydroxylation free energies can be accurately approximated using FFSiOH.

Table 1. Progressive hydroxylation free energies (kJ/mol) of anhydrous nanoclusters calculated using DFT and FFSiOH. Note that all energies are per added water molecule.

Reaction	ΔG_{react} (DFT)	ΔG_{react} (FFSiOH)	$\Delta\Delta G_{react}$ (DFT-FFSiOH)
$Si_4O_8 + H_2O \rightarrow H_2Si_4O_9$	-323.0	-328.5	5.5
$H_2Si_4O_9 + H_2O \rightarrow H_4Si_4O_{10}$	-461.8	-459.7	-2.1
$Si_8O_{16} + 2H_2O \rightarrow H_4Si_8O_{18}$	-392.7	-396.4	3.7
$H_4Si_8O_{18} + 2H_2O \rightarrow H_8Si_8O_{20}$	-244.3	-239.6	-4.7
$Si_{16}O_{32} + 4H_2O \rightarrow H_8Si_{16}O_{36}$	-343.2	-348.0	4.8
$H_8Si_{16}O_{36} + 4H_2O \rightarrow H_{16}Si_{16}O_{40}$	-47.9	-46.6	-1.3

4. Conclusions

We have benchmarked the ability of FFSiOH to reproduce thermodynamic quantities of silica nanoclusters calculated using DFT, with respect to size, degree of hydroxylation, isomer structure, and hydroxylation reactions. In terms of size and degree of hydroxylation (nanocluster set A, Figure 1), the results show that the factor which most affects the thermodynamic properties is the degree of hydroxylation. Since the FFSiOH forcefield was parametrized to reproduce properties of hydroxylated silica surfaces, it is not surprising that it performs slightly worse for the anhydrous nanoclusters, where the presence of ring-strain and defects negatively affect its ability to accurately calculate vibrations. The results also show that FFSiOH performs best for nanoclusters with 25% hydroxylation, which is also the expected average degree of hydroxylation of amorphous silica materials exposed to moisture in the air. This is important, as some of these nanoclusters can be adopted to simulate the hydroxylated silica surfaces either in their crystalline or amorphous state. In terms of energy per unit, the free energy differences between DFT and FFSiOH results are found to reduce with increasing size. These results imply that FFSiOH would be well suited for evaluating free energies of relatively large silica nanoparticles with a moderate degree of hydroxylation. In the B set of low energy $(SiO_2)_{16} \cdot (H_2O)_4$ isomers (see Figure 2), we show that the FFSiOH-DFT differences in free energies between isomers is, in all cases, lower than 4 kJ/mol. This small error (<0.25 kJ/mol per SiO_2) will not thus lead to FFSiOH free energy corrections that could change in U_{elec} stability orderings of isomers (as calculated using DFT), where the typical U_{elec} difference between isomers is >0.8 kJ/mol per SiO_2. For reactions, we find that the free energy of progressive hydroxylation (ΔG_{react}) of the considered silica nanoclusters predicted using FFSiOH is also in very good agreement with the DFT data.

In view of the obtained results, we believe that FFSiOH could be generally employed to obtain thermodynamic corrections to U_{elec} values of hydroxylated nanosilica systems (calculated by DFT or other means) with an accuracy similar to that obtained with DFT methods (at a fraction of the computational cost of the latter). Due to the good performance of FFSiOH for bulk and surface systems [22,23] we also expect our specific methodology to be of use for a wide range of extended hydroxylated silica systems. Overall, given a suitably accurate forcefield, the general approach outlined herein for hydroxylated nanosilica could be promising for a number of other inorganic nanosystems.

Acknowledgments: Support from Spanish MINECO/FEDER grant CTQ2015-64618-R grant and, in part, by Generalitat de Catalunya grants 2014SGR97, XRQTC is acknowledged. We also acknowledge the NOMAD Center of Excellence project (this project has received funding from the European Union's Horizon 2020 research and innovation programme under grant agreement No. 676580). Access to supercomputer resources was provided through grants from the Red Española de Supercomputación. We also thank Alberto Figueroba for assistance in obtaining the point group symmetries of some of the nanoclusters.

Author Contributions: Antoni Macià Escatllar performed all calculations and prepared the original version of the manuscript. Piero Ugliengo discussed the topic of the work and results and helped correct the manuscript. Stefan T. Bromley supervised the work and finalized the manuscript.

References

1.   Kollman, P. Free energy calculations: Applications to chemical and biochemical phenomena. *Chem. Rev.* **1993**, *93*, 2395–2417. [CrossRef]

2.   Van Gunsteren, W.F.; Dolenc, J. Thirty-five years of biomolecular simulation: Development of methodology, force fields and software. *Mol. Simul.* **2012**, *38*, 1271–2181. [CrossRef]

3.   Oostenbrink, C.; Villa, A.; Mark, A.E.; van Gunsteren, W.F. A biomolecular force field based on the free enthalpy of hydration and solvation: The GROMOS force-field parameter sets 53A5 and 53A6. *J. Comput. Chem.* **2004**, *25*, 1656–1676. [CrossRef] [PubMed]

4.   Shivakumar, D.; Williams, J.; Wu, Y.; Damm, W.; Shelley, J.; Sherman, W. Prediction of absolute solvation free energies using molecular dynamics free energy perturbation and the OPLS force field. *J. Chem. Theory Comput.* **2010**, *6*, 1509–1519. [CrossRef] [PubMed]

5.   Becker, C.; Tavazza, F.; Trautt, Z.T.; de Macedo, R.A.B. Considerations for choosing and using force fields and interatomic potentials in materials science and engineering. *Curr. Opin. Solid State Mater. Sci.* **2013**, *17*, 277–283. [CrossRef]

6.   Goumans, T.P.M.; Bromley, S.T. Efficient nucleation of stardust silicates via heteromolecular homogeneous condensation. *Mon. Not. R. Astron. Soc.* **2012**, *420*, 3344–3349. [CrossRef]

7.   Bromley, S.T.; Martin, J.C.G.; Plane, J.M.C. Under what conditions does $(SiO)_N$ nucleation occur? A bottom-up kinetic modelling evaluation. *Phys. Chem. Chem. Phys.* **2016**, *18*, 26913–26922. [CrossRef] [PubMed]

8.   Zachariah, M.R.; Tsang, W. Application of ab initio molecular orbital and reaction rate theories to nucleation kinetics. *Aerosol Sci. Tech.* **1993**, *19*, 499–513. [CrossRef]

9.   Köhler, T.M.; Gail, H.-P.; Sedlmayr, E. MgO dust nucleation in M-Stars: Calculation of cluster properties and nucleation rates. *Astron. Astrophys.* **1997**, *320*, 553–567.

10.   Bhattacharya, S.; Levchenko, S.V.; Ghiringhelli, L.M.; Scheffler, M. Efficient ab initio schemes for finding thermodynamically stable and metastable atomic structures: Benchmark of cascade genetic algorithms. *New J. Phys.* **2014**, *16*. [CrossRef]

11. Bhattacharya, S.; Levchenko, S.V.; Ghiringhelli, L.M.; Scheffler, M. Stability and Metastability of clusters in a Reactive Atmosphere: Theoretical Evidence for Unexpected Stoichiometries of Mg_MO_x. *Phys. Rev. Lett.* **2013**, *111*. [CrossRef] [PubMed]

12. Lepeshkin, S.; Baturin, V.; Tikhonov, E.; Matsko, N.; Uspenskii, Y.; Naumova, A.; Feya, O.; Schoonene, M.A.; Oganov, A.R. Super-oxidation of silicon nanoclusters: Magnetism and reactive oxygen species at the surface. *Nanoscale* **2016**, *8*, 18616–18620. [CrossRef] [PubMed]

13. Mora-Fonz, M.J.; Catlow, C.R.A.; Lewis, D.W. Oligomerization and cyclization processes in the nucleation of microporous silicas. *Angew. Chem. Int. Ed.* **2005**, *44*, 3082–3086. [CrossRef] [PubMed]

14. Trinh, T.T.; Jansen, A.P.J.; van Santen, R.A. Mechanism of oligomerization reactions of silica. *J. Phys. Chem. B* **2006**, *110*, 23099–23106. [CrossRef] [PubMed]

15. Schaffer, C.L.; Thomson, K.T. Density functional theory investigation into structure and reactivity of prenucleation silica species. *J. Phys. Chem. C* **2008**, *112*, 12653–12662. [CrossRef]

16. Jelfs, K.; Flikkema, E.; Bromley, S.T. Evidence for atomic mixing via multiple intermediates during the dynamic interconversion of silicate oligomers in solution. *Chem. Commun.* **2012**, *48*. [CrossRef] [PubMed]

17. Flikkema, E.; Bromley, S.T. A new interatomic potential for nanoscale silica. *Chem. Phys. Lett.* **2003**, *378*, 622–629. [CrossRef]

18. Aguado, A.; Madden, P.A. Fully transferable interatomic potentials for large-scale computer simulations of simple metal oxides: Application to MgO. *Phys. Rev. B* **2004**, *70*. [CrossRef]

19. Raiteri, P.; Demichelis, R.; Gale, J.D. Thermodynamically consistent force field for molecular dynamics simulations of alkaline-earth carbonates and their aqueous speciation. *J. Phys. Chem. C* **2015**, *119*, 24447–24458. [CrossRef]

20. Raiteri, P.; Gale, J.D.; Quigley, D.; Rodger, P.M. Derivation of an accurate force-field for simulating the growth of calcium carbonate from aqueous solution: A new model for the calcite-water interface. *J. Phys. Chem. C* **2010**, *114*, 5997–6010. [CrossRef]

21. Demichelis, R.; Raiteri, P.; Gale, J.D.; Quigley, D.; Gebauer, D. Stable prenucleation mineral clusters are liquid-like ionic polymers. *Nat. Commun.* **2011**, *2*, 590. [CrossRef] [PubMed]

22. Pedone, A.; Malavasi, G.; Menziani, M.C.; Segre, U.; Musso, F.; Corno, M.; Civalleri, B.; Ugliengo, P. FFSiOH: A new force field for silica polymorphs and their hydroxylated surfaces based on periodic B3LYP calculations. *Chem. Mater.* **2008**, *20*, 2522–2531. [CrossRef]

23. Macia, A.; Ugliengo, P.; Bromley, S.T. Modeling hydroxylated nanosilica: Testing the performance of ReaxFF and FFSiOH forcefields. *J. Chem. Phys.* **2017**, *146*. [CrossRef]

24. Iler, R.K. *The Chemistry of Silica: Solubility, Polymerization, Colloid and Surface Properties*; Wiley: New York, NY, USA, 1979.

25. Dove, P.M.; Rimstidt, J.D. *Silica: Physical Behaviour, Geochemistry and Materials Applications*; Heany, P.J., Pewitt, C.T., Gibbs, G.V., Eds.; Mineralogical Society of America: Washington, DC, USA, 1994; Volume 29, pp. 259–301.

26. McQuarrie, D.A.; Simons, J.D. *Molecular Thermodynamics*; University Science Book: Sausalito, CA, USA, 1999.

27. Jelfs, K.E.; Flikkema, E.; Bromley, S.T. Hydroxylation of silica nanoclusters $(SiO_2)_M(H_2O)_N$, M = 4, 8, 16, 24: Stability and structural trends. *Phys. Chem. Chem. Phys.* **2013**, *15*, 20438–20443. [CrossRef] [PubMed]

28. Flikkema, E.; Jelfs, K.E.; Bromley, S.T. Structure and energetics of hydroxylated silica clusters $(SiO_2)_M(H_2O)_N$ M = 8, 16 and N = 1–4: A global optimisation study. *Chem. Phys. Lett.* **2012**, *554*, 117–122. [CrossRef]

29. Flikkema, E.; Bromley, S.T. Dedicated global optimization search for ground state silica nanoclusters: $(SiO_2)_M$ (N = 6–12). *J. Phys. Chem. B* **2004**, *108*, 9638–9645. [CrossRef]

30. Bromley, S.T.; Flikkema, E. Columnar-to-disk structural transition in nanoscale $(SiO_2)_N$ clusters. *Phys. Rev. Lett.* **2005**, *95*, 185505. [CrossRef] [PubMed]

31. Frisch, M.J.; Trucks, G.W.; Schlegel, H.B.; Scuseria, G.E.; Robb, M.A.; Cheeseman, J.R.; Scalmani, G.; Barone, V.; Petersson, G.A.; Nakatsuji, H.; et al. *Gaussian 09*; Revision D.01; Gaussian Inc.: Wallingford, CT, USA, 2013.

32. Stephens, P.J.; Devlin, F.J.; Chabalowski, C.F.; Frisch, M.J. Ab initio calculation of vibrational absorption and circular dichroism spectra using density functional force fields. *J. Phys. Chem.* **1994**, *98*, 11623–11627. [CrossRef]

33. Ugliengo, P.; Sodupe, M.; Musso, F.; Bush, I.J.; Orlando, R.; Dovesi, R. Realistic Models of Hydroxylated Amorphous Silica Surfaces and MCM-41 Mesoporous Material Simulated by Large-scale Periodic B3LYP calculations. *Adv. Mater.* **2008**, *20*, 1–5. [CrossRef]

34. Schlegel, H.B. Optimization of Equilibrium Geometries and Transition Structures. *J. Comput. Chem.* **1982**, *3*, 214–218. [CrossRef]

35. Gale, J.D.; Rohl, A.L. The General Utility Lattice Program (GULP). *Mol. Simul.* **2003**, *29*, 291–341. [CrossRef]

7

Computational Studies on the Selective Polymerization of Lactide Catalyzed by Bifunctional Yttrium NHC Catalyst

Yincheng Wang, Andleeb Mehmood, Yanan Zhao, Jingping Qu and Yi Luo * (iD)

State Key Laboratory of Fine Chemicals, School of Chemical Engineering, Dalian University of Technology, Dalian 116024, China; wangyincheng@mail.dlut.edu.cn (Y.W.); andleeb.mehmood@gmail.com (A.M.); yananzhao92@foxmail.com (Y.Z.); qujp@dlut.edu.cn (J.Q.)
* Correspondence: luoyi@dlut.edu.cn

Academic Editor: Hani Amouri

Abstract: A theoretical investigation of the ring-opening polymerization (ROP) mechanism of *rac*-lactide (LA) with an yttrium complex featuring a *N*-heterocyclic carbine (NHC) tethered moiety is reported. It was found that the carbonyl of lactide is attacked by $N(SiMe_3)_2$ group rather than NHC species at the chain initiation step. The polymerization selectivity was further investigated via two consecutive insertions of lactide monomer molecules. The insertion of the second monomer in different assembly modes indicated that the steric interactions between the last enchained monomer unit and the incoming monomer together with the repulsion between the incoming monomer and the ligand framework are the primary factors determining the stereoselectivity. The interaction energy between the monomer and the metal center could also play an important role in the stereocontrol.

Keywords: yttrium NHC catalyst; lactide polymerization; selectivity; DFT; ring-opening polymerization

1. Introduction

Polylactide (PLA) is a biodegradable and biocompatible material derived from biorenewable feedstock, and it is considered an excellent commercial alternative for conventional petroleum-based materials. It is of great interest for various biomedical as well as ecological applications, including drug-delivery systems and tissue engineering [1–8]. PLA has attracted much attention from researchers in both industry and academia. The most efficient method for the synthesis of PLA is ring-opening polymerization (ROP) of the six-membered cyclic ester lactide with a variety of metal-based complexes [4,9–11]. Organocatalysts also have been employed to more precisely control the synthesis of ring-opening polymerization [12]. Among the metal-based catalysts, rare-earth metal complexes show excellent performance for the ROP of lactides, due to their qualities such as living behaviours, controlled molar weights, and very narrow molar weight distributions [13–16]. Many different lithium complexes have been identified as excellent catalysts for the ROP of lactides in alcohol [17]. Particularly, divalent metal catalysts bearing a monoanionic auxiliary skeleton are advantageous due to their low cost, and group 3 metal complexes are also exquisite initiators in the synthesis of PLAs by the ROP of *rac*-lactide [18,19]. The explicit group 4 metal complexes bearing polydentate ligands have also been proved to be active initiators for the synthesis of PLA [20]. As compared to the other metal-based catalysts for the synthesis of PLAs by ROP, titanium complexes are not high in terms of activity and stereoselectivity, but their low toxicity is an attractive property for the synthesis of PLAs [21]. Regardless of the high cost, Ga(III) and In(III) precursors have captivated an increasing interest for ROP catalysis and are considered to be potentially effective initiators with biocompatible metal centers [22].

As for organocatalysts, the application of *N*-heterocyclic carbene (NHC) and its derivatives to ROP has attracted considerable attention. It can efficiently promote the polymerization of lactide and related monomers under mild conditions [23–26]. These organocatalysts often exhibit higher functional group tolerance [27] than metal-based complexes, and inherently work under milder conditions. Inspired by above features, bifunctional catalysts formed by combining metal and organic catalysis, which may result in a combination of their respective advantages, have been demonstrated to be a very powerful strategy in organic synthesis; these dual systems have attracted considerable interest in the last few years [28–32]. To exploit the potential of NHC–ZnR2 complexes, different approaches have been followed in which the synthesis and characterization of NHC–ZnR2 are carried out, and these complexes have then been used in ROP [33]. Controlled ROP of lactide was carried out by a dual system $Zn(C_6F_5)_2$ combined with an amine or phosphine as organic bases [34]. In this context, Bourissou and co-workers reported a dual catalytic system combining an original cationic zinc complex with a tertiary amine, which was shown to promote efficiently and in a controlled manner the ROP of lactide under mild conditions [31]. In addition, Guillaume and Carpentier used analogous discrete cationic zinc and magnesium complexes for dual organic/organometallic-catalyzed ROP of trimethylene carbonate. The zinc cationic compounds are highly active catalysts for the ROP of TMC in the presence of an exogenous alcohol and addition of a small amount of tertiary amine [32]. However, in this content, bifunctional rare earth metal catalysts had been rarely explored for polymer synthesis.

Arnold had reported that the ROP of lactide catalyzed a yttrium complex featuring an NHC tethered moiety. Such a yttrium complex showed high activity and stereoselectivity toward the ROP of *rac*-lactide, resulting in a heterotactic based PLA (Scheme 1) [35]. In the same work, it was proposed that the polymerization occurs through a bifunctional mechanism involving electrophilic activation of the monomer by Lewis acidic metal centre Y and nucleophilic attack of the labile NHC fragment on the activated monomer (Scheme 2). However, the influence of the bifunctional catalyst on the mechanism of this reaction and the origin of stereocontrol, which is interesting and important for catalyst design, has remained unclear.

Theoretical calculations have shown their ability to efficiently describe and explain the ROP mechanism of cyclic esters, such as lactones and lactide at the molecular level [36–41]. For instance, Hormnirun et al. conducted a DFT study on the *rac*-lactide ring-opening mechanism initiated by a series of bis(pyrrolidene) Schiff-base aluminium complexes to reveal the correlation between the structure of backbone linker and the polymerization activity and stereoselectivity [42]. Rzepa et al. computationally studied both the mechanism of ring-opening and the origin of heterotactic stereocontrol in the β-diketiminate magnesium system [43]. They found that the stereoselectivity is determined by the ring-opening transition state and apparently arises from the minimization of several steric interactions and possibly from weak attractive C–H···π interactions. Maron et al. reported a DFT study of the ROP of lactide induced by a dinuclear indium catalyst [44]; the mechanism is proposed to involve the dinuclear species rather than a mononuclear complex because the dissociation energy of the dimer is too high. When *rac*-lactide is used, no clear preference for (*R,R*)-lactide vs. (*S,S*)-lactide insertion was found, therefore atactic polymer formation is predicted. A thorough DFT study of latide ROP initiated by an NHC zinc complex was also reported and showed the assistance of the second metal center [45]. These calculations provided valuable information on the design and development of homogeneous transition and main-group metal polymerization catalysts. In comparison, there are relatively few theoretical studies on lactide polymerization catalyzed by rare-earth metal complexes. Xu et al. conducted experimental and computational studies on the ROP of *rac*-lactide catalyzed by a novel yttrium bis(phenolate) ether complex, which showed high activity and excellent isotactic selectivity [46]. They found that the formation of an isotactic polymer originated chiefly from interactions between the methyl groups of the monomer units in the chain and the auxiliary ligand.

To the best of our knowledge, there is no report on DFT studies on the polymerization of *rac*-lactide initiated by bifunctional NHC-ligated rare-earth metal catalysts, such as that shown in Scheme 1. Simulated by our previous computational studies on olefin polymerization catalyzed by rare-earth metal complexes [47–50], we became interested in the ROP of cyclic ester catalyzed by rare-earth metal complexes. To obtain better insight into the mechanism of the reaction shown in Scheme 1 as well as to establish the role of the bifunctional catalyst in the ROP of lactide, DFT calculations were performed for the initiation and propagation of lactide ROP catalyzed by an NHC-ligated yttrium complex (Scheme 1). It was found that the carbonyl of lactide is attacked by the $N(SiMe_3)_2$ group rather than the NHC moiety. The steric repulsion between the incoming monomer and the ligand framework plays an important role in the heterotactic selectivity.

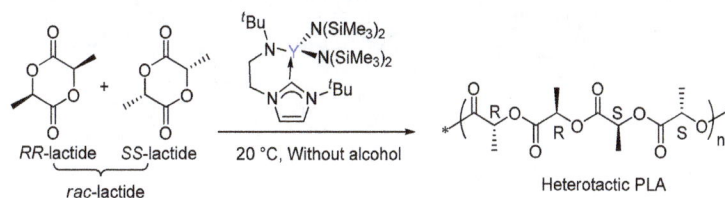

Scheme 1. The ring-opening polymerization of *rac*-lactide by the bifunctional yttrium catalyst.

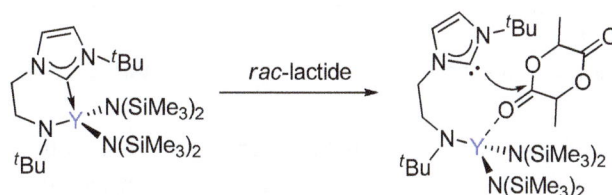

Scheme 2. The previously proposed bifunctional activation mode of *rac*-lactide by the bifunctional yttrium complex.

2. Computational Details

All calculations were performed with Gaussian 09 program [51]. The DFT method of B3PW91 [52,53] was utilized for geometry optimization and subsequent frequency calculations. The 6-31G(d) basis set was used for C, H, O, and N atoms; the LANL2DZ basis set together with the associated effective core potential (ECP) was utilized for Si and Y atoms [54]. One d-polarization function (exponent of 0.284) was augmented for the basis set of Si atoms [55]. Such a computational strategy has been widely used for the study of transition metal-containing systems [56–61]. The transition states were ascertained by a single imaginary frequency for the correct mode. The solvation effects were considered through single-point calculations with the SMD [62] solvation model. These single-point calculations are based on the optimized structures and carried out at the level of B3PW91-D3 (B3PW91with Grimme's DFT-D3 correction [63]. In the single-point calculations, the 6-311g(d,p) basis set for C, H, O, N, and Si atoms in addition to the Stuttgart/Dresden ECP together with the MWB28 basis set for Y were utilized. Toluene (ε = 2.37) was employed as a solvent in the SMD calculations. The free energy (ΔG, 298.15 K, 1 atm) in solution was obtained from the solvation single-point calculation, and the gas-phase Gibbs free energy correction was included. The sum of free energies of the isolated complex and the corresponding free lactide molecule was set to be an energy reference point in the overall reaction. The 3D molecular structure displayed in this paper was drawn by using CYLview [64].

3. Results and Discussion

3.1. The Chain Initiation Step of Lactide ROP Mediated by the Bifunctional Yttrium Complex

The coordination-insertion mechanism, which is generally accepted for the ROP of a cyclic ester, was considered for the current reaction. To address the mechanism of the ROP of *rac*-lactide initiated by bifunctional yttrium NHC catalyst in detail, the insertion and ring-opening of the (*S,S*)-lactide and (*R,R*)-lactide monomer were first examined, respectively. There are two possible different pathways for the chain initiation of the polymerization of *rac*-lactide catalyzed by the bifunctional yttrium catalyst. As shown in Scheme 3, (i) in path A, the lactide coordinates to the metal center, then the nucleophilic attack of the NHC group to the carbonyl C atom of the coordinating lactide takes place, leading to a cyclic insertion product; (ii) in path B, the nucleophilic attack of the $N(SiMe_3)_2$ group to the carbonyl C atom of the coordinating lactide occurs, yielding a linear insertion product.

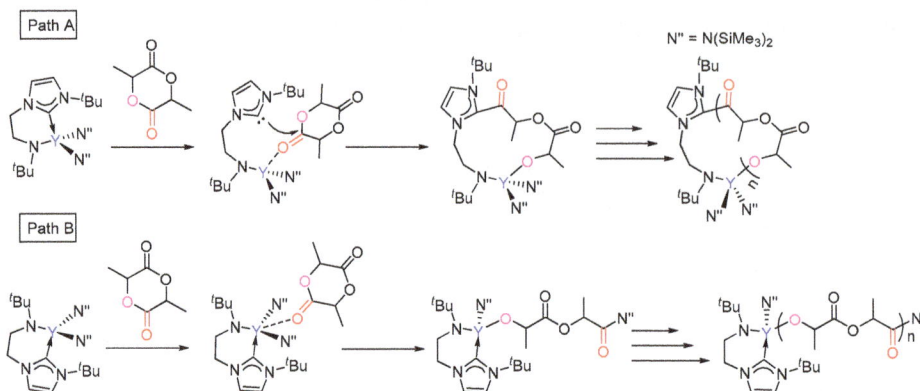

Scheme 3. Possible pathways for the ROP of lactide catalyzed by the bifunctional yttrium complex.

The calculated free energy profile for the chain initiation shown in path A is presented in Figure 1. As shown in this figure, the lactide coordinates to the metal center, which results in the (*S*,*S*)-lactide or (*R*,*R*)-lactide coordinating intermediate **1a** (Y···O = 2.39 and 2.41 Å, respectively), destabilized by 11.8 kcal/mol and 7.6 kcal/mol with respect to the catalyst and lactide monomer. In the insertion transition state **TS$_{1a-2a}$**, the carbonyl C atom of the coordinating lactide undergoes a change in hybridization from sp^2 to sp^3 due to the nucleophilic attack by the NHC species. The activation barriers for this step are 20.4 and 18.1 kcal/mol for the (*S*,*S*)- and (*R*,*R*)-lactide cases, respectively. The intermediate **2a** with a newly formed C–C bond is therefore obtained. The second step proceeds through the transition state **TS$_{2a-3a}$**, in which the lactide acyl–O bond is elongated (2.02 for (*S*,*S*) and 1.94 Å for (*R*,*R*) cases) with respect to the free lactide (1.35 Å) in concomitance with the formation of a new yttrium–alkoxide bond. The carbonyl carbon atom undergoes a change in hybridization from sp^3 to sp^2. This step results in the ring-opening of the lactide moiety and has an activation barrier of 34.6 kcal/mol for (*S*,*S*)-lactide and 27.2 kcal/mol for (*R*,*R*)-lactide. Moreover, the whole initiation step is endergonic by more than 4 kcal/mol. From an energy point of view, such high energy barriers and endergonic character suggest that path A is unlikely to be feasible pathway under the experimental conditions. This drove us to further investigate path B, as shown in Scheme 3.

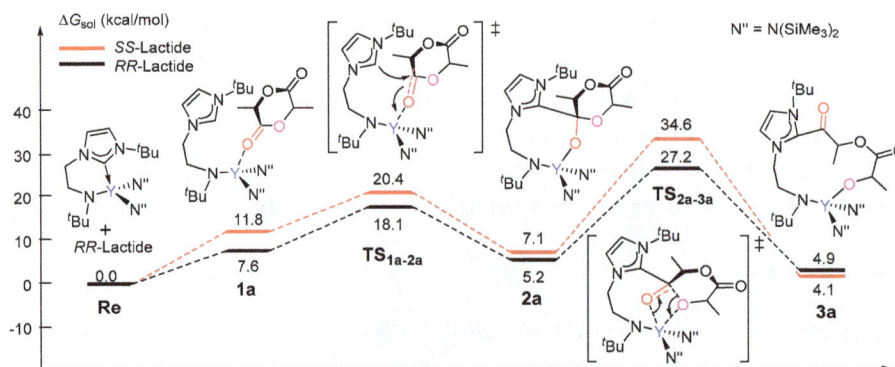

Figure 1. Calculated free energy profile for the chain initiation of the ROP of *rac*-lactide along path A. **TS$_{1a-2a}$**, nucleophilic addition transition state; **TS$_{2a-3a}$**, ring-opening transition state.

Unlike that in path A, the coordination complex of the monomer with the catalyst was not found on the energy profile of path B, possibly due to the coordinative saturation around the metal center (Figure 2). A nucleophilic attack of one $N(SiMe_3)_2$ group on the carbonyl group of (S,S)-lactide and (R,R)-lactide proceeds via the transition state $\mathbf{TS_{Re-1b}}$ with an activation barrier of 26.4 kcal/mol and 25.1 kcal/mol, respectively. This step represents the rate-determining step for the first monomer insertion along with path B. The resulting tetrahedral intermediate $\mathbf{1b}$ rearranges to give intermediate $\mathbf{2b}$. Then, $\mathbf{2b}$ overcomes the transition state $\mathbf{TS_{2b-3b}}$ to accomplish the ring-opening of the lactide moiety, accompanied with the concerted cleavage of the acyl–oxygen bond and the formation of the new metal-alkoxyl bond. After the ring-opening, the chain reorganization occurs, affording a more stable five-membered metallacyclic product $\mathbf{4b}$. The calculated energy barriers for the ring-opening step are less than that of the insertion step. As a whole, the conversion of the reactant into a product is exergonic by 14.1 and 12.3 kcal/mol for (S,S)-lactide and (R,R)-lactide, respectively (Figure 2). A comparison of the energy profiles for paths A and B (Figures 1 and 2) indicates that path B is more kinetically favorable than path A (an energy barrier of 34.6 kcal/mol and 27.2 kcal/mol vs. 26.4 kcal/mol and 25.1 kcal/mol), in addition to being more thermodynamically favorable (4.9 kcal/mol and 4.1 kcal/mol vs. −12.3 kcal/mol and −14.1 kcal/mol). It is also obvious that the reaction of (R,R)-lactide is more kinetically favorable than that of (S,S)-lactide (Figures 1 and 2). These results suggest that the NHC moiety could not directly participate in the insertion and ring-opening processes and the (R,R)-lactide could be favorably involved at the chain initiation step.

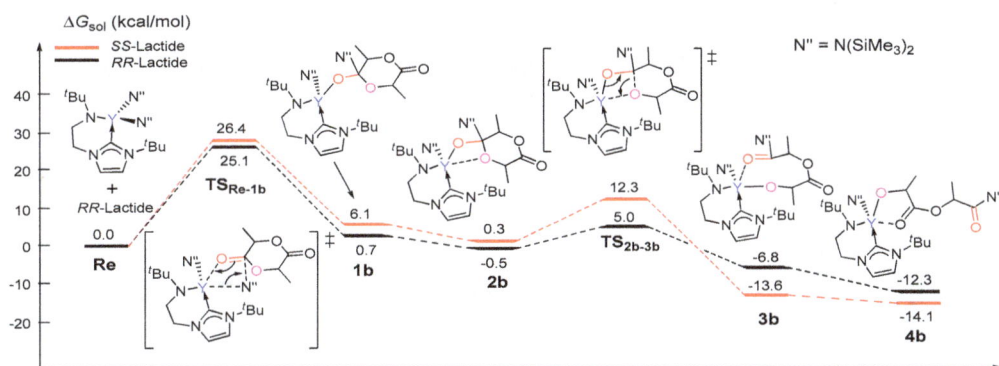

Figure 2. Calculated free energy profile for the chain initiation of the ROP of *rac*-lactide along path B. $\mathbf{TS_{Re-1b}}$, nucleophilic addition transition state; $\mathbf{TS_{2b-3b}}$, ring-opening transition state.

The reasonable chain initiation pathway was studied, as shown above. However, the role of NHC is unclear in the process of the ROP of *rac*-lactide. In order to further investigate the role of NHC in the chain initiation step, we modelled the de-coordination of the NHC moiety from the metal center via a single-bond rotation in the complex. The dissociation energies of a $Y–N(SiMe_3)_2$ bond were cleaved for the monomer insertion as well as the insertion of the first (R,R)-lactide monomer before ($\mathbf{1}$) and after ($\mathbf{2}$) the rotation was calculated (Figures 3 and 4). Obviously, in the case of NHC de-coordination, the dissociation energy of the $Y–N(SiMe_3)_2$ bond ($\Delta E = 107.3$ kcal/mol) is larger than that for the NHC-coordination case ($\Delta E = 96.1$ kcal/mol, Figure 3), because the NHC moiety serves as a good electron donor and could decrease the Lewis acidity of the metal center (NBO (Natural bond orbital) charge of 1.64 vs. 1.72, Figure 3). One may suppose that the increase in the $Y–N(SiMe_3)_2$ bond dissociation energy could increase the insertion energy barrier of the monomer. Actually, the calculated energy profiles support this hypothesis. As shown in Figure 4, in the case of NHC de-coordination, the insertion energy barrier is much higher than that for the coordination case (40.4 vs. 25.1 kcal/mol). Moreover, the former case is significantly endergonic, while the latter is an almost an isoenergetic process. These results suggest that the NHC ligation could accelerate the monomer insertion and therefore improve the polymerization activity.

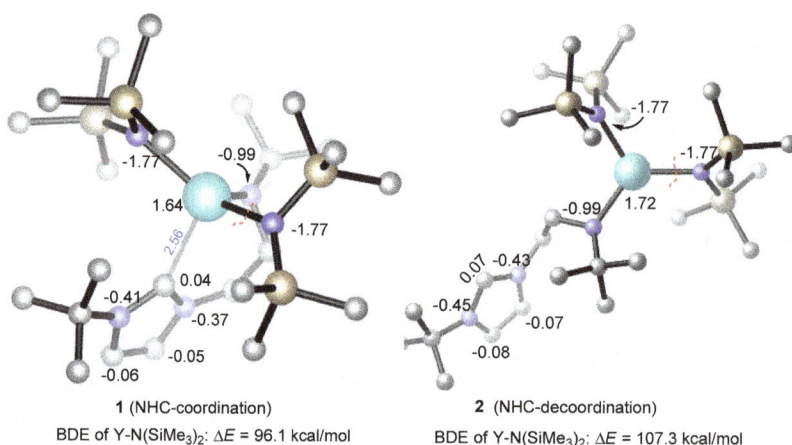

1 (NHC-coordination)
BDE of Y–N(SiMe₃)₂: ΔE = 96.1 kcal/mol

2 (NHC-decoordination)
BDE of Y–N(SiMe₃)₂: ΔE = 107.3 kcal/mol

Figure 3. The bond dissociation energy of Y–N(SiMe$_3$)$_2$ in the coordination and de-coordination of the *N*-heterocyclic carbine (NHC) moiety. NBO (Natural bond orbital) atomic charges are given in black. All H atoms are omitted for clarity.

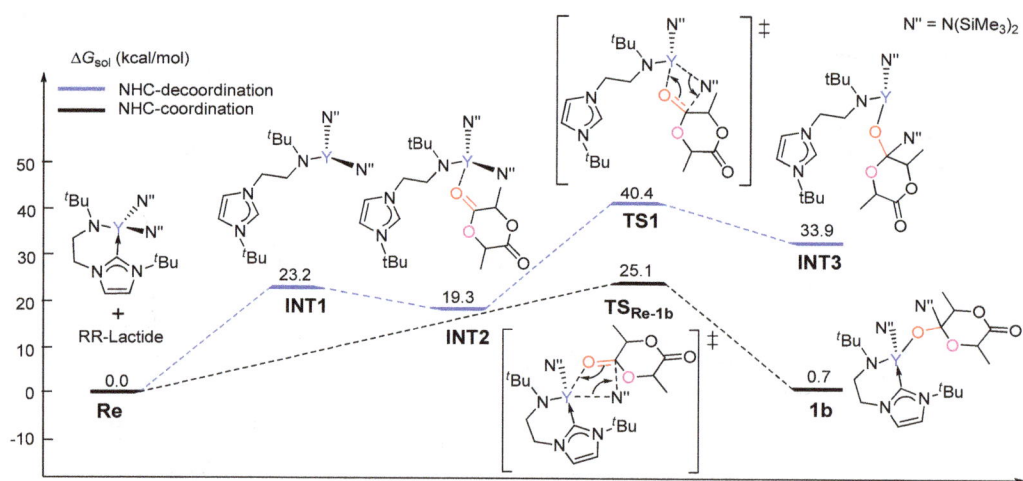

Figure 4. Calculated free energy profiles for the insertion of the (*R*,*R*)-lactide monomer in the cases of NHC de-coordination and coordination.

3.2. Selectivity of rac-Lactide Polymerization

As aforementioned, in the chain initiation step, the reaction of (*R*,*R*)-lactide is more kinetically favorable than that of (*S*,*S*)-lactide. To explore the origin of stereocontrol in the *rac*-lactide polymerization, the insertions of both (*R*,*R*)- and (*S*,*S*)-lactide monomers were calculated in order to model the chain propagation on the basis of the resulting insertion product of the (*R*,*R*)-lactide **4b**. As shown in Figure 5, the incoming monomer coordinates to **4b** to form **5b**, which could undergo a migratory insertion and subsequent ring-opening to finally give **9b**, accompanied by structure isomerizations. A comparison of the coordination complex **4b** indicates that the (*S*,*S*)-lactide is more thermodynamically favorable to coordinate to the metal center than its stereoisomers (relative energies of −10.0 vs. −3.2 kcal/mol). Kinetically, the ring-opening via **TS$_{7b\text{-}8b}$** is the rate-determining step for both cases. Also, the free energy barrier for the case of (*S*,*S*)-lactide is 18.7 kcal/mol relative to **4b** and the free monomer, which is lower than that for the case of its enantiomeric form (20.9 kcal/mol). These results suggest that the reaction of **4b** with (*S*,*S*)-lactide is more favorable compared with (*R*,*R*)-lactide. This results in the (*R*,*R*)-(*S*,*S*) sequence, which is in line with the experimentally observed heterotactic selectivity. Considering that an achiral initiator cannot not lead any enantioselectivity in the chain initiation step, the case of (*S*,*S*)-enantiomer is also used for modeling the chain propagation.

Actually, on the basis of the (S,S)-lactide-initiated product **4b**(S,S), the subsequent reaction of the (R,R)-monomer is also more favorable than (S,S)-lactide, yielding the (S,S)-(R,R) heterotactic sequence (Figure S1).

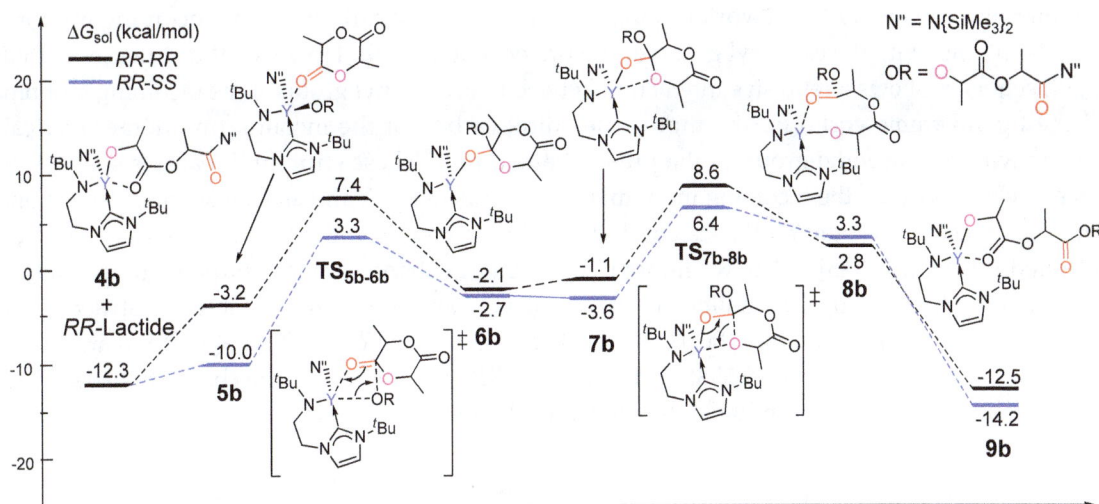

Figure 5. Calculated free energy profile for the ring-opening polymerization of the second *rac*-lactide monomer.

To further elucidate the origin of stereoselectivity, an energy decomposition analysis [65–68] was performed for the coordination complex **5b**, which shows a significant difference in energy between the two stereoisomers. In the energy decomposition analysis, **5b** could be divided into two fragments, viz., lactide moiety (**A**) and the remaining part of the metal complex (**B**). The energies of the fragments **A** and **B** in the geometry of **5b** were evaluated by single-point calculations. Such single-point energies of the fragments and the energy (corrected by the basis set superposition error) of **5b** were used to estimate the interaction energy ΔE_{int}. These energies, together with the energy of each fragment of their optimal geometries, allow for the estimation of the deformation energies of the two fragments, $\Delta E_{def}(\mathbf{A})$ and $\Delta E_{def}(\mathbf{B})$. As the total energy of **5b**, ΔE_{5b} is evaluated with respect to the energy of two separated fragments, and the relation $\Delta E_{5b} = \Delta E_{int} + \Delta E_{def}(\mathbf{A}) + \Delta E_{def}(\mathbf{B})$ holds. As shown in Table 1, the deformation energy of **5b**(RR,SS) is similar to that of **5b**(RR,RR), but the interaction energy is more favorable for **5b**(RR,SS) (−32.1 kcal/mol) as compared to that of (RR,RR) (−26.9 kcal/mol), resulting in the lower ΔE_{5b} for **5b**(RR,SS) (−12.8 kcal/mol) in comparison with that for **5b**(RR,RR) (−7.0 kcal/mol). It is obvious that the smaller ΔE_{int} (more negative) for **5b**(RR,SS) accounts for the lower energy barrier of **5b**(RR,SS) in comparison with **5b**(RR,RR). Therefore, the stability of the coordination complex **5b** is controlled by the interaction energy between the monomer and the metal complex moiety.

Table 1. The energy decomposition of **5b** (kcal/mol).

Structure	ΔE_{int}	$\Delta E_{def}(\mathbf{A})$ [1]	$\Delta E_{def}(\mathbf{B})$ [2]	ΔE_{def}	ΔE_{5b}
5b (RR,RR)	−26.9	0.9	19	19.9	−7
5b (RR,SS)	−32.1	1.2	18.1	19.3	−12.8

[1] Deformation energy of the lactide monomer fragment in **5b**; [2] Deformation energy of the remaining part of **5b**.

Considering that the energies of **TS$_{5b-6b}$** and **TS$_{7b-8b}$** are close (Figure 5), and that the latter is unsuitable for such a fragment division, an energy decomposition analysis was also carried out for **TS$_{5b-6b}$** to add better understanding to the stereoselectivity. It was found that the stronger interaction (more negative ΔE_{int}) between the monomer and the metal could account for the lower energy barrier of TS_{5b-6b} (RR,SS) in comparison with TS_{5b-6b} (RR,RR). This suggests that the electronic factor plays an important role during the formation of the heterotactic sequence (Table S1).

A closer examination of the structure of the ring-opening transition state (TS_{7b-8b}) indicates that the (RR,RR)-sequenced structure shows a strong repulsion between the methyl of the incoming monomer and the ligand framework (Figure 6). The two chiral carbons of the incoming monomer sterically interact with the carbonyl group of the pre-enchained unit. However, the structure with the (RR,SS) sequence shows a repulsive interaction between the carbonyl group of the incoming monomer and the ligand framework. Furthermore, one chiral carbon of the incoming monomer sterically interacts with the carbonyl group of the pre-enchained unit. These structural features suggest that the steric clash between the incoming monomer and the ancillary ligand as well as the pre-enchained monomer unit could account for the stereoselectivity, in accordance with the ligand-assisted chain-end controlled mechanism [43]. Also, we know that the charge dispersion is conducive to the stability of a structure. In general, the smaller the values of $|Q|$ and S, the more stable the structure [69]. The charge analyses indicate that the $|Q| = 1.035$ and S = 0.526 for the TS_{7b-8b} (RR,SS) are smaller than those ($|Q| = 1.038$ and S = 0.558) for the TS_{7b-8b} (RR,RR). Therefore, the charge dispersion also suggests that the former is more stable than the later (Figure S2).

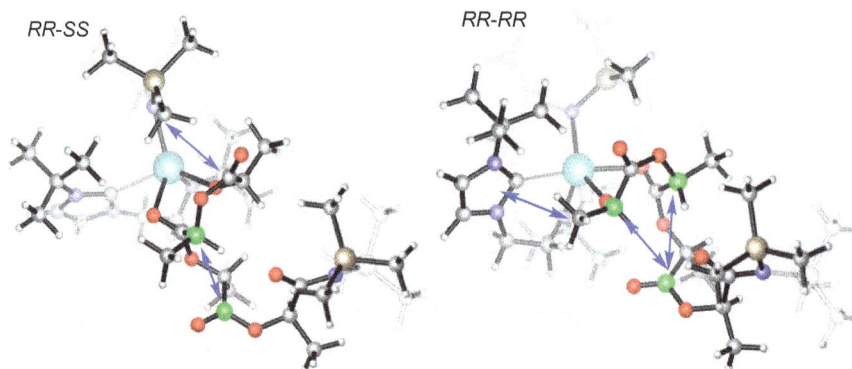

Figure 6. Optimized geometries of TS_{7b-8b}; the arrows show repulsive interaction.

4. Conclusions

The stereoselective polymerization of *rac*-lactide by bifunctional yttrium NHC catalyst LY[N(SiMe₃)₂]₂ (L = ^{t}BuNCH₂CH₂(1-C{NCHCHN^{t}Bu})) was studied by using density functional theory. At the chain initiation stage, it was found that the NHC group could not be directly involved in the polymerization process and the N(SiMe₃)₂ group nucleophilically attacks the carbonyl C atom of the coordinating lactide instead. However, computational modelling suggests that the NHC moiety to be a good electron-donor that could accelerate the carbonyl insertion and thus improve the polymerization activity. The subsequent reaction of the second monomer was also calculated in order to model the chain propagation. Having achieved an agreement between the experimental and theoretical results, the origin of the observed stereoselectivity was further investigated. It was found that the steric repulsion between the incoming monomer and ancillary ligand as well as the pre-enchained monomer unit could account for the stereoselectivity, in accordance with the ligand-assisted chain-end controlled mechanism. In addition, the electronic factor may play an important role in the monomer insertion reaction to form a heterotactic sequence. The results reported in this work are expected to shed light on the development of stereoselective catalysts for the ring-opening polymerization of lactides.

Supplementary Materials:
Figure S1: Calculated free energy profile for the ring-opening polymerization of the second *rac*-lcatide monomer on the basis of the resulting insertion product of the (S,S)-lactide **4b**; Figure S2: the charge dispersion analyses of TS_{7b-8b}; Table S1: The energy decomposition of TS_{5b-6b}. Optimized Cartesian coordinates (XYZ) with the self-consistent field (SCF) energies and the imaginary frequencies of transition states.

Acknowledgments: This work was partially supported by the NSFC (Nos. 21429201, 21674014). The authors also thank the Fundamental Research Funds for the Central Universities (DUT2016TB08) and the Network and Information Center of the Dalian University of Technology for part of computational resources.

Author Contributions: Yincheng Wang and Yi Luo conceived and designed the calculations; Yincheng Wang performed the calculations; Yincheng Wang and Yi Luo analyzed the data and wrote the paper, with contributions from Yanan Zhao; Andleeb Mehmood polished the English sentence and grammar; Jingping Qu participated in the discussion; Yi Luo directed the project.

References

1. Uhrich, K.E.; Cannizzaro, S.M.; Langer, R.S. Polymeric systems for controlled drug release. *Chem. Rev.* **1999**, *99*, 3181–3198. [CrossRef] [PubMed]

2. Gupta, A.P.; Kumar, V. New emerging trends in synthetic biodegradable polymers—Polylactide: A critique. *Eur. Polym. J.* **2007**, *43*, 4053–4074. [CrossRef]

3. Albertsson, A.C.; Varma, I.K. Recent developments in ring opening polymerization of lactones for biomedical applications. *Biomacromolecules* **2003**, *4*, 1466–1486. [CrossRef] [PubMed]

4. Cui, D.; Liu, X.; Shang, X. Achiral lanthanide alkyl complexes bearing N,O multidentate ligands. Synthesis and catalysis of highly heteroselective ring-opening polymerization of *rac*-lactide. *Organometallics* **2007**, *26*, 2747–2757.

5. Ragauskas, A.J.; Williams, C.K.; Davison, B.H.; Britovsek, G.; Cairney, J.; Eckert, C.A.; Frederick, W.J., Jr.; Hallett, J.P.; Leak, D.J.; Liotta, C.L. The path forward for biofuels and biomaterials. *Science* **2006**, *311*, 484–489. [CrossRef] [PubMed]

6. Platel, R.H.; Hodgson, L.M.; Williams, C.K. Biocompatible initiators for lactide polymerization. *Polym. Rev.* **2008**, *48*, 11–63. [CrossRef]

7. Stanford, M.J.; Dove, A.P. Stereocontrolled ring-opening polymerisation of lactide. *Chem. Soc. Rev.* **2010**, *39*, 486–494. [CrossRef] [PubMed]

8. Rosen, T.; Goldberg, I.; Venditto, K.M. Tailor-made stereoblock copolymers of poly(lactic acid) by a truly living polymerization catalyst. *J. Am. Chem. Soc.* **2016**, *138*, 12041–12044. [CrossRef] [PubMed]

9. Robert, C.; Schmid, T.E.; Richard, V.; Haquette, P.; Raman, S.K.; Rager, M.N.; Gauvin, R.M.; Morin, Y.; Trivelli, X.; Guerineau, V. Mechanistic aspects of the polymerization of lactide using a highly efficient aluminum(III) catalytic system. *J. Am. Chem. Soc.* **2017**, *139*, 6217–6225. [CrossRef] [PubMed]

10. Thomas, C.M. Stereocontrolled ring-opening polymerization of cyclic esters: Synthesis of new polyester microstructures. *Chem. Soc. Rev.* **2010**, *39*, 165–173. [CrossRef] [PubMed]

11. Brown, H.A.; Crisci, A.G.D.; Hedrick, J.L.; Waymouth, R.M. Amidine-mediated zwitterionic polymerization of lactide. *ACS Macro Lett.* **2012**, *1*, 1113–1115. [CrossRef]

12. Dove, A.P. Organic Catalysis for Ring-Opening Polymerization. *ACS Macro Lett.* **2012**, *1*, 1409–1412. [CrossRef]

13. Dechy-Cabaret, O.; Martin-Vaca, B.; Bourissou, D. Controlled ring-opening polymerization of lactide and glycolide. *Chem. Rev.* **2004**, *104*, 6147–6176. [CrossRef] [PubMed]

14. Dove, A.P. Controlled ring-opening polymerisation of cyclic esters: Polymer blocks in self. *Chem. Commun.* **2008**, *48*, 6446–6470. [CrossRef] [PubMed]

15. Ajellal, N.; Carpentier, J.F.; Guillaume, C.; Guillaume, S.M.; Heloua, M.; Poiriera, V.; Sarazina, Y.; Trifonovb, A. Metal-catalyzed immortal ring-opening polymerization of lactones, lactides and cyclic carbonates. *Dalton Trans.* **2010**, *39*, 8363–8376. [CrossRef] [PubMed]

16. Platel, R.H.; White, A.J.P.; Williams, C.K. Bis(phosphinic)diamido yttrium amide, alkoxide, and aryloxide complexes: An evaluation of lactide ring-opening polymerization initiator efficiency. *Inorg. Chem.* **2011**, *50*, 7718–7728. [CrossRef] [PubMed]

17. Wu, J.; Yu, T.L.; Chen, C.T.; Lin, C.C. Recent developments in main group metal complexes catalyzed/initiated polymerization of lactides and related cyclic esters. *Coord. Chem. Rev.* **2006**, *250*, 602–626. [CrossRef]

18. Wheaton, C.A.; Hayes, P.G.; Ireland, B.J. Complexes of Mg, Ca and Zn as homogeneous catalysts for lactide polymerization. *Dalton Trans.* **2009**, 4832–4846. [CrossRef] [PubMed]

19. Amgoune, A.; Thomas, A.M.; Carpentier, J.F. Controlled rong-opening polymerization of lactide by group 3 metal complexes. *Pure Appl. Chem.* **2007**, *79*, 2013–2030. [CrossRef]

20. Sauer, A.; Kapelski, A.; Fliedel, C.; Dagorne, S.; Kol, M.; Okuda, J. Structurally well-defined group 4 metal complexesas initiators for ring-opening polymerization of lactide monomers. *Dalton Trans.* **2013**, *42*, 9007–9023. [CrossRef] [PubMed]

21. Le Roux, E. Recent advances on tailor-made titanium catalysts for biopolymer synthesis. *Coord. Chem. Rev.* **2016**, *306*, 65–85. [CrossRef]

22. Dagorne, S.; Normand, M.; Kirillov, E.; Carpentier, J.F. Gallium and Indium complexes for ring-opening polymerization of cyclic ethers, esters and carbonates. *Coord. Chem. Rev.* **2013**, *257*, 1869–1886. [CrossRef]

23. Connor, E.F.; Nyce, G.W.; Myers, M.; Möck, K.; Hedrick, J.L. First example of *N*-heterocyclic carbenes as catalysts for living polymerization: Organocatalytic ring-opening polymerization of cyclic esters. *J. Am. Chem. Soc.* **2002**, *124*, 914–915. [CrossRef] [PubMed]

24. Kamber, N.E.; Jeong, W.; Waymouth, R.M. Organocatalytic ring-opening polymerization. *Chem. Rev.* **2007**, *107*, 5813–5840. [CrossRef] [PubMed]

25. Kiesewetter, M.K.; Shin, E.J.; Hedrick, J.L.; Waymouth, R.M. Organocatalysis: Opportunities and challenges for polymer synthesis. *Macromolecules* **2010**, *43*, 2093–2107. [CrossRef]

26. Acharya, A.K.; Chang, Y.A.; Jones, G.O.; Rice, J.E.; Hedrick, J.L.; Horn, H.W.; Waymouth, R.M. Experimental and computational studies on the mechanism of zwitterionic ring-opening polymerization of δ-valerolactone with *N*-heterocyclic carbenes. *J. Phys. Chem. B* **2014**, *118*, 6553–6560. [CrossRef] [PubMed]

27. Suriano, F.; Coulembier, O.; Hedrick, J.L.; Dubois, P. Functionalized cyclic carbonates: From synthesis and metal-free catalyzed ring-opening polymerization to applications. *Polym. Chem.* **2011**, *2*, 528–533. [CrossRef]

28. Piedra-Arroni, E.; Amgoune, A.; Bourissou, D. Dual catalysis: New approaches for the polymerization of lactones and polar olefins. *Dalton Trans.* **2013**, *42*, 9024–9029. [CrossRef] [PubMed]

29. Shao, Z.; Zhang, H. Combining transition metal catalysis and organocatalysis: A broad new concept for catalysis. *Chem. Soc. Rev.* **2009**, *40*, 2745–2755. [CrossRef] [PubMed]

30. Zhong, C.; Shi, X. When organocatalysis meets transition-metal catalysis. *Eur. J. Org. Chem.* **2010**, *16*, 2999–3025. [CrossRef]

31. Piedra-Arroni, E.; Brignou, P.; Amgoune, A.; Guillaume, S.M.; Carpentier, J.F.; Bourissou, D. A dual organic/organometallic approach for catalytic ring-opening polymerization. *Chem. Commun.* **2011**, *47*, 9828–9830. [CrossRef] [PubMed]

32. Brignou, P.; Guillaume, S.M.; Roisnel, T.; Bourissou, D.; Carpentier, J.-F. Discrete cationic zinc and magnesium complexes for dual organic/organometallic-catalyzed ring-opening polymerization of trimethylene carbonate. *Chem. Eur. J.* **2012**, *18*, 9360–9370. [CrossRef] [PubMed]

33. Schnee, G.; Fliedel, C.; Aviles, T.; Dagrone, S. Neutral and cationic *N*-heterocyclic carbene zinc adducts and the BnOH/Zn(C$_6$F$_5$)$_2$ binary mixture characterization and use in the ring-opening polymerization of β-Butyrolactone, lactides and trimethylene carbonate. *Eur. J. Inorg. Chem.* **2013**, 3699–3709. [CrossRef]

34. Piedra-Arroni, E.; Ladaviere, C.; Amgoune, A.; Bourissou, D. Ring-opening polymerization with Zn(C$_6$F$_5$)$_2$-based lewis pairs: Original and efficient approach to cyclic polyesters. *J. Am. Chem. Soc.* **2013**, *135*, 13306–13309. [CrossRef] [PubMed]

35. Patel, D.; Liddle, S.T.; Mungur, S.A.; Rodden, M.; Blake, A.J.; Arnold, P.L. Bifunctional yttrium(III) and titanium(IV) NHC catalysts for lactide polymerization. *Chem. Commun.* **2006**, 1124–1126. [CrossRef] [PubMed]

36. Vieira, I.S.; Whitelaw, E.L.; Jones, M.D.; Pawlis, S.H. Synergistic empirical and theoretical study on the stereoselective mechanism for the aluminum salalen complex mediated polymerization of *rac*-lactide. *Chem. Eur. J.* **2013**, *19*, 4712–4716. [CrossRef] [PubMed]

37. Dyer, H.E.; Huijser, S.; Susperregui, N.; Bonnet, F.; Schwarz, A.D.; Duchateau, R.; Maron, L.; Mountford, P. Ring-opening polymerization of *rac*-Lactide by bis(phenolate)amine-supported samarium borohydride complexes: An experimental and DFT Study. *Organometallics* **2010**, *29*, 3602–3621. [CrossRef]

38. Broderick, E.M.; Guo, N.; Wu, T.; Vogel, C.S.; Xu, C.; Sutter, J.; Miller, J.T.; Meyer, K.; Cantat, T.; Diaconescu, P.L. Redox control of a polymerization catalyst by changing the oxidation state of the metal center. *Chem. Commun.* **2011**, *47*, 9897–9899. [CrossRef] [PubMed]

39. Fang, J.; Walshe, A.; Maron, L.; Baker, R.J. Ring-opening polymerization of epoxides catalyzed by uranyl complexes: An experimental and theoretical study of the reaction mechanism. *Inorg. Chem.* **2012**, *51*, 9132–9140. [CrossRef] [PubMed]

40. Fang, J.; Tschan, M.J.-L.; Roisnel, T.; Trivelli, X.; Gauvin, R.M.; Thomas, C.M.; Maron, L. Yttrium catalysts for syndioselective b-butyrolactone polymerization: On the origin of ligand-induced stereoselectivity. *Polym. Chem.* **2013**, *4*, 360–367. [CrossRef]

41. Rosal, I.D.; Brignou, P.; Guillaume, S.M.; Carpentier, J.F.; Maron, L. DFT investigations on the ring-opening polymerization of substituted cyclic carbonates catalyzed by zinc-{β-diketiminate} complexes. *Polym. Chem.* **2015**, *6*, 3336–3352. [CrossRef]

42. Tabthong, S.; Nanok, T.; Sumrit, P.; Kongsaeree, P.; Prabpai, S.; Chuawong, P.; Hormnirun, P. Bis(pyrrolidene) schiff base aluminum complexes as isoselective-biased initiators for the controlled ring-opening polymerization of *rac*-lactide: Experimental and theoretical Studies. *Macromolecules* **2015**, *48*, 6846–6861. [CrossRef]

43. Marshall, E.L.; Gibson, V.C.; Rzepa, H.S. A computational analysis of the ring-opening polymerization of rac-lactide initiated by single-site β-diketiminate metal complexes: Defining the mechanistic pathway and the origin of stereocontrol. *J. Am. Chem. Soc.* **2005**, *127*, 6048–6051. [CrossRef] [PubMed]

44. Fang, J.; Yu, I.; Mehrkhodavandi, P.; Maron, L. Theoretical investigation of lactide ring-opening polymerization induced by a dinuclear indium catalyst. *Organometallics* **2013**, *32*, 6950–6956. [CrossRef]

45. Fliedel, C.; Vila-Vicosa, D.; Calhorda, M.J.; Dagorne, S.; Aviles, T. Dinuclear Zinc–N-Heterocyclic carbine complexes either the controlled ring-opening polymerization of lactide or the controlled degradation of polylactide under mild conditions. *ChemCatChem* **2014**, *6*, 1357–1367. [CrossRef]

46. Xu, T.Q.; Yang, G.W.; Liu, C.; Lu, X.B. Highly robust yttrium bis(phenolate) ether catalysts for excellent isoselective ring-opening polymerization of racemic lactide. *Macromolecules* **2017**, *50*, 515–522. [CrossRef]

47. Kang, X.H.; Song, Y.M.; Luo, Y.; Li, G.; Hou, Z.M.; Qu, J.P. Computational studies on isospecific polymerization of 1-hexene catalyzed by cationic rare-earth metal alkyl complex bearing a C3 iPr-trisox ligand. *Macromolecules* **2012**, *45*, 640–651. [CrossRef]

48. Kang, X.H.; Yamamoto, A.; Nishiura, M.; Luo, Y.; Hou, Z.M. Computational analyses of the effect of lewis bases on styrene polymerization catalyzed by cationic scandium half-sandwich complexes. *Organometallics* **2015**, *34*, 5540–5548. [CrossRef]

49. Kang, X.H.; Zhou, G.L.; Wang, X.B.; Luo, Y.; Hou, Z.M.; Qu, J.P. Alkyl effects on the chain initiation efficiency of olefin polymerization by cationic half-sandwich scandium catalysts: A DFT study. *Organometallics* **2016**, *35*, 913–920. [CrossRef]

50. Kang, X.H.; Luo, Y.; Zhou, G.L.; Wang, X.B.; Yu, X.R.; Hou, Z.M.; Qu, J.P. Theoretical mechanistic studies on the *trans*-1,4-specific polymerization of isoprene catalyzed by a cationic La–Al binuclear complex. *Macromolecules* **2014**, *47*, 4596–4606. [CrossRef]

51. Frisch, M.J.; Trucks, G.W.; Schlegel, H.B.; Scuseria, G.E.; Robb, M.A.; Cheeseman, J.R.; Scalmani, G.; Barone, Y.; Mennucci, B.; Petersson, G.A. *Gaussian 09, Revision A.02*; Gaussian Inc.: Wallingford, CT, USA, 2009.

52. Becke, A.D. Density-functional thermochemistry. III. The role of exact exchange. *J. Chem. Phys.* **1993**, *98*, 5648–5653. [CrossRef]

53. Perdew, J.P.; Wang, Y. Accurate and simple analytic representation of the electron-gas correlation energy. *Phys. Rev. B* **1992**, *45*, 13244–13249. [CrossRef]

54. Yang, Y.; Weaver, M.N.; Merz, K.M., Jr. Assessment of the "6-31 + G** + LANL2DZ" mixed basis set coupled with density functional theory methods and the effective core potential: Prediction of heats of formation and ionization potentials for first-row-transition-metal complexes. *J. Phys. Chem. A* **2009**, *113*, 9843–9851. [CrossRef] [PubMed]

55. Hoellwarth, A.; Boehme, M.; Dapprich, S.; Ehlers, A.W.; Gobbi, A.; Jonas, V.; Köhler, K.F.; Stegmann, R.; Veldkamp, A.; Frenking, G. A set of d-polarization functions for pseudo-potential basis sets of the main group elements Al–Bi and f-type polarization functions for Zn, Cd, Hg. *Chem. Phys. Lett.* **1993**, *208*, 237–240.

56. Liu, Y.; Liu, Y.; Drew, M.G.B. Correlation between regioselectivity and site charge in propene polymerisation catalysed by metallocene. *Struct. Chem.* **2010**, *21*, 21–28. [CrossRef]

57. Zhang, C.; Yu, S.; Zhang, L.; Li, H.Y.; Wang, Z.X. DFT mechanistic study of the H_2-assisted chain transfer copolymerization of propylene and *p*-methylstyrene catalyzed by zirconocene complex. *J. Polym. Sci. Part A Polym. Chem.* **2015**, *53*, 576–585. [CrossRef]

58. Valente, A.; Zinck, P.; Mortreux, A.; Visseauxa, M.; Mendes, P.J.G.; Silva, T.J.L.; Garcia, M.H. Polymerization of ε-caprolactone using ruthenium(II) mixed metallocene catalysts and isopropyl alcohol: Living character and mechanistic study. *J. Mol. Catal. A Chem.* **2011**, *346*, 102–110. [CrossRef]

59. Jitonnom, J.; Molloy, R.; Punyodom, W.; Meelua, W. Theoretical studies on aluminum trialkoxide-initiated lactone ring-opening polymerizations: Roles of alkoxide substituent and monomer ring structure. *Comput. Theor. Chem.* **2016**, *1097*, 25–32. [CrossRef]

60. Jitonnom, J.; Meelua, W. Effects of silicon-bridge and π-ligands on the electronic structures and related properties of dimethyl zirconocene polymerization catalysts: A comparative theoretical study. *Chiang Ma J. Sci.* **2014**, *41*, 1220–1229.

61. Jitonnom, J.; Sontag, C. Catalytic oxidation of glucose with hydrogen peroxide and colloidal gold as pseudo-homogenous catalyst: A combined experimental and theoretical investigation. *Chiang Mai J. Sci.* **2016**, *43*, 825–833.

62. Marenich, A.V.; Cramer, C.J.; Truhlar, D.G. Universal solvation model based on solute electron density and on a continuum model of the solvent defined by the bulk dielectric constant and atomic surface tensions. *J. Phys. Chem. B* **2009**, *113*, 6378–6396. [CrossRef] [PubMed]

63. Grimme, S.; Antony, J.; Ehrlich, S.; Krieg, H. A consistent and accurate ab initio parametrization of density functional dispersion correction (DFT-D) for the 94 elements H–Pu. *J. Chem. Phys.* **2010**, *132*, 154104. [CrossRef] [PubMed]

64. Legault, C.Y. *CLYview, Version 1.0b*; University of California: Los Angeles, CA, USA, 2007.

65. Kitaura, K.; Morokuma, K. A new energy decomposition scheme for molecular interactions within the Hartree-Fock approximation. *Int. J. Quantum Chem.* **1976**, *10*, 325–340. [CrossRef]

66. Pan, Y.; Xu, X.; Wei, N.N.; Hao, C.; Zhu, X.D.; He, G.H. DFT study on 1,7-octadiene polymerization catalyzed by non-bridged half-titanocene system. *RSC Adv.* **2016**, *6*, 69939–69946. [CrossRef]

67. Li, Y.; Qi, X.; Lei, Y.; Lan, Y. Mechanism and selectivity for zinc-mediated cycloaddition of azides with alkynes: A computational study. *RSC Adv.* **2015**, *5*, 49802–49808. [CrossRef]

68. Bhattacharjee, R.; Nijamudheen, A.; Datta, A. Mechanistic insights into the synergistic catalysis by Au(I), Ga(III), and counterions in the nakamura reaction. *Org. Biomol. Chem.* **2015**, *13*, 7412–7420. [CrossRef] [PubMed]

69. Liu, F.; Luo, G.; Hou, Z.M.; Luo, Y. Mechanistic insights into scandium-catalyzed hydroaminoalkylation of olefins with amines: Origin of regioselectivity and charge-based prediction model. *Organometallics* **2017**, *36*, 1557–1565. [CrossRef]

8

Modification of Cooperativity and Critical Temperatures on a Hofmann-Like Template Structure by Modular Substituent

Takashi Kosone [1,*], Takeshi Kawasaki [2], Itaru Tomori [2], Jun Okabayashi [3] and
Takafumi Kitazawa [2,4,*] (iD)

[1]   Department of Creative Technology Engineering Course of Chemical Engineering, Anan College, 265 Aoki,
       Minobayashi, Anan, Tokushima 774-0017, Japan
[2]   Department of Chemistry, Faculty of Science, Toho University, 2-2-1 Miyama, Funabashi, Chiba 274-8510,
       Japan; takeshi.kawasaki@sci.toho-u.ac.jp (T.K.); synapse_yf@yahoo.co.jp (I.T.)
[3]   Research Center for Spectrochemistry, University of Tokyo, Bunkyo-ku, Tokyo 113-0033, Japan;
       jun@chem.s.u-tokyo.ac.jp
[4]   Research Centre for Materials with Integrated Properties, Toho University, 2-2-1 Miyama, Funabashi,
       Chiba 274-8510, Japan
*   Correspondence: kosone@anan-nct.ac.jp (T.K.)

Abstract: In a series of Hofmann-like spin crossover complexes, two new compounds, {Fe(3-F-4-Methyl-py)$_2$[Au(CN)$_2$]$_2$} (**1**) and {Fe(3-Methyl-py)$_2$[Au(CN)$_2$]$_2$} (**2**) (py = pyridine) are described. The series maintains a uniform 2-dimentional (2-D) layer structure of {Fe[Au(CN)$_2$]$_2$}. The layers are combined with another layer by strong aurophilic interactions, which results in a bilayer structure. Both coordination compounds **1** and **2** at 293 K crystallize in the centrosymmetric space groups $P2_1/c$. The asymmetric unit contains two pyridine derivative ligands, one type of Fe^{2+}, and two types of crystallographically distinct [Au(CN)$_2$]$^-$ units. Compound **1** undergoes a complete two-step spin transition. On the other hand, **2** maintains the characteristic of the high-spin state. The present compounds and other closely related bilayer compounds are compared and discussed in terms of the cooperativity and critical temperature. The bilayer structure is able to be further linked by substituent-substituent contact resulting in 3-dimentional (3-D) network cooperativity.

Keywords: coordination polymer; cooperative interaction; crystal engineering; spin crossover

1. Introduction

An essential part for designing spin crossover (SCO) materials is to control and optimize the crystal structure [1]. Especially, construction of the strong cooperative intermolecular interactions in the whole of a structure leads to steep transition behavior with a wide hysteresis loop, which is important for the practical materials [2]. The cooperativity is now being investigated by a variety of coordination polymers. Coordination polymers are one of the interesting materials for constructing supramolecular networks. However, systematic designing of the networks is hard because of the structural diversity of coordination polymers.

Since we reported the first Hofmann like SCO coordination polymer {Fe(py)$_2$[Ni(CN)$_4$]}$_n$ (py = pyridine) [3], many derived types of {Fe^{II}(L)$_{1\sim2}$[M^{II}(CN)$_4$]}$_n$ [4–9] and {Fe^{II}(L)$_{1\sim2}$[M^I(CN)$_2$]$_2$}$_n$ [10–21] (M^I = Cu, Ag, or Au, M^{II} = Ni, Pd, or Pt, L = pyridine derivatives) have been developed. These compounds show a template 2-dimentional (2-D) sheet structure because of their strongly determinate self-assembly process in which they link octahedral metal centers through the N atoms of the bidentate [Au(CN)$_2$]$^-$ unit. Therefore, this structural system can be modified only at the axial ligands, L, for designing the cooperative networks. Here we report and discuss new Hofmann-like 2-D compounds of the general formula {Fe(X-py)$_2$[Au(CN)$_2$]$_2$} (X = 3-F-4-Methyl (**1**) or 3-Methyl (**2**) as shown in Scheme 1).

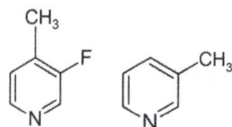

Scheme 1. Molecular structure of the ligands of 3-Furuoro-4-methyl-py and 3-Methyl-py.

2. Results

2.1. X-ray Structural Analysis

2.1.1. Structure of Compound **1** (T = 293 K)

Compound **1** at 293 K crystallizes in the monoclinic centrosymmetric space group $P2_1/c$. The asymmetric unit of the complex consists of the hetero-metal Fe^{II}Au^I unit (Figure 1a). The Fe^{II} ion is octahedrally coordinated by six N atoms. The Fe–N$_{py}$ bond lengths (Fe(1)–N(1) = 2.219(6) Å, Fe(1)–N(2) = 2.220(6) Å) are longer than the Fe–N$_{CN}$ bond lengths (Fe(1)–N(3) = 2.142(7) Å, Fe(1)–N(4) = 2.153(6) Å, Fe(1)–N(5) = 2.160(7) Å, Fe(1)–N(6) = 2.148(6) Å). The average lengths of Fe–N$_{py}$ = 2.220 Å and Fe–N$_{CN}$ = 2.151 Å (total average length of Fe–N = 2.185 Å are estimated). All Au^I atoms have linear coordination geometries with the CN substituents binding to the Fe^{II} ions. While the F(1) in the 3-F-4-Me-py ligand is disordered, the F(2) F(3) in the other 3-F-4-Me-py ligand are not disordered. Thus, the two 3-F-4-Me-py ligands in [Fe^{II}(3-F-4-Me-py)$_2$][Au^I(CN)$_2$] are not equivalent and coexist in *transoid* and *cisoid* conformations for Fe(1). The bidentate [Au^I(CN)$_2$] linear units give rise to an infinite corrugated 2-D mesh-layer formed by the assembly of –Au–N–C–Fe–C–N–Au– infinite chains (Figure 1b). In addition, the layers interact by pairs to define bilayers which stem from strong aurophilic interactions (Figure 1c). The average Au···Au distance in the bilayers is 3.142 Å, less than the sum of the van der Waals radii of Au (3.60 Å). The nearest aromatic rings form almost face-to-face superposition (dihedral angles = 8.55°). It constructs weak π-stacking interactions. The closet C$_{py}$···C$_{py}$ distances between py rings [C(2)···C(8) = 3.548(13) Å] are smaller than the sum of the van der Waals radius (ca. 3.70 Å).

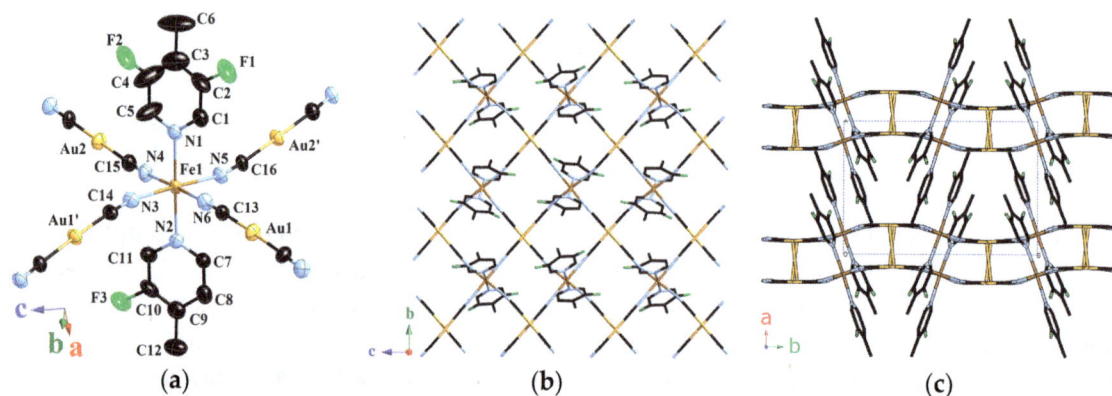

Figure 1. (**a**) Coordination structure of compound **1** containing its asymmetric unit at 293 K; (**b**) View of the 2-D layer structure of **1**; (**c**) Stacking of four consecutive layers of **1** at 293 K. In these pictures, hydrogen atoms are omitted for clarity.

2.1.2. Structure of Compound **1** (T = 180 K)

The crystal structure of this state is similar to that for 293 and 90 K. However, as compared to the fully high spin (HS) state, the single crystals lose their qualities due to the narrow temperature region of the half transition phase. As a result, the crystal data at 180 K is not good enough ($R = 0.1058$ [$I > 2\sigma (I)$]) for determining the correct structure (see Table S1). So, here we discuss in detail the structure at 293 K (fully high spin (HS) state) and 90 K (fully low spin (LS) state).

2.1.3. Structure of Compound 1 (T = 90 K)

The crystal structure of **1** at 90 K is almost identical to that observed at 293 K. The Fe–N_{py} bond lengths (Fe(1)–N(1) = 2.013(4) Å, Fe(1)–N(2) = 2.002(4) Å) are longer than the Fe–N_{CN} bond lengths [Fe(1)–N(3) = 1.942(4) Å, Fe(1)–N(4) = 1.944(4) Å, Fe(1)–N(5) = 1.934(4) Å, Fe(1)–N(6) = 1.936(4) Å]. Total average length is estimated as Fe–N = 1.962 Å. The change of the average length upon spin transition is 0.223 Å, which is almost identical with the expected values for the Fe^{II} 100% LS state. The rings form more parallel superposition than that of the HS state (dihedral angles = 2.57°).

2.1.4. Structure of Compound 2 (T = 293 K)

The crystal structure of **2** at 293 K is almost similar to that of **1** at 293 K which also crystallizes in the monoclinic centrosymmetric space group $P2_1/c$. The 2-D bilayer structure is also almost same shape (Au···Au distance is 3.174 Å) (Figure 2). However, local differences are observed. The rings array is slightly more unparallel than that of **1** (dihedral angles = 11.97°). However, weak π-stacking interactions are also observed (C(1)···C(10) = 3.542(10) Å). There is a *trans* orientation of 3-Me substituents with respect to the N(1)–Fe–N(2) axis. The Fe–N_{py} bond lengths (Fe(1)–N(1) = 2.247(4) Å, Fe(1)–N(2) = 2.244(4) Å) are longer than the Fe–N_{CN} bond lengths [Fe(1)–N(3) = 2.164(4) Å, Fe(1)–N(4) = 2.162(4) Å, Fe(1)–N(5) = 2.159(4) Å, Fe(1)–N(6) = 2.158(4) Å]. The average lengths of Fe–N_{py} = 2.246 Å and Fe–N_{CN} = 2.162 Å (total average length of Fe–N = 2.204 Å are also estimated).

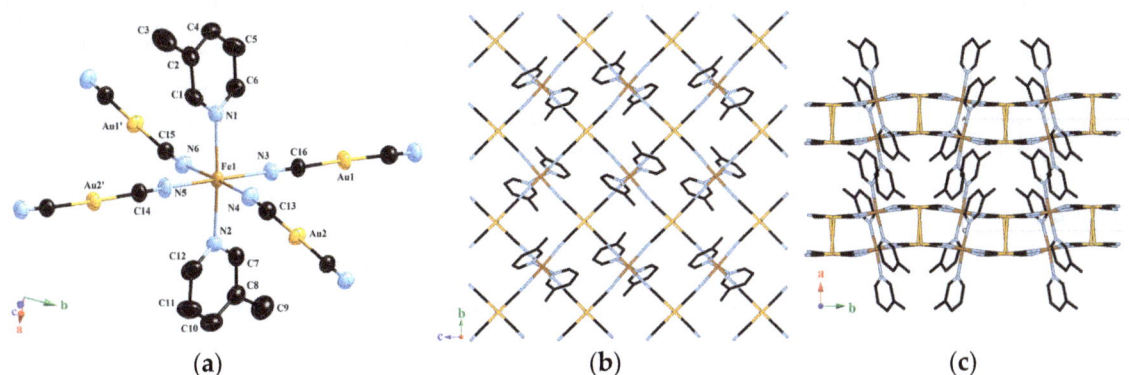

(a)   (b)   (c)

Figure 2. (a) Coordination structure of compound **2** containing its asymmetric unit at 293 K; (b) View of the 2-D layer structure of **2**; (c) Stacking of four consecutive layers of **2** at 293 K. In these pictures, hydrogen atoms are omitted for clarity.

2.2. Magnetic Properties

2.2.1. Thermal Dependence Magnetic Behavior of Compound 1

Figure 3a shows the thermal dependence of $\chi_M T$ for **1** with χ_M being the molar magnetic susceptibility and T the temperature. At room temperature, $\chi_M T$ is 3.65 $cm^3 \cdot K \cdot mol^{-1}$. This value is slightly higher than usual for paramagnetic Fe^{II} compounds, possibly due to oxidation of the complex. Upon cooling, $\chi_M T$ remains almost constant down to 195 K; below this temperature, $\chi_M T$ undergoes a sharp decrease. The complex displays two-step spin transition with a characteristic plateau centered at around 50% conversion, and the warming mode reveals the occurrence of a hysteresis loop (second step). The SCO for this complex causes a reversible change of color from blue (HS) to purple (LS). The critical temperature (T_c) in the first step (T_c^1) is 188.5 K and the cooling ($T_c^{2\,down}$) and warming

$(T_c^{2\,\text{up}})$ modes in the second step are 164.5 and 171.5 K, respectively, giving an approximately 7.0 K width hysteresis loop.

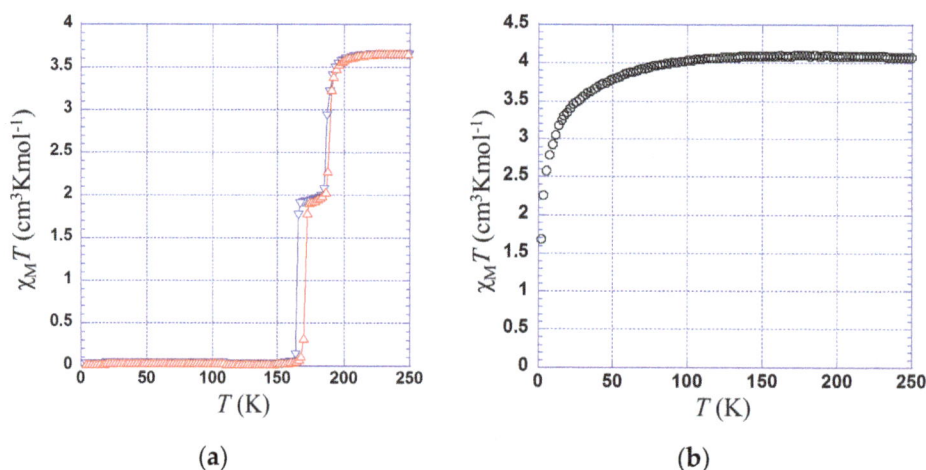

Figure 3. (**a**) Thermal dependence of $\chi_M T$ for compound **1**. The sample was cooled from 300 to 2 K (blue) and then warmed from 2 to 300 K (red). (**b**) Thermal dependence of $\chi_M T$ for compound **2**.

2.2.2. Thermal Dependence Magnetic Behavior of Compound 2

$\chi_M T$ versus T plots for **2** are shown in Figure 3b. At room temperature, the magnetic behavior of the complex **2** is characteristic of Fe(II) compounds in the HS state, $\chi_M T = 4.05\ \text{cm}^3\cdot\text{K}\cdot\text{mol}^{-1}$. The value is slightly higher than that of a pure spin only system, whereas the value is similar to the values of other Hofmann-like SCO Fe^{II} compounds. The $\chi_M T$ value is nearly constant in the range of 80–300 K. However, it decreases steeply as the temperature is lowered to less than 30 K. The decrease in the value of $\chi_M T$ at lower temperatures is due to a typical behavior of zero-field splitting (ZFS) effects of the metallic Fe^{II} centers in the residual HS (S = 2) species [22].

3. Discussion

The bilayer structures of **1** and **2** are almost identical with the former reported compounds of $\{Fe^{II}(L)_2[Au^I(CN)_2]_2\}$ (L = pyridine derivatives). In the previous papers, the synthesis and characterization of the closely related bilayer structures of $\{Fe(X\text{-py})_2[Au(CN)_2]_2\}$ (X = 3-F (**3**) [12,13,15], 4-Methyl (**4**) [17], 3-Br (**5**) [12], and 3-Br-4-Methyl (**6**) [16]) have been reported. These analogous compounds are also discussed in this paper.

Cell volumes, SCO behavior type, and T_c for **1–6** are summarized in Table 1. It is noted that the cell volume clearly shows an expansion with the increase of the substituent bulk. The compounds from smallest to largest are as follows: 3-F-py (1915.7 Å^3) < 3-Methyl-py (1976.3 Å^3) < 3-Br-py (1990.9 Å^3) < 4-Methyl-py (2112.5 Å^3) < 3-F-4-Methyl-py (2155.2 Å^3) < 3-Br-4-Methyl-py (2209.7 Å^3). In spite of the lattice expansion, these compounds completely maintain a bilayer structure.

In terms of the magnetic properties of **1** and **2**, **1** displays a steep two-step spin conversion with a hysteresis loop (second step), while **2** is fully HS in the whole range of temperatures. This result suggests that the T_c of **2** might be so low that no spin transition is observed in an ambient pressure. There is also no spin conversion observed for X = 3-Br. In this series, T_c increases in the following way: 3-Me-py, 3-Br-py (no transition) << 3-Br-4-Me-py (109.1 K) < 3-F-py (first step 147.9 K) < 3-F-4-Me-py (first step 188.5 K) < 4-Me-py (first step 210 K). The order of T_c must result from the difference in the ligand field strength Δ_o. It seems that the 4-Me substituent is the most effective for decreasing Δ_o (see Scheme S1). This trend of T_c is in the same order for $[Fe(L)_2Ag(CN)_2]$ (L = 3-F-py, 3-Br-py, 3-Me-py, and 4-Me-py) and is as follows: 3-Me-py [18], 3-Br-py (no transition) [11] << 3-F-py (first step

146.3 K) [11] < 4-Mepy (first step 189 K) [18]. These orders of T_c are opposite to the expected electronic effect according to the Hammet constants [23].

Table 1. Spin crossover (SCO) behavior types, cell volumes, and critical temperatures of $\{Fe^{II}(X\text{-py})_2[Au^I(CN)_2]_2\}$ at 293 K or room temperature.

Substituents	Cell Volume (Å^3)	SCO Behavior Type	Critical Temperature T_c (K)	Ref.
3-Methyl (**2**)	1976.30 Å^3	None	None	–
4-Methyl (HS) (**4**)	2112.5 Å^3	Steep (3-step)	210 K (first step)	[17]
3-Fuluoro (HS) (**3**)	1915.7 Å^3	Steep (2-step)	147.9 K (first step)	[12,13,15]
3-Fuluoro-4-Methyl (HS) (**1**)	2155.2 Å^3	Steep (2-step)	188.5 K (first step)	–
3-Bromo (**5**)	1990.9 Å^3	None	None	[12]
3-Bromo-4-Methyl (HS) (**6**)	2204.5 Å^3	Steep (1-step, 50% transition)	109.1 K	[16]

In a previous related study, Real and co-workers discussed the effect of substituted pyridines for the analogous series of $[Fe(X\text{-py})_2Ag(CN)_2]$ (X = halogen atoms) [11]. The research says that T_c may be dominantly influenced by crystal packing factor and polymeric structure. We interpret the former research as meaning that the Fe–N bond length is strongly influenced by the interlayer spaces. The steric effect from substituent bulk makes the space tight. In fact, 4-position is vertical to the layer. Thus, a 4-Methyl substituent gives rise to higher chemical pressure in the Fe–N bond. Consequently, the trend of T_c can be explained by the substituent bulk. Electronic effects righteously coexist and compete with steric effects.

Both **1** and **3** have significantly similar transition behavior. However, each behavior has a different hysteresis width. In compound **3**, intermolecular distance between F···F is significantly shorter as compared to that of **1** (Scheme 2). The crystal structure of **3** shows the lowest cell volume which makes the interlayer spaces narrow. It results in shorter F···F contact (closest distance of F···F (HS) = 3.104 Å) between bilayers. Specifically, the half spin transition state (HS–LS) has extremely short F···F contact (2.955 Å). Therefore, this complex must have higher cooperative networks which stabilize the HS–LS state. On the other hand, the F···F contact of **1** is much longer than that of **3** (closest distance of F···F (HS) = 3.717 Å). This longer distance of **1** is due to the 4-Me substituent which serves as bulk for the expansion along the *a* axis between the layers. Thus it generates weakened cooperativity. In fact, the thermal trapping of the HS–LS state for **3** is observed (see Figure S1), while this process cannot be seen for **1** at the same cooling/warming rate. This result is also strongly supported by the difference in the cooperativity of **1** and **3**. SCO behavior types, hysteresis width, closest approach F···F distance, and presence or absence of thermal trapping for **1** and **3** are summarized in Table 2.

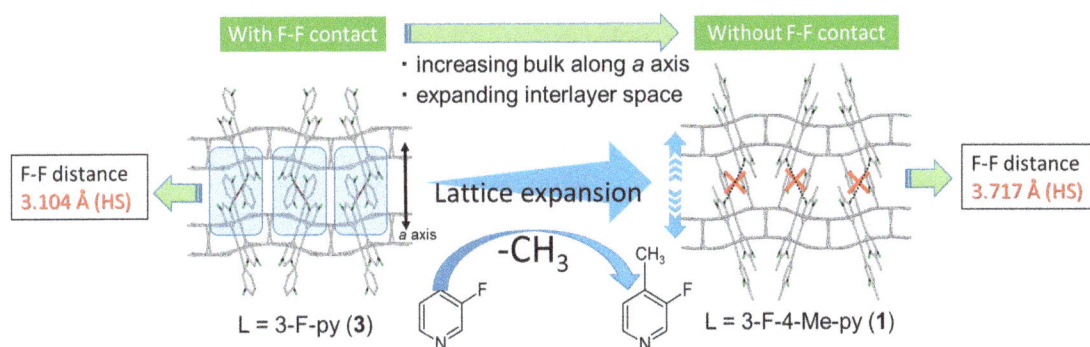

Scheme 2. Representation of the lattice expansion with the addition of the 4-position substituent. Blue highlighted box shows the F···F interactions (black and red dotted line).

Table 2. Comparative hysteresis width, closest approach F···F distances, and presence or absence of thermal trapping for compounds **1** and **3**. SI = supplemental information.

Fe(X-py)$_2$[Au(CN)$_2$]$_2$	Hysteresis Width (K)	Closest Approach F···F Distances [Å]	Thermal Trapping After Cooling at 2 K/min
X = 3-F (**3**)	ca. 40 K	F(1)···F(1): 3.104 Å (HS) F(1)···F(1): 2.955 Å (HS–LS) F(3)···F(3): 3.264 Å (LS)	Observed (see SI)
X = 3-F-4-Me (**1**)	ca. 7 K	F(3)···F(3): 3.717 Å (HS) F(2)···F(3): 3.305 Å (LS)	None

4. Materials and Methods

4.1. Materials

All the chemicals were purchased from commercial sources and used without any further purification.

4.2. Preparation of Compounds 1 and 2

To 10 mL of water, FeSO$_4$·(NH$_4$)$_2$SO$_4$·6H$_2$O (0.10 mmol, 39.2 mg) and K[Au(CN)$_2$] (0.1 mmol, 28.8 mg) were dissolved. The vapor diffusion method using 3-F-4-Me-py or 3-Me-py as a source of ligand molecules provided colorless single crystals suitable for X-ray diffraction of **1–2** over a period of two days. Elem. Anal. Calcd. for C$_{16}$H$_{12}$N$_6$F$_2$Au$_2$Fe (**1**): C, 25.76; H, 1.56; N, 10.83. Found: C, 25.98; H, 1.58; N, 10.90. Calcd. for C$_{16}$H$_{14}$N$_6$Au$_2$Fe (**2**): C, 25.97; H, 1.91; N, 11.36. Found: C, 25.78; H, 1.92; N, 11.35.

4.3. Magnetic Measurements

Measurements of the temperature dependence of the magnetic susceptibility of the complexes **1** and **2** of the powdered samples in the temperature range of 2–300 K with a cooling and heating rate of 2 K·min^{-1} in a 1 kOe field were measured on a MPMS-XL Quantum Design SQUID magnetometer in the Cryogenic Research Center, the University of Tokyo. The diamagnetism of the samples and sample holders were taken into account.

4.4. X-ray Crystallography

Data collection was performed on a BRUKER APEX SMART CCD area-detector diffractometer at 293, 180 K, and 90 K for **1** and **2** with monochromated Mo–Kα radiation (λ = 0.71073 Å) (Bruker, Billerica, MA, USA). A selected single crystal was carefully mounted on a thin glass capillaly and immediately placed under a liquid N$_2$ cooled N$_2$ stream in each case. Crystal structures of the complexes **1** and **2** were determined using a BRUKER APEX SMART CCD area-detector diffractometer with monochromed Mo Kα radiation (λ = 0.71073 Å). The diffraction data were treated using SMART and SAINT, and absorption correction was performed using SADABS [24]. The structures were solved by using direct methods with *SHELXTL* [25]. All non-hydrogen atoms were refined anisotropically, and the hydrogen atoms were generated geometrically. Pertinent crystallographic parameters and selected metric parameters for **1** and **2** are displayed in Tables S1 and S2. Crystallographic data have been deposited with Cambridge Crystallographic Data Centre: Deposition numbers CCDC-1559723 for compound **1** (293 K), CCDC-1559724 for **1** (90 K), and CCDC-1559832 for **2** (293 K). These data can be obtained free of charge via http://www.ccdc.cam.ac.uk/conts/retrieving.html.

5. Conclusions

In this paper, the 2-D bilayer structures of Hofmann-like coordination polymers {Fe^{II}(X-py)$_2$[Au^I(CN)$_2$]$_2$} were synthesized. It is noted that the substituents of pyridine derivatives have two steric effects in the whole of the crystal structure. One is the blocking interlayer spaces resulting in modification of the cooperativity. The other is the chemical pressure to the Fe^{II} centers resulting in modification of the critical temperature. The two effects are opposite forces to each other, like action-reaction forces with a wall.

In general, the 3-D coordination bonding types, such as $[Fe^{II}(pyrazine)[M^{II}(CN)_4]]$ (M = Ni, Pd, and Pt) display stable discontinuous spin transitions with strong cooperativity. On the other hand, most of the 2-D coordination types had weaker cooperativity than that of the 3-D type.

However, the present 2-D system shows a strong cooperativity that is as strong as the 3-D type due to the substituent-substituent interaction.

The 2-D coordination structure can be used in designs due to the unique properties, such as gate-opening behavior [26]. Therefore, the 2-D structure would create novel and unique SCO material with strong cooperativity for practical applications.

Supplementary Materials:
Scheme S1: Critical temperature (T_c) changes depending on the substituent of $\{Fe(X-py)_2[Au(CN)_2]_2\}$ (X = 3-F-4-Methyl (**1**), 3-Methyl (**2**), 3-F (**3**), 4-Methyl (**4**), 3-Br (**5**) and 3-Br-4-Methyl (**6**)), Figure S1: Thermal dependence of spin transition curve for compound **3**. The sample was cooled from 300 to 2 K at a rate of 2 K·min^{-1} and then warmed again to 300 K at a rate of 2 K·min^{-1}. Table S1: Crystal data and structure refinement for compounds **1** and **2**, Table S2: Selected bond lengths for compounds **1** and **2**. For CIF, check CIF files.

Acknowledgments: A part of this work was supported by a Ministry of Education, Culture, Sports, Science, and Technology, Japan (MEXT)-Supported program for the Strategic Research Foundation at Private Universities 2012–2016. This work was also supported by JSPS KAKENHI Grant Number 15K05485.

Author Contributions: Takashi Kosone, Itaru Tomori, and Takeshi Kawasaki carried out the synthesis, XRD measurements with the structural analysis, and Jun Okabayashi carried out the magnetic measurements. Takashi Kosone and Takafumi Kitazawa interpreted and discussed the result.

References

1. Real, J.A.; Andrés, E.; Munoz, M.C.; Julve, M.; Granier, T.; Bousseksou, A.; Varret, F. Spin crossover in a catenane supramolecular system. *Science* **1995**, *268*, 265–268. [CrossRef] [PubMed]

2. Gütlich, P.; Garcia, Y.; Goodwin, H.A. Spin crossover phenomena in Fe(II) complexes. *Chem. Soc. Rev.* **2000**, *29*, 419–427. [CrossRef]

3. Kitazawa, T.; Gomi, Y.; Takahashi, M.; Takeda, M.; Enomoto, M.; Miyazaki, A.; Enoki, T. Spin-crossover behaviour of the coordination polymer $Fe^{II}(C_5H_5N)_2Ni^{II}(CN)_4$. *J. Mater. Chem.* **1996**, *6*, 119–121. [CrossRef]

4. Niel, V.; Martinez-Agudo, J.M.; Muñoz, M.C.; Gaspar, A.B.; Real, J.A. Cooperative Spin Crossover Behavior in Cyanide-Bridged Fe(II)–M(II) Bimetallic 3D Hofmann-like Networks (M = Ni, Pd, and Pt). *Inorg. Chem.* **2001**, *40*, 3838–3839. [CrossRef] [PubMed]

5. Agusti, G.; Cobo, S.; Gaspar, A.B.; Molnár, G.; Moussa, N.O.; Szilágyi, P.Á.; Pálfi, V.; Vieu, C.; Carmen Muñoz, M.; Real, J.A.; et al. Thermal and light-induced spin crossover phenomena in new 3D Hofmann-like microporous metalorganic frameworks produced as bulk materials and nanopatterned thin films. *Chem. Mater.* **2008**, *20*, 6721–6732. [CrossRef]

6. Martínez, V.; Gaspar, A.B.; Muñoz, M.C.; Bukin, G.V.; Levchenko, G.; Real, J.A. Synthesis and Characterisation of a New Series of Bistable Iron(II) Spin-Crossover 2D Metal–Organic Frameworks. *Chem. Eur. J.* **2009**, *15*, 10960–10971. [CrossRef] [PubMed]

7. Bartual-Murgui, C.; Ortega-Villar, N.A.; Shepherd, H.J.; Muñoz, M.C.; Salmon, L.; Molnár, G.; Bousseksou, A.; Real, J.A. Enhanced porosity in a new 3D Hofmann-like network exhibiting humidity sensitive cooperative spin transitions at room temperature. *J. Mater. Chem.* **2011**, *21*, 7217. [CrossRef]

8. Ohtani, R.; Arai, M.; Hori, A.; Takata, M.; Kitao, S.; Seto, M.; Kitagawa, S.; Ohba, M. Modulation of Spin-Crossover Behavior in an Elongated and Flexible Hofmann-Type Porous Coordination Polymer. *J. Inorg. Organomet. Polym. Mater.* **2013**, *23*, 104–110. [CrossRef]

9. Sciortino, N.F.; Zenere, K.A.; Corrigan, M.E.; Halder, G.J.; Chastanet, G.; Létard, J.-F.; Kepert, C.J.; Neville, S.M. Four-step iron(II) spin state cascade driven by antagonistic solid state interactions. *Chem. Sci.* **2017**, *8*, 701–707. [CrossRef] [PubMed]

10. Galet, A.; Muñoz, M.C.; Martinez, V.; Real, J.A. Supramolecular isomerism in spin crossover networks with aurophilic interactions. *Chem. Commun.* **2004**, 2268–2269. [CrossRef] [PubMed]

11. Muñoz, M.C.; Gaspar, A.B.; Galet, A.; Real, J.A. Spin-Crossover Behavior in Cyanide-Bridged Iron(II)–Silver(I) Bimetallic 2D Hofmann-like Metal–Organic Frameworks. *Inorg. Chem.* **2007**, *46*, 8182–8192. [CrossRef] [PubMed]

12. Agustí, G.; Muñoz, M.C.; Gaspar, A.B.; Real, J.A. Spin-Crossover Behavior in Cyanide-bridged Iron(II)–Gold(I) Bimetallic 2D Hofmann-like Metal–Organic Frameworks. *Inorg. Chem.* **2008**, *47*, 2552–2561. [CrossRef] [PubMed]

13. Kosone, T.; Kachi-Terajima, C.; Kanadani, C.; Saito, T.; Kitazawa, T. A two-step and hysteretic spin-crossover transition in new cyano-bridged hetero-metal $Fe^{II}Au^{I}$ 2-dimensional assemblage. *Chem. Lett.* **2008**, *37*, 422–423. [CrossRef]

14. Kosone, T.; Kachi-Terajima, C.; Kanadani, C.; Saito, T.; Kitazawa, T. Isotope Effect on Spin-crossover Transition in a New Two-dimensional Coordination Polymer $[Fe^{II}(C_5H_5N)_2][Au^{I}(CN)_2]_2$, $[Fe^{II}(C_5D_5N)_2][Au^{I}(CN)_2]_2$, and $[Fe^{II}(C_5H_5{}^{15}N)_2][Au^{I}(CN)_2]_2$. *Chem. Lett.* **2008**, *37*, 754–755. [CrossRef]

15. Kosone, T.; Kanadani, C.; Saito, T.; Kitazawa, T. Synthesis, crystal structures, magnetic properties and fluorescent emissions of two-dimensional bimetallic coordination frameworks Fe^{II}(3-fluoropyridine)$_2$ $[Au^{I}(CN)_2]_2$ and Mn^{II}(3-fluoropyridine)$_2[Au^{I}(CN)_2]_2$. *Polyhedron* **2009**, *28*, 1930–1934. [CrossRef]

16. Kosone, T.; Kanadani, C.; Saito, T.; Kitazawa, T. Spin crossover behavior in two-dimensional bimetallic coordination polymer Fe^{II}(3-bromo-4-picoline)$_2[Au^{I}(CN)_2]_2$: Synthesis, crystal structures, and magnetic properties. *Polyhedron* **2009**, *28*, 1991–1995. [CrossRef]

17. Kosone, T.; Tomori, I.; Kanadani, C.; Saito, T.; Mochida, T.; Kitazawa, T. Unprecedented three-step spin-crossover transition in new 2-dimensional coordination polymer $\{Fe^{II}$(4-methylpyridine)$_2[Au^{I}(CN)_2]_2\}$. *Dalton Trans.* **2010**, *39*, 1719–1721. [CrossRef] [PubMed]

18. Rodríguez-Velamazán, J.A.; Carbonera, C.; Castro, M.; Palacios, E.; Kitazawa, T.; Létard, J.-F.; Burriel, R. Two-Step Thermal Spin Transition and LIESST Relaxation of the Polymeric Spin-Crossover Compounds Fe(X-py)$_2$[Ag(CN)$_2$]$_2$ (X=H, 3-methyl, 4-methyl, 3,4-dimethyl, 3-Cl). *Chem. Eur. J.* **2010**, *16*, 8785–8796. [CrossRef] [PubMed]

19. Kosone, T.; Kitazawa, T. Guest-dependent spin transition with long range intermediate state for 2-dimensional Hofmann-like coordination polymer. *Inorg. Chim. Acta* **2016**, *439*, 159–163. [CrossRef]

20. Chiruta, D.; Linares, J.; Garcia, Y.; Dimian, M.; Dahoo, P.R. Analysis of multi-step transitions in spin crossover nanochains. *Phys. B Condens. Matter* **2014**, *434*, 134–138. [CrossRef]

21. Okabayashi, J.; Ueno, S.; Kawasaki, T.; Kitazawa, T. Ligand 4-X pyridine (X = Cl, Br, I) dependence in Hofmann-type spin crossover complexes: Fe(4-Xpyridine)$_2$[Au(CN)$_2$]$_2$. *Inorg. Chim. Acta* **2016**, *445*, 17–21. [CrossRef]

22. Krzystek, J.; Ozarowski, A.; Telser, J. Multi-frequency, high-field EPR as a powerful tool to accurately determine zero-field splitting in high-spin transition metal coordination complexes. *Coord. Chem. Rev.* **2006**, *250*, 2308–2324. [CrossRef]

23. Nakano, K.; Suemura, N.; Yoneda, K.; Kawata, S.; Kaizaki, S. Substituent effect of the coordinated pyridine in a series of pyrazolato bridged dinuclear diiron(II) complexes on the spin-crossover behavior. *Dalton Trans.* **2005**, 740. [CrossRef] [PubMed]

24. Sheldrick, G.M. *SADABS, Program for Empirical Absorption Correction for Area Detector Data*; University of Göttingen: Göttingen, Germany, 1996.

25. Sheldrick, G.M. *SHELXL, Program for the Solution of Crystal Structures*; University of Göttingen: Göttingen, Germany, 1997.

26. Sakaida, S.; Otsubo, K.; Sakata, O.; Song, C.; Fujiwara, A.; Takata, M.; Kitagawa, H. Crystalline coordination framework endowed with dynamic gate-opening behaviour by being downsized to a thin film. *Nat. Chem.* **2016**, *8*, 377–383. [CrossRef] [PubMed]

9

Improved Cluster Structure Optimization: Hybridizing Evolutionary Algorithms with Local Heat Pulses

Johannes M. Dieterich [1,2] and Bernd Hartke [2,*]

[1] Mechanical and Aerospace Engineering, Princeton University, Princeton, New Jersey, NJ 08544-5263, USA; dieterich@ogolem.org

[2] Institute for Physical Chemistry, Christian-Albrechts-University, 24098 Kiel, Germany

* Correspondence: hartke@pctc.uni-kiel.de

Abstract: Cluster structure optimization (CSO) refers to finding the globally minimal cluster structure with respect to a specific model and quality criterion, and is a computationally extraordinarily hard problem. Here we report a successful hybridization of evolutionary algorithms (EAs) with local heat pulses (LHPs). We describe the algorithm's implementation and assess its performance with hard benchmark CSO cases. EA-LHP showed superior performance compared to regular EAs. Additionally, the EA-LHP hybrid is an unbiased, general CSO algorithm requiring no system-specific solution knowledge. These are compelling arguments for a wider future use of EA-LHP in CSO.

Keywords: evolutionary algorithms; genetic algorithms; heat pulses; global optimization; cluster structure optimization

---

1. Introduction

Evolutionary algorithms (EAs) remain one of the prime strategies used to solve global optimization problems such as global cluster structure optimization (CSO). Global CSO attempts to find atomic or molecular cluster conformers with globally minimal energy, or with globally minimal deviations from measured properties [1]. This is of immediate interest in many areas—both in fundamental science (e.g., non-continuum nucleation [2] and solvation [3]) and in applied studies (e.g., aerosols [4], designed materials [5], combustion/astrophysics [6,7], and semiconductors [8–11], involving carbon clusters, silicon clusters, and many other systems). CSO is considered a non-deterministic polynomial-hard (NP-hard) problem; i.e., the number of minima scales exponentially with the dimensionality of the problem. Breaking through this exponential wall is hence the first challenge and main obstacle for any solution strategy. One of us has shown [12] that EAs can be made to locate global minima of Lennard-Jones-type atomic clusters in roughly cubic time complexity. This is a significant scaling reduction compared to the exponential scaling of brute-force enumerative solutions. However, we have shown these same EAs to solve many of the standard analytical benchmark functions [13] in almost linear (i.e., optimal) time complexity, which so far has not been achieved for CSO. CSO is therefore a difficult challenge for the scaling reduction ability of global optimization strategies.

Second, if we are to connect our simulation data with experiment—as should be our target as scientists—we require the hypersurface model to be accurate enough for this purpose. Unfortunately, this requirement means we must typically employ computationally expensive models for the hypersurface itself. For a given cluster size (i.e., independent of the size scaling), this generates a high prefactor on the computational cost, which seriously limits search space exploration and/or enforces restriction of the search to small clusters. This constitutes the second challenge for global

CSO: prefactor reduction. Current cases in point are CSO studies performed directly at the density functional theory (DFT) level [14], which are limited to very small clusters despite the use of national supercomputer resources.

Efficiency improvements to the model (e.g., reduced-scaling techniques) and efficiency improvements to the search itself (e.g., those of the present work) alleviate some of the computational expense and hence can extend direct-DFT CSO to larger systems. However, given the huge prefactor of first-principles methods compared to force-field calculations (4–6 orders of magnitude), equally huge efficiency improvements would be necessary.

Prefactor reduction can be approached in fundamentally different ways. A very successful avenue is the inclusion of system-specific deterministic knowledge into the otherwise generic heuristic EA. However, this simpler approach relies on the similarity of the problem to be solved to the reference from which the deterministic knowledge was extracted. Beyond simple systems, such similarity may be hard to ensure, hence biasing the solution strategy in the wrong direction and potentially jeopardizing its ability to locate the correct global minimum.

A theoretically superior but practically harder avenue is the analysis of a given global optimization strategy for its strengths and weaknesses in hypersurface exploration and exploitation, followed by algorithmic improvements designed to eliminate weaknesses while maintaining strengths. This more generic approach ensures that the similarity requirement is less strict. Again, various options arise. We have contributed, e.g., niches [12] and graph-based technologies [15], and others have employed order parameters [16,17] (which are similar to niches, in guiding the search by criteria other than the property being optimized), bond transpositions in Monte-Carlo graph search [18] and graph partitioning [19].

Over the course of the last two decades, researchers in the field have hypothesized that hybridizing EAs with other solution strategies of complementary strength profiles would result in overall superior behavior. Proofs-of-concept to this extent were demonstrated. Attempts to evolve beyond this stage were mainly unsuccessful. For example, in unpublished work, both of the present authors independently of each other attempted to hybridize EAs with Monte-Carlo with minimizations (MCM [20]; later known as basin-hopping [21]), but failed to arrive at a hybrid that performed significantly better than EAs alone. Augmenting MCM with short molecular dynamics trajectories has been successfully employed in two studies of transition metal nanoalloys as an indiscriminant heating of the entire cluster [22,23].

In this publication, we will introduce a successful attempt to hybridize EAs with another global optimization strategy—local heat pulses (LHPs). We will first concisely introduce our EA implementation and our implementation of LHPs, with an emphasis on changes of the latter compared to the original. Subsequently, we will discuss our hybridization strategy in detail. We then present benchmark results for a selection of hard CSO problems and conclude with an outlook on how this hybrid strategy will enable us to connect theory and experiment in the future with a smaller computational footprint.

In 1997, Möbius, Hoffmann, and Schreiber [24] proposed "thermal cycling" as an improvement over simulated annealing. Briefly, thermal cycling consists of two steps: (a) "locally heating" the system by perturbing a small part of it (e.g., via a short series of Metropolis Monte-Carlo steps at a given temperature T, or by applying small changes in a randomly selected region), followed by (b) a quench (e.g., via local optimization). Steps (a) and (b) are repeated, while T is lowered gradually. For several traveling salesman instances, substantial improvements over standard simulated annealing were found.

In 2011, Möbius and Schön [25] extended this concept to continuous problems, now under the name of "local heat pulses", and applied it to structure determination of the periodic bulk system $Mg_{10}Al_4Ge_2Si_8O_{36}$, using a force field containing Buckingham and three-body terms. Again, they could demonstrate the superiority of LHP compared to standard simulated annealing, and also to multistart local search.

Clearly, LHP is historically and conceptually related to simulated annealing. However, it is also similar to MCM (aka basin-hopping): basically, both consist of cycling two steps, namely getting out of the current minimum and quenching into the next one. The difference is in the first step: in MCM, as well as in typical Monte-Carlo mutation operators of genetic algorithm (GA)/EA optimization, this "stepping-out" is achieved by randomly mutating randomly chosen particles. In contrast, in LHP these mutations are not entirely random; instead, both the choice of particles and the mutation extent is spatially localized. In this sense, the mutation step of LHP can be interpreted as a "phenotype" version of a Monte-Carlo mutation, in which prior spatial neighborhoods of the cluster particles are preserved instead of being potentially destroyed. In GA/EA applications to CSO, the phenotype crossover by Deaven and Ho [12,26–28] turned out to be superior to the traditional genotype crossover, because the latter did not take spatial neighborhoods into account. Similarly, with the present work we show that the phenotype character of LHP tends to make it superior to traditional MC mutation that acts in a spatially random fashion.

We will in the following first introduce EAs and our hybridization strategy of them with LHP. We will discuss details of our LHP implementation in light of algorithmic requirements posed by hybridization. Subsequently, we will test the EA-LHP hybrid in comparison to the pure EA and LHP approaches on extremely hard CSO test cases and analyze its performance. We conclude with a summary and outlook of future applications of EA-LHP.

2. Algorithms

2.1. Evolutionary Algorithms

EAs encompass a large set of algorithms and implementations. In the following, we will restrict ourselves to discuss our implementation employed here and highlight important features of it of relevance to this work. More exhaustive discussions can be found in References [29,30] for our implementation and References [31–34] for EAs in general.

EAs have a history of using biology-inspired terms to describe algorithmic steps. A convincing argument has been made that this is needlessly obfuscating [35,36]. In the following, we will hence limit our use of biological terms to a minimum.

Our pool-based EA maintains a set of energetically low-lying isomers or conformers (henceforth referred to as candidates) throughout the optimization. Each candidate contains an encoding of its solution string. For CSO, these are the nuclear coordinates as a mix of absolute Cartesian center of mass positions, Euler angles, and internal Cartesians. At startup, this pool is filled with initial solution candidates, based on randomly packing molecules and relaxing the resulting clusters. This is already a first step in gathering knowledge about possible solutions (e.g., on which particle pair-distances are presumably optimal). In fact, for very small clusters, the global minimum may already be in this initial pool. For larger clusters this quickly becomes unlikely, and in such a case simply increasing the initial pool is not a good strategy, since series of local optimizations from random starting structures will not suffice [37]. Instead, exchange of partial knowledge embodied in the candidates and further variations of these candidates is much more profitable. Hence, we next select two known candidates—one based on a Gaussian-weighted fitness distribution, the other randomly—from the pool for further operations. First, solution knowledge is exchanged between the old candidates to form two new candidates (crossover). This exchange may either happen based on the numeric representation (genotype) or the physical form (phenotype). We have found phenotype exchanges [26–28] by cutting through clusters and exchanging the parts above the cutting plane to be superior to other options [12]. Subsequently, the two new solution candidates can be augmented with new solution knowledge to explore the solution space further (mutation with a configured probability). This operation may take the form of selecting and moving a few random atoms through a Monte Carlo (MC) type move, a graph-based approach [15] where least connected molecules are identified and directed into a more connected position, or system-specific manipulations. In this contribution, we use

LHP as a means to manipulate the cluster structure for better conformational space exploration. We will discuss our implementation of LHP and its importance for conformational space exploration below.

After these operations are completed, the candidate structures are exposed to fast quality filters. In the case of CSO, graph-based filters assert that the candidates do not have colliding nuclei or parts of the cluster dissociated from the rest. This step ensures that only candidates of a minimal quality enter the next, computationally most intensive step: local optimization. Each candidate geometry is quenched to a local minimum based on the method description chosen. We are employing classical force fields in this work (see below)—the fastest option available. Nevertheless, the local optimization steps account for the vast majority of computational expense.

This described hybridization employs LHP as a mutation step. Alternatively, we can integrate LHP to replace the local optimizations; i.e., after regular crossover and mutation, instead of a simple local optimization, LHP is employed for a few iterations to sample the vicinity of the candidates.

We then check the total energy of the two candidates after optimization, pick the lower energy (i.e., better individual) and attempt to add it to the pool of candidates. Here, the pool only adds the candidate if it can ensure a minimal diversity with respect to other candidates in the pool and if the energy of the candidate to be added is better than the one of the worst individuals in the pool. This minimal diversity is either achieved through primitive comparison of total energies or through more advanced technologies such as niching [12]. If the candidate is found to qualify for addition, the pool size is kept below a configured maximal size through removal of worse individual(s).

Subsequently, the next two candidates are picked from the pool and the cycle repeats until either a maximal number of steps is reached or the best candidate in the pool has an energy less than or equal to a configured cutoff. Obviously, the cycles can be parallelized with great efficiency, as the only serial step is the addition to the pool of candidates, which is computationally significantly cheaper than local optimization, even if force fields and non-trivial diversity measures are used.

Lastly, for the sake of comparison, we have also enabled the option to use LHP in its intended purpose as a standalone global optimization. We want to note explicitly that we do not make the claim that our implementation is an optimal LHP for this purpose. Indeed, our semi-local tuning may be detrimental to its efficiency as a global optimization. Instead, we simply want to compare our implementation as part of the hybrid EA-LHP approach to it alone. If optimal LHP performance is required, we currently refer to the original implementations. The LHP also supports resetting to either a known good geometry or reinitializing with a randomized solution candidate. This breaks the Markov chain of events of LHP and hence serves as a good benchmark for LHP's performance without considering accumulated system knowledge.

2.2. Local Heat Pulses and EA-LHP Hybrid

Our implementation of LHP follows the ideas of Möbius et al., but differs in key aspects. At the most basic level, local heat pulses go through a repeated protocol of selecting a set of atoms to perturb, perturbing them (i.e., heating them), and quenching the so-perturbed geometry by means of a local optimization. Finally, a basic criterion is applied whether or not to accept the quenched geometry and do the next local heat cycle based on it or revert to the previously used geometry instead.

As we stated before, integration with our existing EA protocol is straightforward: we simply use LHP as the solution augmentation step, in contrast to its original use as a complete global optimization strategy. The requirements for these two usages are somewhat different: a global optimization protocol must be able to efficiently explore the conformational space, exploit semi-local regions/funnels to find low-lying minima, and ensure enough diversity in solution candidates to progress. On the other hand, as a substep of the overall optimization protocol, the solution augmentation step has a reduced responsibility. It only needs to explore and exploit semi-local regions of conformational space. Other parts of the EA protocol have other responsibilities: the exchange of solution knowledge ensures long-range exploration, the use of local optimization ensures local exploitation, and the pool with its diversity criteria and potential niching ensures that progress will be made to locate the global minimum.

Hence, we can tune the LHP implementation for this intended purpose. LHP's perturbation of a set of atoms lends itself from the onset to be used as a semi-local exploitation step. We can further strengthen this aspect by how the to-be-perturbed nuclei are selected. Instead of randomly choosing atoms, we try to localize our selection operators. Beyond trivial global selectors for a fixed number or percentage of the number of nuclei, we have implemented semi-local phenotype selectors that create a random pulse origin either inside or on the surface of the cluster and then pick nuclei (or center of masses of molecules) within a specified radius around this point for perturbation. The following perturbation is also localized by weighing the strength of move with a three-dimensional Gaussian centered at the pulse origin.

After the quenching local optimization, we either apply a strict criterion where only steps decreasing the energy are accepted and used for the next step or a standard Metropolis criterion using an artificial temperature.

All implementations described in this publication are available free of charge under a BSD 4-clause license from https://www.ogolem.org as part of the OGOLEM framework.

2.3. Benchmark System Models

As explained in the introduction, one aim of the present work is to arrive at a CSO algorithm that offers improved performance compared to standard EAs but without relying on system-specific information; i.e., performance improvements should incur for very different application cases. Hence, we have tested our EA-LHP-hybrid both with atomic clusters and with molecular ones, and also for clusters without and with directional interactions between the particles. These desirable characteristics can be covered with just two pure neutral cluster systems as test cases: Lennard-Jones (LJ) clusters and water clusters. Notably, the energy landscape of TIP4P (transferable intermolecular potential, 4 points) water clusters can reasonably be expected to differ strongly from that of LJ clusters, since TIP4P supports strongly directional molecular interactions and reasonably captures the propensity of several water molecules to form ordered H-bonded chains and rings.

2.4. Lennard-Jones Clusters

LJ clusters are defined via total cluster energies E_{LJ} given by

$$E_{LJ} = \sum_{i<j} 4\epsilon_{ij} \left(\left(\frac{\sigma_{ij}}{r_{ij}} \right)^{12} - \left(\frac{\sigma_{ij}}{r_{ij}} \right)^{6} \right) \tag{1}$$

with pair well depth ϵ and pair equilibrium distance $r_e = \sqrt[6]{2}\,\sigma$, interatomic distance r, and indices i, j enumerating the atoms. The LJ model is a reasonable first-order approximation for systems dominated by van-der-Waals interactions (e.g., rare gases), but even for this purpose it is not quantitatively accurate [38]. Of course, we use the LJ model here not for its (limited) application accuracy, but for two other reasons: it is a well-established benchmark system for CSO, and by construction LJ particles have exclusively pairwise and perfectly isotropic particle–particle interactions.

For different cluster sizes n, homogeneous LJ_n clusters offer a wide range of CSO challenges, from very simple single-funnel landscapes to difficult deceptive ones in which the global minimum-energy structure resides in narrow funnel separated by high barriers from the remaining search region which is dominated by significantly different structures. These situations have been analyzed in detail [39]. From the well-known most challenging cases below $n = 150$ we pick $n = 98$, since in that case not just two structural types (icosahedral and decahedral) compete with each other in a close race, but the true global minimum is of yet another structural type with T_d symmetry [40]. Hence, for LJ_{98}, a successful CSO run not only has to get beyond the icosahedral structures dominating most of the search landscape, but also has to differentiate between decahedral and T_d as two elusive alternatives. Cartesian coordinates for these three minima are provided as Supplementary Materials.

2.5. Water Clusters

Water clusters differ from LJ clusters not only by being molecular, but they also feature strongly directional interactions (hydrogen bonds) that strongly favor far less than 12 nearest neighbors. Hence, search landscapes that strongly differ from those exhibited by LJ clusters can be expected. Compared to these basic features, the highly non-trivial choice of which water model to choose is of secondary importance. Here, we chose the TIP4P model [41].

TIP4P water clusters are defined in atomic units by

$$E_{\text{TIP4P}} = \frac{A}{r_{mn}} - \frac{B}{r_{mn}} + \sum_{i \in m} \sum_{j \in n} \frac{q_i q_j}{r_{ij}} \tag{2}$$

where indices m, n, with $m < n$, enumerate individual water molecules and indices i, j indicate charged sites within each water molecule. Water molecules are rigid, with an OH bond length of 0.9572 Å and a bond angle of 104.52 degrees. Partial charges q of +0.52 sit on all H-atoms, while a partial charge of −1.04 sits on the so-called M-site on the HOH angle bisector, 0.15 Å away from the O-atom. The first two LJ-like terms are centered on the O-atoms themselves and prevent water molecules from collapsing upon each other. The two parameter values $A = 6 \times 10^5$ Å^{12} kcal/mol and $B = 610$ Å^6 kcal/mol complete the TIP4P definition. TIP4P water exhibits reasonable H-bond patterns and (in contrast to, e.g., TIP3P) a phase diagram that at least vaguely resembles the true one. Again, we use TIP4P here not for modeling true water clusters, but because it is also frequently employed as CSO benchmark.

Homogeneous neutral water clusters modeled with the TIP4P potential are simple for CSO algorithms until $n = 15$. Beyond this limit, not only differences between different water potentials start to have qualitative effects [42], but the search landscapes also become significantly more challenging than for clusters consisting of isotropic atoms without directional bonding propensities. A case in point is $n = 21$ with TIP4P, where a low-energy local minimum with a reasonable-looking geometry is easy to find, while the true global minimum structure is very different and difficult to locate [43,44]—in other words, this is another deceptive search landscape. Cartesian coordinates for these two minima are provided as Supplementary Materials.

3. Results

3.1. LJ$_{98}$

For LJ$_{98}$, we have compared an EA setup with LHP as a main mutation ingredient with another EA setup in which LHP is replaced by a standard Monte-Carlo mutation. In both cases, a pool size of 2000 individuals was chosen, and LJ neighborhood niching [45] was employed, as well as a mix of several phenotype crossover operators. Twenty-five percent of the mutations were done with our graph-based directed mutation [15]. In the remaining 75% of the mutation events, either LHP-mutation or MC-mutation were used. In both cases, the number of mutated particle positions was Gaussian-distributed, with a similar mean and width of the distribution, such that in most cases between two and five particle positions were changed.

In Table 1, numbers of global optimization steps are compared that are needed until the true T_d global minimum is found, starting from random seeds.

To cope with the typically large random scatter of non-deterministic search, 50 runs were performed in each case, and both average and median results are shown. Note that in each run, the maximum number of global optimization steps was set to 4.2 million, and unsuccessful runs were counted as successful after 4.2 million steps. Therefore, the results in Table 1 contain a bias, making bad results better than they actually are. Despite this disadvantage, LHP mutation is clearly better than conventional Monte-Carlo mutation. We have used the phenotype-style spatial localization of LHP "heating" here (concerning both the selection of atoms to be moved and their amount of movement), and this is the decisive difference between LHP and MC mutation in these runs. Therefore, it can be concluded that this phenotype mutation is beneficial—at least in this very hard CSO test case.

Table 1. Numbers of global optimization steps until the true T_d global minimum is found, comparing local heat pulse (LHP) and standard Monte-Carlo (MC) as main mutation operators.

Mutation	Number of Steps Until T_d	
	Median	Mean
MC mutation	257,759.5	1,357,727
LHP	131,326.0	735,428

3.2. $(H_2O)_{21}$

For $(H_2O)_{21}$ in the TIP4P model, we have again compared LHP mutation with conventional Monte-Carlo mutation. Additionally, we have also tested LHP as a pre-stage in local optimization within our standard global–local hybrid EA scheme, and LHP as a stand-alone global optimization algorithm, without any EA ingredients.

For the EA cases, we used a pool size of 84 individuals and 2 million global optimization steps. As for the LJ clusters, we employed a mix of two mutation operators here: 25% of the mutations were always done with our graph-based directed mutation [15]. The remaining 75% were either LHP-mutation or a standard Monte-Carlo mutation, here operating directly in the six-dimensional external-coordinate space for each water molecule (three positional coordinates of the center of mass and three Euler angles determining the rotational orientation), denoted extMC in the following. For LHP-mutation, the number of mutated molecules was chosen randomly via a Gaussian probability distribution; below we show results for three different choices of this Gaussian width.

In contrast to the LJ clusters, we used only a straightforward phenotype crossover operator (OGOLEM's sweden); this brings our CSO runs in a range where the effects of LHP are most easily visible.

Normally in our EA runs, each crossover/mutation global optimization step was finished by subjecting the resulting new clusters to a local geometry optimization. Here, we have performed additional extMC-EA-runs in which LHP was applied after crossover/mutation, but before local optimization. Hence, in these cases, LHP was added on top of a fully functional EA-based CSO.

In contrast to all this, we also used LHP closer to the original intention; i.e., in a stand-alone manner, without any EA ingredients.

The strongly deceptive search landscape of $(H_2O)_{21}$ in the TIP4P model leads to an even wider spread of the number of global optimization steps needed to find the true global minimum than for LJ_{98}. Hence, in Table 2, for the cases mentioned above we compare a different measure of success—namely the percentage of runs in which the true global minimum is found within the prescribed 2 million steps.

Table 2. Percentage of runs finding the true global minimum within 2 million steps for different cluster structure optimization (CSO) setups, as discussed in the text.

CSO Setup	Percentage of Successful Runs
LHP as mutation, wide ($\sigma = 9$)	27.5
LHP as mutation, medium ($\sigma = 8$)	38.0
LHP as mutation, narrow ($\sigma = 7$)	16.6
LHP before local optimization	66.0
extMC	30.0
LHP standalone	15.0
LHP standalone, randomized	6.6

Note that in the "wide" case of the LHP-as-mutation distribution, there is a significantly non-zero chance that most or even all molecules are affected by an LHP instance: 19 of 21 molecules are mutated with a chance of 2%, and all molecules with a chance of 0.007%. Conversely, for the "narrow" case, these probabilities are much smaller (zero percent for 16 molecules and more, 1% for 12 molecules). However, as Table 2 indicates, this significantly reduces the effectiveness of LHP. Partly, this re-emphasizes the visual finding that the true global minimum and the more dominant region of the search landscape correspond to very different structures, and hence large changes in cluster structures are needed to "jump" from one funnel to the other. Nevertheless, when doing such large structure changes with LHP, they can be less disrupting to the local particle neighborhoods (in this case also to the H-bond network), since in our LHP implementation the amount of positional and orientational mutations is spatially ordered and localized—in contrast to the completely randomized Monte-Carlo mutation ("extMC"). For this reason, with proper tuning, LHP can be better than extMC, as documented by Table 2.

When LHP was inserted as an additional step before local optimization and after normal crossover/mutation, two thirds of all runs were successful, which is roughly twice the success rate of extMC or LHP acting alone. At first sight, this is not very surprising, since then both extMC and LHP are applied in each global search step, so it seems reasonable that the combined success is the sum of the isolated successes. However, this also means that (at least in this case) LHP and extMC are largely "orthogonal" to each other; i.e., one search mode can achieve changes that the other cannot, and vice versa. In any case, this test shows that adding LHP to an otherwise standard EA scheme indeed does improve the overall search power significantly. However, it has to be remembered that in this test case we used a rather simplistic crossover/mutation set in our EA. With more sophisticated EA ingredients, the impact of adding LHP may well be smaller.

On the other hand, operating LHP by itself without the usual EA ingredients (a pool of solutions being improved in parallel, via crossover/mutation and selection) is clearly worse than even the simplistic EA used here, without or with LHP as mutation operator. Unexpectedly, resetting standalone-LHP to new random structures periodically—in an effort to enhance exploration—does not help, but even lowers the success rate markedly.

4. Discussion and Conclusions

With hard and deceptive CSO application examples of very different characteristics, we have shown that EAs do benefit from hybridization with LHP, producing a combined search algorithm that is better than both of these ingredients on their own. This is a non-trivial result, since both ingredients alone had been previously shown to solve hard global optimization problems and since no system-specific knowledge was employed in their combination.

Extending the perspective beyond CSO towards other global optimization problems, however, it should be emphasized that both EA and LHP—as implemented and used here—do profit from a more general a-priori knowledge about CSO, namely locality in 3D space. Whatever the nature of the interactions between cluster particles may be (covalent bonds, hydrogen bonds, van-der-Waals, or Coulomb interactions, etc.), all of them decay to zero with increasing distance. For all but the smallest clusters (for which CSO is not problematic anyway), this guarantees that localized changes in one region of the cluster do not strongly couple with changes in other far-away regions. This may well be different in other global optimization problems. Hence, even for LHP-EA, by construction there still is no free lunch [46,47].

Acknowledgments: Johannes M. Dieterich thanks Arnulf Möbius for convincing him of LHP as a technology during long hikes in Colorado's Rocky Mountains. He wishes to thank Dean Emily Carter for her ongoing support of his other scientific endeavors. He expresses his gratitude to Ricardo Mata for sending him to the Telluride Energy Landscapes workshop. Bernd Hartke gratefully acknowledges financial support by the German Science Foundation DFG, via grant Ha2498/16-1.

Author Contributions: Johannes M. Dieterich conceived and implemented the EA-LHP hybrid. Bernd Hartke suggested some additional LHP features and did the performance tests. Both contributed equally to writing the paper.

Abbreviations

The following abbreviations are used in this manuscript:

CSO   Cluster Structure Optimization
DFT   Density Functional Theory
EA   Evolutionary Algorithm
GA   Genetic Algorithm
LHP   Local Heat Pulse
LJ   Lennard-Jones
MC   Monte Carlo
MCM   Monte Carlo with Minimizations
TIP4P transferable intermolecular potential, 4 points

References

1. Dieterich, J.M.; Hartke, B. Observable-targeting global cluster structure optimization. *Phys. Chem. Chem. Phys.* **2015**, *17*, 11958–11961, doi:10.1039/C5CP01910A.

2. Angélil, R.; Diemand, J.; Tanaka, K.K.; Tanaka, H. Properties of liquid clusters in large-scale molecular dynamics nucleation simulations. *J. Chem. Phys.* **2014**, *140*, 074303, doi:10.1063/1.4865256.

3. Sunoj, R.B.; Anand, M. Microsolvated transition state models for improved insight into chemical properties and reaction mechanisms. *Phys. Chem. Chem. Phys.* **2012**, *14*, 12715, doi:10.1039/c2cp41719g.

4. Bonačić-Koutecky, V.; Bernhardt, T.M. Structure and reactivity of small particles: From clusters to aerosols. *Phys. Chem. Chem. Phys.* **2012**, *14*, 9252, doi:10.1039/C2CP90066A.

5. Lu, Z.; Yin, Y. Colloidal nanoparticle clusters: Functional materials by design. *Chem. Soc. Rev.* **2012**, *41*, 6874, doi:10.1039/C2CS35197H.

6. Weltner, W., Jr.; van Zee, R.J. Carbon Ions, Molecules, and Clusters. *Chem. Rev.* **1989**, *89*, 1713.

7. Von Helden, G.; Hsu, M.-T.; Kemper, P.R.; Bowers, M.T. Structures of carbon cluster ions from 3 to 60 atoms: Linears to rings to fullerenes. *J. Chem. Phys.* **1991**, *95*, 3835, doi:10.1063/1.460783.

8. Ho, K.-M.; Shvartsburg, A.A.; Pan, B.; Lu, Z.-Y.; Wang, C.-Z.; Wacker, J.G.; Fye, J.L.; Jarrold, M.F. Structures of medium-sized silicon clusters. *Nature* **1998**, *392*, 582.

9. Hiura, H.; Miyazaki, T.; Kanayama, T. Formation of Metal-Encapsulating Si Cage Clusters. *Phys. Rev. Lett.* **2001**, *86*, 1733, doi:10.1103/PhysRevLett.86.1733.

10. Koyasu, K.; Akutsu, M.; Mitsui, M.; Nakajima, A. Selective Formation of MSi_{16} (M = Sc, Ti, and V). *J. Am. Chem. Soc.* **2005**, *127*, 4998, doi:10.1021/ja045380t.

11. Li, Y.; Tam, N.M.; Claes, P.; Woodham, A.P.; Lyon, J.T.; Ngan, V.T.; Nguyen, M.T.; Lievens, P.; Fielicke, A.; Janssens, E. Structure Assignment, Electronic Properties, and Magnetism Quenching of Endohedrally Doped Neutral Silicon Clusters, Si_nCo (n = 10–12). *J. Phys. Chem. A* **2014**, *118*, 8198, doi:10.1021/jp500928t.

12. Hartke, B. Global cluster geometry optimization by a phenotype algorithm with niches: Location of elusive minima, and low-order scaling with cluster size. *J. Comput. Chem.* **1999**, *20*, 1752–1759.

13. Dieterich, J.M.; Hartke, B. Empirical review of standard benchmark functions using evolutionary global optimization. *Appl. Math.* **2012**, *3*, 1552–1564, doi:10.4236/am.2012.330215.

14. Aslan, M.; Davis, J.B.A.; Johnston, R.L. Global optimization of small bimetallic Pd–Co binary nanoalloy clusters: A genetic algorithm approach at the DFT level. *Phys. Chem. Chem. Phys.* **2016**, *18*, 6676–6682, doi:10.1039/c6cp00342g.

15. Dieterich, J.M.; Hartke, B. A graph-based short-cut to low-energy structures. *J. Comput. Chem.* **2014**, *35*, 1618, doi:10.1002/jcc.23669.

16. Rossi, G.; Ferrando, R. Searching for low-energy structures of nanoparticles: A comparison of different methods and algorithms. *J. Phys. Condens. Matter* **2009**, *21*, 084208, doi:10.1088/0953-8984/21/8/084208.

17. Barcaro, G.; Sementa, L.; Fortunelli, A. A grouping approach to homotop global optimization in alloy nanoclusters. *Phys. Chem. Chem. Phys.* **2014**, *16*, 24256, doi:10.1039/c4cp03745f.

18. Flikkema, E.; Bromley, S.T. Defective to fully coordinated crossover in complex directionally bonded nanoclusters. *Phys. Rev. B* **2009**, *80*, 035402, doi:10.1103/PhysRevB.80.035402.

19. Schebarchov, D.; Wales, D.J. A new paradigm for structure prediction in multicomponent systems. *J. Chem. Phys.* **2013**, *139*, 221101, doi:10.1063/1.4843956.

20. Li, Z.; Scheraga, H.A. Monte Carlo-minimization approach to the multiple-minima problem in protein folding. *Proc. Natl. Acad. Sci.* **1987**, *84*, 6611–6615.

21. Wales, D.J.; Doye, J.P.K. Global Optimization by Basin-Hopping and the Lowest Energy Structures of Lennard-Jones Clusters Containing up to 110 Atoms. *J. Phys. Chem. A* **1997**, *101*, 5111, doi:10.1021/jp970984n.

22. Bochicchio, D.; Ferrando, R. Size-Dependent Transition to High-Symmetry Chiral Structures in AgCu, AgCo, AgNi, and AuNi Nanoalloys. *Nano Lett.* **2010**, *10*, 4211, doi:10.1021/nl102588p.

23. Rossi, G.; Ferrando, R. Shape-changing with Exchange Moves in the Optimization of Nanoalloys. *Comput. Theor. Chem.* **2017**, *1107*, 66, doi:10.1016/j.comptc.2017.01.002.

24. Möbius, A.; Neklioudov, A.; Díaz-Sánchez, A.; Hoffmann, K.H.; Fachat, A.; Schreiber, M. Optimization by thermal cycling. *Phys. Rev. Lett.* **1997**, *79*, 4297, doi:10.1103/PhysRevLett.79.4297.

25. Möbius, A.; Schön, J.C. Periodic Structure Optimization via Local Heat Pulses. Available online: https://www.researchgate.net/publication/258596434 and http://www.physik.uni-leipzig.de/~janke/CompPhys11/Folien/moebius.pdf (accessed on 25 March 2017).

26. Deaven, D.M.; Ho, K.M. Molecular Geometry Optimization with a Genetic Algorithm. *Phys. Rev. Lett.* **1995**, *75*, 288, doi:10.1103/PhysRevLett.75.288.

27. Deaven, D.M.; Tit, N.; Morris, J.R.; Ho, K.M. Structural optimization of Lennard-Jones clusters by a genetic algorithm. *Chem. Phys. Lett.* **1996**, *256*, 195, doi:10.1016/0009-2614(96)00406-X.

28. Pullan, W.J. Genetic operators for the atomic cluster problem. *Comput. Phys. Commun.* **1997**, *107*, 137, doi:10.1016/S0010-4655(97)00092-1.

29. Dieterich, J.M.; Hartke, B. OGOLEM: Global cluster structure optimisation for arbitrary mixtures of flexible molecules. A multiscaling, object-oriented approach. *Mol. Phys.* **2010**, *108*, 279, doi:10.1080/00268970903446756.

30. Dieterich, J.M.; Hartke, B. Error-safe, portable and efficient evolutionary algorithms implementation with high scalability. *J. Chem. Theory Comput.* **2016**, *12*, 5226, doi:10.1021/acs.jctc.6b00716.

31. Cartwright, H.M. An introduction to evolutionary computation and evolutionary algorithms. *Struct. Bonding* **2004**, *110*, 1, doi:10.1007/b13931.

32. Hartke, B. Global optimization. *WIREs Comput. Mol. Sci.* **2011**, *1*, 879, doi:10.1002/wcms.70.

33. Weise, T. Global Optimization Algorithms: Theory and Application. Available online: http://www.it-weise.de/projects/book.pdf (accessed on 8 September 2017).

34. Weise, T.; Chiong, R.; Tang, K. Evolutionary Optimization: Pitfalls and Booby Traps. *J. Comput. Sci. Technol.* **2012**, *27*, 907, doi:10.1007/s11390-012-1274-4.

35. Weise, T. Why Research in Computational Intelligence Should Be Less Inspired. Available online: http://www.it-weise.de/thoughts/text/ecInspiration.html (accessed on 9 May 2016).

36. Sörensen, K. Metaheuristics—The metaphor exposed. *Int. Trans. Oper. Res.* **2015**, *22*, 3, doi:10.1111/itor.12001.

37. Avaltroni, F.; Corminboeuf, C. Efficiency of random search procedures along the silicon cluster series. *J. Comput. Chem.* **2011**, *32*, 1869, doi:10.1002/jcc.21769.

38. Dieterich, J.M.; Hartke, B. Composition-induced structural transitions in mixed LJ clusters: Global reparametrization and optimization. *J. Comput. Chem.* **2011**, *32*, 1377–1385, doi:10.1002/jcc.21721.

39. Doye, J.P.K.; Miller, M.A.; Wales, D.J. Evolution of the potential energy surface with size for Lennard-Jones clusters. *J. Chem. Phys.* **1999**, *111*, 8417, doi:10.1063/1.480217.

40. Leary, R.H.; Doye, J.P.K. Tetrahedral global minimum for the 98-atom Lennard-Jones cluster. *Phys. Rev. E* **1999**, *60*, R6320, doi:10.1103/PhysRevE.60.R6320.

41. Jorgensen, W.L.; Chandresekhar, J.; Madura, J.D.; Impey, R.W.; Klein, M.L. Comparison of simple potential functions for simulating liquid water. *J. Chem. Phys.* **1983**, *79*, 926.

42. Hartke, B. Size-dependent transition from all-surface to interior-molecule structures for pure neutral water clusters. *Phys. Chem. Chem. Phys.* **2003**, *5*, 275–284, doi:10.1039/b209966g.

43. Wales, D.J.; Hodges, M.P. Global minima of water clusters $(H_2O)_n$, $n \leq 21$, described by an empirical potential. *Chem. Phys. Lett.* **1998**, *286*, 65, doi:10.1016/S0009-2614(98)00065-7.

44. Hartke, B. Global geometry optimization of molecular clusters: TIP4P water. *Z. Phys. Chem.* **2000**, *214*, 9, doi:10.1524/zpch.2000.214.9.1251.

45. Dittner, M.; Hartke, B. Conquering the hard cases of Lennard-Jones clusters with simple recipes. *Comput. Theor. Chem.* **2017**, *1107*, 7–13, doi:10.1016/j.comptc.2016.09.032.

46. Ho, Y.C.; Pepyne, D.L. Simple explanation of the no-free-lunch theorem and its implications. *J. Optim. Theory Appl.* **2002**, *115*, 549, doi:10.1023/A:1021251113462.

47. Gómez, D.; Rojas, A. An empirical overview of the no free lunch theorem and its effect on real-world machine learning classification. *Neural Comput.* **2016**, *28*, 216, doi:10.1162/NECO_a_00793.

10

Optimization of Electrochemical Performance of LiFePO$_4$/C by Indium Doping and High Temperature Annealing

Ajay Kumar [1], Parisa Bashiri [1], Balaji P. Mandal [1], Kulwinder S. Dhindsa [1], Khadije Bazzi [1], Ambesh Dixit [2] ⓘ, Maryam Nazri [1], Zhixian Zhou [1], Vijayendra K. Garg [3], Aderbal C. Oliveira [3], Prem P. Vaishnava [4], Vaman M. Naik [5], Gholam-Abbas Nazri [1,6] and Ratna Naik [1,*]

[1] Department of Physics and Astronomy, Wayne State University, Detroit, MI 48202, USA; ajay.kumar3@wayne.edu (A.K.); parisa.bashiri@wayne.edu (P.B.); bpmandal80@gmail.com (B.P.M.); ee3087@wayne.edu (K.S.D.); eb2920@wayne.edu (K.B.); maryam.nazri@wayne.edu (M.N.); dw0795@wayne.edu (Z.Z.); nazri@wayne.edu (G.-A.Z.)

[2] Indian Institute of Technology, Jodhpur 342011, India; adixit2@gmail.com

[3] Universidade de Brasilia, Instituto de Fisica, Brasilia, DF 70919-970, Brazil; vijgarg@gmail.com (V.K.G.); aderbal47@gmail.com (A.C.O.)

[4] Department of Physics, Kettering University, Flint, MI 48504, USA; pvaishna@kettering.edu

[5] Department of Natural Sciences, University of Michigan-Dearborn, Dearborn, MI 48128, USA; vmnaik@umich.edu

[6] Electrical and Computer Engineering, Wayne State University, Detroit, MI 48202, USA

* Correspondence: rnaik@wayne.edu

Abstract: We have prepared nano-structured In-doped (1 mol %) LiFePO$_4$/C samples by sol–gel method followed by a selective high temperature (600 and 700 °C) annealing in a reducing environment of flowing Ar/H$_2$ atmosphere. The crystal structure, particle size, morphology, and magnetic properties of nano-composites were characterized by X-ray diffraction (XRD), scanning electron microsopy (SEM), transmission electron microscopy (TEM), and ^{57}Fe Mössbauer spectroscopy. The Rietveld refinement of XRD patterns of the nano-composites were indexed to the olivine crystal structure of LiFePO$_4$ with space group *Pnma*, showing minor impurities of Fe$_2$P and Li$_3$PO$_4$ due to decomposition of LiFePO$_4$. We found that the doping of In in LiFePO$_4$/C nanocomposites affects the amount of decomposed products, when compared to the un-doped ones treated under similar conditions. An optimum amount of Fe$_2$P present in the In-doped samples enhances the electronic conductivity to achieve a much improved electrochemical performance. The galvanostatic charge/discharge curves show a significant improvement in the electrochemical performance of 700 °C annealed In-doped-LiFePO$_4$/C sample with a discharge capacity of 142 mAh·g^{-1} at 1 C rate, better rate capability (~128 mAh·g^{-1} at 10 C rate, ~75% of the theoretical capacity) and excellent cyclic stability (96% retention after 250 cycles) compared to other samples. This enhancement in electrochemical performance is consistent with the results of our electrochemical impedance spectroscopy measurements showing decreased charge-transfer resistance and high exchange current density.

Keywords: Lithium iron phosphate; conductive Fe$_2$P; indium doping

1. Introduction

LiFePO$_4$ has become one of the most viable commercial cathode materials after the ground breaking work of Padhi et al. [1]. This material has received an extensive attention due to its high thermal and electrochemical safety, lower cost compared to mixed oxide cathode materials, low toxicity, stable voltage range even at overcharge condition, and long cycle life. However, the poor electronic conductivity and slow diffusion of lithium ion in bulk $LiFePO_4$ have been major challenges requiring new electrode material engineering. To improve electronic conductivity and reduce lithium ion diffusion length, many approaches, such as reducing the particle size to nanoscale [2–5], coating the particles with conductive carbon [6–12], and doping $LiFePO_4$ with various cations [13–20] have been proposed. In addition, $LiFePO_4$ decomposes above 700 °C leading to in-situ formation of conductive iron phosphides (Fe_2P, FeP, Fe_3P), and compounds with superior lithium-ion diffusion coefficients, such as, Li_3PO_4 and $Li_2FeP_2O_7$ [21–26]. Although initial formation of conductive iron phosphides at the grain boundaries improves electrochemical performance, these phases are not electrochemically active and excessive decomposition of $LiFePO_4$ reduces the active material leading to reduced specific capacity of the sample. Therefore, careful annealing temperature and addition of proper amount of dopants that reduce the decomposition of active material is crucial for preparation of high performance $LiFePO_4$ cathode materials. In our previous work [27], we studied the formation Fe_2P and Li_3PO_4 by the decomposition of $LiFePO_4/C$ as a function of annealing temperature between 600–900 °C in a reducing environment and found that the amount of Fe_2P increases very steeply from 5 to 38 wt % with the annealing temperature. Li_3PO_4 may contribute to the high ionic conduction at the electrode/electrolyte interface when preferentially deposited on the surface at the grain boundaries [28]. The presence of conductive Fe_2P significantly improved the electronic conductivity of the samples which varied from 2×10^{-3} S·cm^{-1} (600 °C) to 2×10^{-1} S·cm^{-1} (900 °C). Of all the samples studied, the $LiFePO_4/C$ sample calcined at 700 °C which consists of 14 wt % of Fe_2P exhibited a better electrochemical performance with a discharge capacity of ~136 mAh·g^{-1} at 1 C, ~121 mAh·g^{-1} at 10 C (70% of the theoretical capacity of $LiFePO_4$), and excellent cycleability. The observed steep decrease in the discharge capacity of samples annealed at higher temperatures was attributed to the increased amount of inactive decomposed products in the electrode. Hence, our previous work suggests that the synthesis environment can be controlled to optimize the amount of Fe_2P to obtaining the best discharge capacity of $LiFePO_4/C$ nanocomposites.

In addition, cation doping at Li and Fe sites in $LiFePO_4$ have been investigated by several researchers [13–20] to improve the electrochemical properties of $LiFePO_4$. Substitution of Mg, Al, Na at Li sites [14,15,19] have been shown to improve the overall electrochemical properties of $LiFePO_4$. Theoretical calculations by Islam et al. [20] have suggested, on energetic grounds, that $LiFePO_4$, is favorable for divalent dopants (e.g., Mg, Mn, Co), but not tolerant to aliovalent doping (e.g., Nd, La, In) on either Li (M1) or Fe (M2) sites. Nevertheless, a few experimental studies have investigated the effects of substituting aliovalent ions, such as Gd, Nd, La, at Fe sites in $LiFePO_4$ [16–18]. For example, 1% La-doped $LiFePO_4/C$ sample showed the best electrochemical behavior with a discharge capacity of 156 mAh·g^{-1} at a rate of 0.2 C [18]. However, there have been no experimental studies available in the literature to see the effect of In-doping in $LiFePO_4$. There are multiple beneficial effects expected with In-doping. The redox potential of indium in nonaqueous electrolyte has shown that In remains in 3-oxidation state (In^{III}) at voltages above 1.5 V vs. lithium. In addition, indium oxide has superior electronic conductivity compared to the $LiFePO_4$ that may lead to an improved electronic conductivity, particularly when it resides on the surface of the sample. On the other hand, if some of the iron sites are occupied by indium ions, it may increase the concentration of charge carrier in the sample as indium has a high thermodynamic tendency to remain as In^{III} cation, while Fe in the original $LiFePO_4$ material is at Fe^{II} state. Furthermore, the indium ion is a more polarizable, softer and diffuse ion than the hard sphere Fe^{III}. Therefore, Indium doping may reduce ionic lattice energy and energy barrier for Li-ion hopping between available sites.

In this work, we have studied the effect of In (1 mol %)-doping on the formation of Fe_2P due to decomposition of $LiFePO_4/C$ nanocomposites when annealed at two different temperatures of 600 and 700 °C in a reducing environment. We find that the In-doped-$LiFePO_4/C$ sample annealed at 700 °C which consists of 11 wt % of Fe_2P showed an improved discharge capacity (142 mAh·g^{-1} at 1 C rate), better rate capability at higher rates (~128 mAh·g^{-1} at 10 C rate, ~75% of the theoretical capacity) and excellent cyclic stability compared to that of un-doped sample annealed under similar conditions. By combining In-doping with high temperature (700 $\pm$ 50 °C) annealing, the electrochemical performance of $LiFePO_4/C$ can be further improved by optimizing the amount of Fe_2P in the nanocomposites for good electronic conductivity without sacrificing the active material.

2. Results and Discussion

2.1. X-ray Diffraction

The In-doped-$LiFePO_4/C$ samples were analyzed by XRD to verify both the crystallinity and phase purity. Their XRD patterns (Figure 1) were indexed to an orthorhombic $LiFePO_4$ phase with space group *Pnma*, according to the standard pattern of JCPDF 83-2092, indicating that an olivine-type structure is well maintained upon doping with 1 mol % of In. We do notice the presence of minor impurity phases that are indexed to iron phosphide (Fe_2P) and lithium phosphate (Li_3PO_4), which are formed, particularly, in In-LFP-700 sample. We have performed Rietveld analysis of XRD patterns using GSAS (General Structure Analysis System, Los Alamos National Laboratory Report LAUR 86-748 (2000)), software implemented with EXPGUI interface, to estimate the amount of Fe_2P and Li_3PO_4 in In-LFP-600 and In-LFP-700 samples (Figure 1), and the estimated amounts are listed in Table 1. For comparison, we have also including the data for LFP-600 and LFP-700 samples from our previous study [27]. The threshold temperature for forming Fe_2P from the decomposition of $LiFePO_4/C$ appears to be around 700 °C for both un-doped and doped samples. It is interesting to note the amount of Fe_2P formed is less in In-LFP-700 compared to LFP-700 sample. In-doping seems to reduce the formation of Fe_2P or the rate of decomposition of $LiFePO_4/C$ at 700 °C. As Fe_2P is conducting, it affects the conductivity of the In-doped-$LiFePO_4/C$ and hence its electrochemical properties. The reduction of Fe_2P with in In-doping in In-LFP-700, as indicated by XRD Rietveld analysis, is also confirmed by Mössbauer spectroscopy measurement, as discussed in a later section. The effect of In-doping on the crystallite size was also investigated using Rietveld fitting (GSAS software package) of the XRD patterns. The In-doped samples seem to have slightly smaller crystallite size compared to un-doped samples.

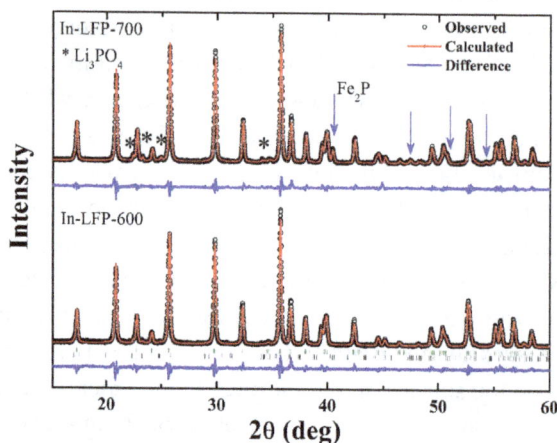

Figure 1. XRD patterns and Rietveld refinement of In-LFP-600 and In-LFP-700 samples.

Table 1. Values of $LiFePO_4$, Fe_2P and Li_3PO_4 estimated from Rietveld analysis of XRD patterns.

Sample	$LiFePO_4$ (wt %)	* Fe_2P (wt %)	Li_3PO_4 (wt %)	Crystallite Size (nm)
LFP-600 *	97.1	0	2.9	99
In-LFP-600	98.6	0	1.4	97
LFP-700 *	93.2	3.6	3.2	102
In-LFP-700	94.6	2.2	3.2	94

* see reference [27].

2.2. Electrical Conductivity

The room temperature electrical conductivity was measured for the samples using Van der Pauw method. The electronic conductivity for the LFP-600, In-LFP-600, LFP-700 and In-LFP-700 are 2×10^{-3}, 8×10^{-3}, 8×10^{-2} and 1×10^{-2} S·cm^{-1}, respectively. These results indicate that electronic conductivity of un-doped samples increases with the annealing temperature which is attributed to the formation of conductive Fe_2P phase at higher temperatures. We will show from our Mössbauer data analysis that the electrical conductivity correlates with the amount of Fe_2P (crystalline or sub-nanocrystalline) present in these samples [27].

2.3. Morphology and Microstructure

The morphology of the In-LFP-600 and In-LFP-700 samples was analyzed by SEM and they are shown in Figure 2a,b. The samples show a uniform distribution of nearly spherical particles with some agglomerated particles very similar to the un-doped samples [27], and 1 mol % In-doping does not affect the morphology significantly. This could be due to the fact that once the particles are carbon coated, the particle growth and the formation of aggregates are suppressed. In addition, the presence of carbon prevents the oxidation of Fe^{2+} to Fe^{3+}. Thus, the addition of the surfactant, lauric acid, is believed to play a crucial role in controlling the particle size and morphology of samples. We also investigated the particle size distribution using TEM as shown in Figure 2c,d for In-LFP-600 and In-LFP-700 samples. Again, the size distribution is very similar to the corresponding un-doped samples (~80–100 nm), with a rough morphology due to decomposition of LiFePO$_4$ into Fe_2P and Li_3PO_4. Our previous work [27] also showed that the particle surface of un-doped samples (for example, LFP-700) reveals sub-nano (2–4 nm) regions of the decomposed products, which may not be detected by XRD. The TEM the results are consistent with the average particle size calculated using XRD patterns.

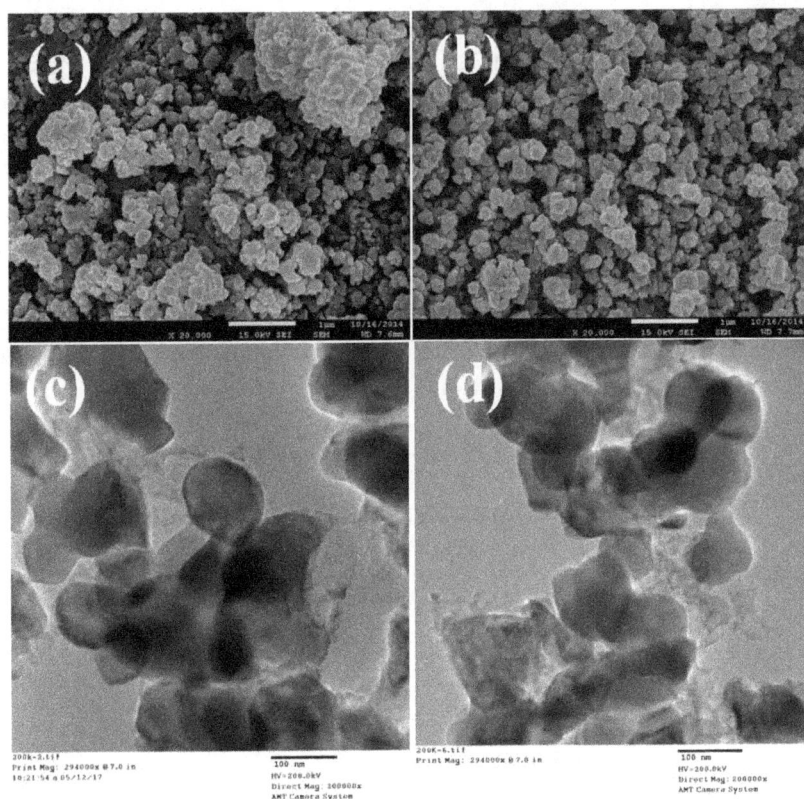

Figure 2. SEM images of (**a**) In-LFP-600 and (**b**) In-LFP-700 samples and their corresponding TEM images (**c,d**). The scale bars in (**a,b**) represent 1 μm and 100 nm in (**c,d**).

2.4. X-ray Photoelectron Spectroscopy

X-ray photoelectron spectroscopy allows us to access the local environment of atoms and their oxidation states. The technique, therefore, is used to differentiate between Fe^{3+} and Fe^{2+}. The Fe elemental XPS spectra of In-LFP-600 and In-LFP-700 samples are shown in Figure 3. All the spectra were fitted with three peaks, two at 710 and 714.5 eV are due to Fe^{2+} ions in $LiFePO_4$ and the third one at 712 eV due to Fe^{3+} originate mainly from $FePO_4$ and/or Fe_2P in agreement with the literature values [29,30]. The amount of ferric iron in In-LFP-600 was determined to be lower (~10%) compared to In-LFP-700 (~13%) sample, which is consistent with the Mössbauer spectroscopy measurements as described in Section 2.5.

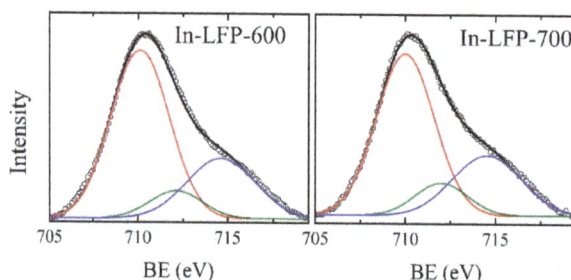

Figure 3. XPS spectra of Fe of In-LFP-600 and In-LFP-700 samples.

2.5. ^{57}Fe Mossbauer Spectroscopy

The room temperature ^{57}Fe Mössbauer spectra for the In-LFP-600 and In-LFP-700 samples are shown in Figure 4 to confirm the presence of Fe_2P. A summary of the Mössbauer parameters are given in Table 2. The ^{57}Fe Mössbauer spectrum of $LiFePO_4$ with Fe_2P consists of three quadrupole doublets. The dominant symmetric doublet with an isomer shift (IS) of 1.22 mm/s and quadrupole splitting (QS) of 2.97 mm/s arises from the high spin Fe^{2+} configuration of the $3d$ electrons and the distorted environment at the Fe atom in $LiFePO_4$ [31,32]. The other two doublets in the Mössbauer spectrum arise from two different favorable sites for Fe^{3+}, namely, tetrahedral with four nearest neighbor P atoms ($3f$ site) and pyramidal with five nearest neighbor P atoms ($3g$ site) in the structure of Fe_2P [31]. The second doublet with an IS of 0.61 mm/s and a QS of 0.43 mm/s is assigned to Fe^{3+} occupying $3f$ site and the third doublet with an IS of 0.19 mm/s and a QS of 0.1 mm/s is assigned to Fe^{3+} occupying $3g$ site in Fe_2P in the samples annealed at 700 °C [32]. The amount of Fe^{2+} and Fe^{3+} have been estimated using relative area under the corresponding peaks in the Mössbauer spectra. Table 2 lists the percentage of Fe^{2+} and Fe^{3+} phases and Table 3 lists the corresponding mol % and wt % of $LiFePO_4$, Fe_2P and Li_3PO_4 calculated using Equation (1). We have also listed the data for un-doped samples, LFP-600 and LFP-700, for a comparison.

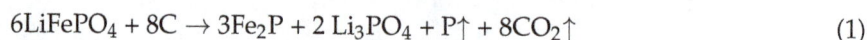

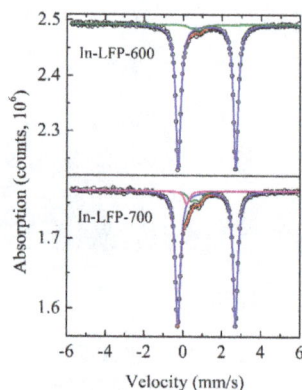

$$6LiFePO_4 + 8C \rightarrow 3Fe_2P + 2\,Li_3PO_4 + P\uparrow + 8CO_2\uparrow \qquad (1)$$

Figure 4. Mössbauer spectra of In-LFP-600 and In-LFP-700 samples measured at room temperature.

Table 2. Room temperature Mossbauer parameters of In-doped $LiFePO_4$/C samples annealed at 600 °C and 700 °C.

Sample	Doublet 1			Doublet 2			Doublet 3			Total
	IS	QS	%	IS	QS	%	IS	QS	%	Fe_2P (%)
In-LFP-600	1.22	2.97	92.2	0.61	0.43	7.8	-	-	-	7.8
In-LFP-700	1.22	2.97	86.7	0.61	0.43	8.7	0.19	0.10	4.6	13.3
	Fe^{2+}			Fe(I) site of Fe_2P			Fe(II) site of Fe_2P			

We note that the total amount of Fe_2P determined by Mössbauer spectroscopy do not agree with the estimated values by Rietveld refinement of the XRD data (Table 1). As discussed in our earlier work [27], this is due to presence of amorphous or sub-nanoregions of Fe_2P and Li_3PO_4.

Table 3. Percentage of $LiFePO_4$, Fe_2P and Li_3PO_4 in LFP and In-LFP samples annealed at 600 °C and 700 °C deduced from Mössbauer measurements.

Sample	$LiFePO_4$		Fe_2P		Li_3PO_4		Capacity (mAh·g^{-1})	
	mol %	wt %	mol %	wt %	mol %	wt %	Expected [a]	Measured at 1 C [b]
LFP-600 *	91.5	92.8	5.1	4.7	3.4	2.5	158	120
In-LFP-600	87.7	89.5	7.4	6.8	4.9	3.7	152	134
LFP-700 *	75.8	78.9	14.5	13.7	9.7	7.4	134	136
In-LFP-700	79.6	82.4	12.2	11.4	8.2	6.2	140	142

[a] 170 mAh·g^{-1} × wt % of $LiFePO_4$; [b] ± % due to uncertainty in mass determination; * see reference [27].

2.6. Electrochemical Measurements

Galvanostatic charge/discharge curves of the coin cells prepared with un-doped and indium doped $LiFePO_4$/C cathodes were measured between 2.2–4.2 V versus lithium at different rates. Charge/discharge curves for In-LFP-600 and In-LFP-700 samples at 1 C are depicted in Figure 5. Typical two-phase nature of the lithium extraction and insertion reactions between $LiFePO_4$ and $FePO_4$ is implied by the flat nature of the charge-discharge potential curves around ~3.4 V [33]. The steep rise and fall in the profiles at the large specific capacity values refer to the charge transfer activation and concentration polarizations with contribution from limited miscibility between the $LiFePO_4$ and $FePO_4$. The expected capacities calculated by taking into account the amount of observed Fe_2P and Li_3PO_4 masses from the Mössbauer measurements, and the measured capacities for the samples are listed in Table 3. The data clearly shows that the expected (~158 mAh·g^{-1} and 152 mAh·g^{-1}) and measured capacities (~120 mAh·g^{-1} and ~136 mAh·g^{-1}) for LFP-600 and In-LFP-600 differ significantly, although the latter sample shows significant improvement due to improved electronic conductivity. At higher annealing temperature of 700 °C, the expected and measured capacities are very close to each other, and the capacity of In-LFP-700 (142 mAh·g^{-1}) is larger than the corresponding un-doped sample. We observe that at both the annealing temperatures, the measured capacity increases upon adding the indium dopant.

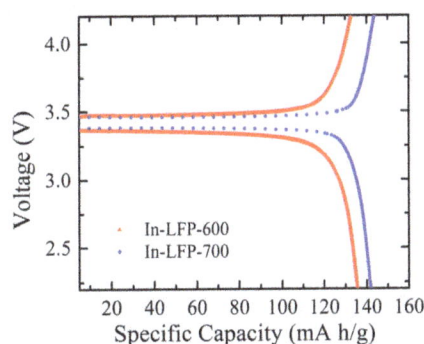

Figure 5. Charge/discharge profiles of In-LFP-600 and In-LFP-700 samples measured at 1 C rate.

The capacity of the samples at various charge/discharge rates are shown in Figure 6, including the data for un-doped LFP-600 and LFP-700 samples for a comparison. At higher rate, for example at 10 C, the supply of electrons from the interface electrochemical reaction becomes a problem leading to a lower specific capacity for un-doped sample annealed at 600 °C. However, addition of In increases its capacity because of its enhanced electronic conductivity. As seen in Figure 6, even at a high charge/discharge of 10 C the capacity of un-doped $LiFePO_4$/C annealed at 600 °C increases from 84 mAh·g^{-1} to 114 mAh·g^{-1} upon doping with In. When the doped sample is annealed at 700 °C, the performance of the In-LFP-700 improves slightly at all rates, and at 10C rate it approached 128 mAh·g^{-1} (75% of the theoretical capacity). The inset in Figure 6 shows the cycling performance of the doped and un-doped samples annealed at 700 °C for 250 cycles. Clearly, In-LFP-700 shows better charge/discharge stability compared to the corresponding un-doped sample, and even after 250 cycles at 10 C rate the sample retains 96% of its initial capacity. Our results suggest that a combination of In-doping and annealing at high temperatures (700 ± 50 °C), the electrochemical performance of $LiFePO_4$/C can be further improved by optimizing the amount of Fe_2P for good electronic conductivity without sacrificing the active material.

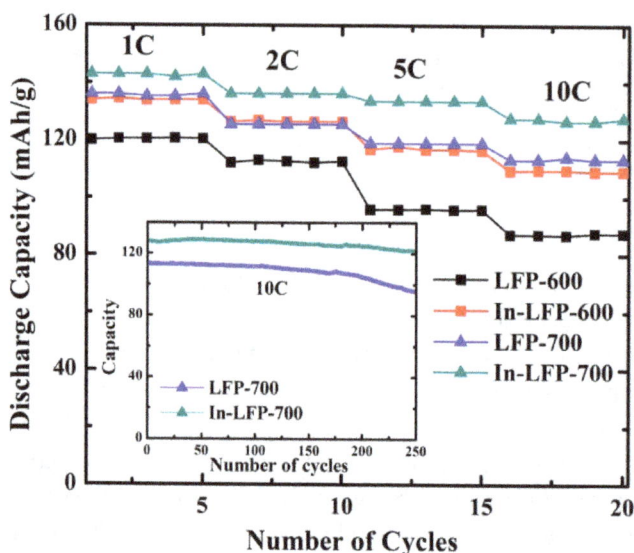

Figure 6. Rate capability curves of LFP-600, In-LFP-600, LFP-700, and In-LFP-700 samples during continuous cycling at different charging rates. The inset shows the capacity retention for LFP-700 and In-LFP-700 sample at 10 C rate.

Electrical impedance spectroscopy (EIS) measurements were also performed to understand the effects of In-doping on electrode impedance. The impedance spectra (Nyquist plots) of un-doped and In-doped $LiFePO_4$ samples are shown in Figure 7a are characteristic of electrochemical cells. The initial intercept of the semi-circle at highest frequency indicates resistance (R_s) associated to the electrolyte. The intercept of the semicircle in the intermediate frequency region corresponds to the charge transfer resistance (R_{ct}) in the bulk of electrode material, and the inclined line in the low frequency range represents the Warburg resistance (R_w), which is associated with lithium-ion diffusion. The data can be fitted to a Randles circuit (see insert in Figure 7a) in consisting of a constant phase element (CPE) representing the double layer capacitance and passivation film capacitance [34]. It has been observed that R_s values for the cells are very close to each other because the same electrolyte (1M $LiPF_6$ in EC/DMC 50:50 solvent) is used in all the cells. On the other hand, R_{ct} is lower in case of In-doped samples compared to the pure samples. This lower impedance of the In-doped sample may help to overcome the kinetic activation over potential for the Fe^{2+}/Fe^{3+} redox reaction during the charge–discharge process, and improve the capacity and cycling performance of the material.

We have determined the diffusion coefficient of lithium ion (D_{Li}) by using Z' dependence on ω in the low frequency region, which is described by [35],

$$Z' = R_s + R_{ct} + \sigma\omega^{-1/2} \tag{2}$$

where, σ is the Warburg coefficient, R_s and R_{ct} are the solution and the charge transfer resistances. σ is related to D_{Li} by

$$D_{Li} = R^2 T^2 / 2A^2 n^4 F^4 C_{Li}^2 \sigma^2 \tag{3}$$

where, R is the gas constant, T is the temperature in Kelvins, n is the number of electrons per molecule during oxidation, A is the surface area of the cathode (0.28 cm^2 in our case), F is the Faraday constant, and C_{Li} is the concentration of lithium ion (0.0228 mol/cm^3 in this case). As expected, a plot of Z' vesus $\omega^{-1/2}$ shows (Figure 7b) a linear relationship which yields σ. An apparent exchange current density (I_o) [35], has been calculated to measure the enhanced reaction rate of electrodes, which is a measure of kinetics of an electrochemical reaction.

$$I_o = RT / nR_{ct}F \tag{4}$$

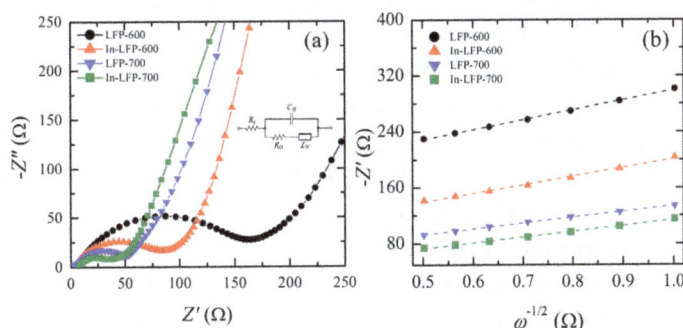

Figure 7. (a) Nyquist plots of LFP-600, In-LFP-600, LFP-700, and In-LFP-700 samples and (b) plot of the Z' vs. $\omega^{-1/2}$ in the low frequency region.

The calculated charge transfer resistance, lithium diffusion cofficent and apparent exchange current density along with other relevant parameters for the samples are given in Table 4. Indium doped LiFePO4/C samples have a lower charge transfer resistance of 77 Ω and 32 Ω for the In-LFP-600 and In-LFP-700 samples compared to 158 Ω and 72 Ω for the un-doped LFP-600 and LFP-700 samples. There is no drastic improvement in the lithium diffusion coefficient with In doping. However, lithium diffusion coefficient for In-LFP-700 is about a factor of two higher than the In-LFP-600 sample thus implying that annealing at 700 °C is desirable to improve the electrochemical properties of this material. Overall, the electrochemical measurements show that In-LFP-700 has the least charge transfer resistance, relatively higher Li-ion diffusion coefficient, and large exchange current density, which are consistent with its superior electrochemical performance in terms capacity and cycleability.

Table 4. Charge transfer resistance, Lithium diffusion coefficient, and exchange current density LFP-600 and LFP-700 compared with In-LFP-600 and In-LFP-700 samples.

Sample	R_{ct} (Ω)	σ (Ω s$^{1/2}$)	D_{Li} (cm$^2 \cdot$s^{-1})	I_o (mA$\cdot$g^{-1})
LFP-600 *	158	142	4.5×10^{-14}	163
In-LFP-600	77	124	6.0×10^{-14}	334
LFP-700 *	52	83	1.3×10^{-13}	494
In-LFP-700	32	82	1.4×10^{-13}	802

* see reference [27].

3. Materials and Methods

3.1. Synthesis Procedure

In-doped LiFePO$_4$/C samples were prepared by sol–gel technique, using CH$_3$CO$_2$Li·2H$_2$O, FeCl$_2$·4H$_2$O, P$_2$O$_5$ and InCl$_3$ as starting raw materials. These chemicals were mixed in stoichiometric ratio in dry ethanol and stirred for three hours, followed by the addition of 0.75 M lauric acid as carbon source to the mixture. After three hours of mixing the sol was dried under atmospheric conditions. The In-doped-LiFePO$_4$/C sample was prepared by adding 1 mol % of InCl$_3$ during the first step so that Fe:In ratio remains 99:1. The dried powders were ground and annealed under reduced environment of H$_2$ (10%) and Ar (90%) for 10 h. Two temperatures, 600 °C and 700 °C, were used to investigate the effects of annealing temperature with heating and cooling rate of 1 °C/min. In what follows, In-doped-LiFePO$_4$/C samples annealed at 600 °C and 700 °C will be referred as In-LFP-600, and In-LFP-700. The un-doped LiFePO$_4$ samples annealed at 600 °C and 700 °C, investigated in our previous study [27] will be referred as LFP-600 and LFP-700.

3.2. Characterization

Powder X-ray diffraction (XRD) patterns were obtained with a Rigaku Minflex-600 diffractometer (Osaka, Japan) using Cu Kα (λ = 1.54 Å) X-rays. Carbon content of the samples was measured by CHN analyzer, and found to be ~4.5%. The room temperature conductivity of the sample pellets were measured using 4-probe Van der Pauw method. The morphology of the samples was investigated using a JSM-6510-LV-LGS scanning electron microscope (SEM) (Tokyo, Japan) and a JEOL 2010 transmission microscope (TEM) (Tokyo, Japan). ^{57}Fe Mössbauer spectra were measured and fitted to obtain Mössbauer parameters using equipment and procedure described in our earlier work [36]. XPS measurements were performed using a Perkin-Elmer XPS systems (Waltham, UK), equipped with a cylindrical mirror analyzer and a highly monochromatic Al Kα (1486.6 eV) X-ray source. The observed binding energies of each element were identified with Perkin-Elmer database and an internal carbon source.

3.3. Electrochemical Measurements

The electrochemical characterization of the samples was performed in a standard coin cell geometry, using a Gamry electrochemical measurement system, in the frequency range of 0.1 Hz–100 kHz with an ac amplitude of 10 mV, as described in Ref. [12].

4. Conclusions

In (1 mol %)-doped-C-LiFePO$_4$ samples were prepared successfully by sol–gel method using lauric acid as surfactant to coat the particles under high temperature (600–700 °C) annealing. The carbon supplied by the decomposition of fatty acid not only provides reducing environment for maintaining Fe^{2+} in the LiFePO$_4$, but also restricts the growth of particle size of LiFePO$_4$. The XRD patterns of the samples indicates that In-doping does not affect the olivine crystal structure of LiFePO$_4$/C nanocomposites, but affects the amount of minority impurity phases (Fe$_2$P and Li$_3$PO$_4$) formed due to decomposition LiFePO$_4$ at higher annealing temperatures under a reducing environment. The presence of Fe$_2$P in the samples significantly enhances the electronic conductivity and hence affects its electrochemical properties. Of all the samples studied (doped and un-doped), we found that the In-doped-LiFePO$_4$/C nanocomposite annealed at 700 °C, containing 11 wt % Fe$_2$P showed the highest specific discharge capacity of ~142 mAh·g^{-1} at 1 C, ~128 mAh·g^{-1} at 10 C rate with a retention of 96% after 250 cycles of charging/discharging. However, our results demonstrate that by combining In-doping with high temperature (700 $\pm$ 50 °C) annealing, the electrochemical performance of LiFePO$_4$/C can be further improved by optimizing the amount of Fe$_2$P in the nanocomposites for good electronic conductivity without sacrificing the active material. It is important to note that carbon coating alone would not enhance the performance of LiFePO$_4$, but simultaneous indium doping and carbon coating is a feasible way to improve electrochemical performance of $LiFePO_4$ for high power applications.

Acknowledgments: We thank the Richard Barber Foundation for financial support for this work.

Author Contributions: Gholam-Abbas Nazri, Ratna Naik, Vaman M. Naik, and Zhixian Zhou conceived and supervised the project and wrote the manuscript. Ajay Kumar, Parisa Bashiri, Balaji P. Mandal, Kulwinder S. Dhindsa, Khadije Bazzi synthesized the materials and did the XRD, SEM, TEM and electrochemical characterization and analysis. Maryam Nazri helped in the fabrication of coin cells, and Ambesh Dixit did the XPS measurements and analysis. Vijayendra K. Garg, Aderbal C. Oliveira, and Prem P. Vaishnava did the Mössbauer measurements and analyses.

References

1. Padhi, A.K.; Nanjundaswamy, K.S.; Goodenough, J.B. Phospho-olivines as Positive-Electrode Materials for Rechargeable Lithium Batteries. *J. Electrochem. Soc.* **1997**, *144*, 1188–1194. [CrossRef]

2. Lim, S.; Yoon, C.S.; Cho, J. Synthesis of nanowire and hollow $LiFePO_4$ cathodes for high-performance lithium batteries. *Chem. Mater.* **2008**, *20*, 4560–4564. [CrossRef]

3. Delmas, C.; Maccario, M.; Croguennec, L.; Cras, F.L.; Weill, F. Lithium deintercalation in $LiFePO_4$ nanoparticles via a domino-cascade model. *Nat. Mater.* **2008**, *7*, 665–671. [CrossRef] [PubMed]

4. Gibot, P.; Casas-Cabanas, M.; Laffont, L.; Levasseur, S.; Carlach, P.; Hamelet, S.; Tarascon, J.-M.; Masquelier, C. Room-temperature single-phase Li insertion/extraction in nanoscale Li_xFePO_4. *Nat. Mater.* **2008**, *7*, 741–747. [CrossRef] [PubMed]

5. Hsu, K.-F.; Tsay, S.-Y.; Hwang, B.-J. Synthesis and characterization of nano-sized $LiFePO_4$ cathode materials prepared by a citric acid-based sol–gel route. *J. Mater. Chem.* **2004**, *14*, 2690–2695. [CrossRef]

6. Huang, Y.-H.; Goodenough, J.B. High-Rate $LiFePO_4$ Lithium Rechargeable Battery Promoted by Electrochemically Active Polymers. *Chem. Mater.* **2008**, *20*, 7237–7241. [CrossRef]

7. Chen, Z.; Dahn, J.R. Reducing carbon in $LiFePO_4$/C composite electrodes to maximize specific energy, volumetric energy, and tap density. *J. Electrochem. Soc.* **2002**, *149*, A1184–A1189. [CrossRef]

8. Doeff, M.M.; Wilcox, J.D.; Kostecki, R.; Lau, G. Optimization of carbon coatings on $LiFePO_4$. *J. Power Sources* **2006**, *163*, 180–184. [CrossRef]

9. Dominko, R.; Bele, M.; Gaberscek, M.; Remskar, M.; Hanzel, D.; Pejovnik, S.; Jamnik, J. Impact of the Carbon coating thickness on the electrochemical performance of $LiFePO_4$/C composites. *J. Electrochem. Soc.* **2005**, *152*, A607–A610. [CrossRef]

10. Dominko, R.; Bele, M.; Goupil, J.-M.; Gaberscek, M.; Hanzel, D.; Arcon, I.; Jamnik, J. Wired porous cathode materials: A novel concept for synthesis of $LiFePO_4$. *Chem. Mater.* **2007**, *19*, 2960–2969. [CrossRef]

11. Huang, H.; Yin, S.-C.; Nazar, L.F. Approaching theoretical capacity of $LiFePO_4$ at room temperature at high rates. *Electrochem. Solid State Lett.* **2001**, *4*, A170–A172. [CrossRef]

12. Bazzi, K.; Mandal, B.P.; Nazri, M.; Naik, V.M.; Garg, V.K.; Oliveira, A.C.; Vaishnava, P.P.; Nazri, G.A.; Naik, R. Effect of Surfactants on the Electrochemical Behavior of $LiFePO_4$ Cathode Material for Lithium Ion Batteries. *J. Power Sources* **2014**, *265*, 67–74. [CrossRef]

13. Wagemaker, M.; Ellis, B.L.; Lützenkirchen-Hecht, D.; Mulder, F.M.; Nazar, L.F. Proof of Supervalent Doping in Olivine $LiFePO_4$. *Chem. Mater.* **2008**, *20*, 6313–6315. [CrossRef]

14. Ou, X.; Liang, G.; Wang, L.; Xu, S.; Zhao, X. Effect of Magnesium doping on electronic conductivity and electrochemical properties of $LiFePO_4$. *J. Power Sources* **2008**, *184*, 543–547. [CrossRef]

15. Yin, X.; Huang, K.; Liu, S.; Wang, H.; Wang, H. Preparation and characterization of Na-doped $LiFePO_4$/C composites as cathode materials for lithium-ion batteries. *J. Power Sources* **2010**, *195*, 4308–4312. [CrossRef]

16. Pang, L.; Zhao, M.; Zhao, X.; Chai, Y. Preparation and electrochemical performance of Gd-doped $LiFePO_4$/C composites. *J. Power Sources* **2012**, *201*, 253–258. [CrossRef]

17. Zhao, X.; Tang, X.; Zhang, L.; Zhao, M.; Zhai, J. Effects of neodymium aliovalent substitution on the structure and electrochemical performance of $LiFePO_4$. *Electrochim. Acta* **2010**, *55*, 5899–5904. [CrossRef]

18. Cho, Y.-D.; Fey, G.T.-K.; Kao, H.-M. Physical and electrochemical properties of La-doped $LiFePO_4$/C composites as cathode materials for lithium-ion batteries. *J. Solid State Electrochem.* **2008**, *12*, 815–823. [CrossRef]

19. Chung, S.-Y.; Bloking, J.T.; Chiang, Y.-M. Electronically conductive phospho-olivines as lithium storage electrodes. *Nat. Mater.* **2002**, *1*, 123–128. [CrossRef] [PubMed]

20. Islam, M.S.; Driscoll, D.J.; Fisher, C.A.; Slater, P.R. Atomic-Scale Investigation of Defects, Dopants, and Lithium Transport in the $LiFePO_4$ Olivine-Type Battery Material. *Chem. Mater.* **2005**, *17*, 5085–5092. [CrossRef]

21. Kim, C.W.; Park, J.S.; Lee, K.S. Effect of Fe_2P on the electron conductivity and electrochemical performance of $LiFePO_4$ synthesized by mechanical alloying using Fe^{3+} raw material. *J. Power Sources* **2006**, *163*, 144–150. [CrossRef]

22. Qiu, Y.; Geng, Y.; Yu, J.; Zuo, X. High-capacity cathode for lithium-ion battery from $LiFePO_4/(C + Fe_2P)$ composite nanofibers by electrospinning. *J. Mater. Sci.* **2014**, *49*, 504–509. [CrossRef]

23. Xu, Y.; Lu, Y.; Yan, L.; Yang, Z.; Yang, R. Synthesis and effect of forming Fe_2P phase on the physics and electrochemical properties of $LiFePO_4/C$ materials. *J. Power Sources* **2006**, *160*, 570–576. [CrossRef]

24. Herle, P.S.; Ellis, B.; Coombs, N.; Nazar, L.F. Nano-network electronic conduction in iron and nickel olivine phosphates. *Nat. Mater.* **2004**, *3*, 143–152. [CrossRef] [PubMed]

25. Rho, Y.-H.; Nazar, L.F.; Perry, L.; Ryan, D. Surface Chemistry of $LiFePO_4$ Studied by Mössbauer and X-ray Photoelectron Spectroscopy and Its Effect on Electrochemical Properties. *J. Electrochem. Soc.* **2007**, *154*, A283–A289. [CrossRef]

26. Lin, Y.; Gao, M.; Zhu, D.; Liu, Y.; Pan, H. Effects of carbon coating and iron phosphides on the electrochemical properties of $LiFePO_4/C$. *J. Power Sources* **2008**, *184*, 444–448. [CrossRef]

27. Dhindsa, K.; Kumar, A.; Nazri, G.; Naik, V.; Garg, V.; Oliveira, A.; Vaishnava, P.; Zhou, Z.; Naik, R. Enhanced electrochemical performance of $LiFePO_4/C$ nanocomposites due to in situ formation of Fe_2P impurities. *J. Solid State Electrochem.* **2016**, *20*, 2275–2282. [CrossRef]

28. Kang, B.; Ceder, G. Battery materials for ultrafast charging and discharging. *Nature* **2009**, *458*, 190–193. [CrossRef] [PubMed]

29. Castro, L.; Dedryvere, R.; El Khalifi, M.; Lippens, P.-E.; Bréger, J.; Tessier, C.; Gonbeau, D. The Spin-Polarized Electronic Structure of $LiFePO_4$ and $FePO_4$ Evidenced by in-Lab XPS. *J. Phys. Chem. C* **2010**, *114*, 17995–18000. [CrossRef]

30. Castro, L.; Dedryvère, R.; Ledeuil, J.-B.; Bréger, J.; Tessier, C.; Gonbeau, D. Aging Mechanisms of $LiFePO_4$//Graphite Cells Studied by XPS: Redox Reaction and Electrode/Electrolyte Interfaces. *J. Electrochem. Soc.* **2012**, *159*, A357–A363. [CrossRef]

31. Yamada, A.; Chung, S.-C.; Hinokuma, K. Optimized $LiFePO_4$ for Lithium Battery Cathodes. *J. Electrochem. Soc.* **2001**, *148*, A224–A229. [CrossRef]

32. Prince, A.; Mylswamy, S.; Chan, T.; Liu, R.; Hannoyer, B.; Jean, M.; Shen, C.; Huang, S.; Lee, J.; Wang, G. Investigation of Fe valence in $LiFePO_4$ by Mössbauer and XANES spectroscopic techniques. *Solid State Commun.* **2004**, *132*, 455–458. [CrossRef]

33. Liu, Y.; Cao, C.; Li, J. Enhanced electrochemical performance of carbon nanospheres–$LiFePO_4$ composite by PEG based sol–gel synthesis. *Electrochim. Acta* **2010**, *55*, 3921–3926. [CrossRef]

34. Wang, G.; Yang, L.; Chen, Y.; Wang, J.; Bewlay, S.; Liu, H. An investigation of polypyrrole-$LiFePO_4$ composite cathode materials for lithium-ion batteries. *Electrochim. Acta* **2005**, *50*, 4649–4654. [CrossRef]

35. Bard, A.J.; Faulkner, L.R.; Leddy, J.; Zoski, C.G. *Electrochemical Methods: Fundamentals and Applications*; Wiley: New York, NY, USA, 1980; Volume 2.

36. Dhindsa, K.; Mandal, B.P.; Bazzi, K.; Lin, M.; Nazri, M.; Nazri, G.; Naik, V.; Garg, V.; Oliveira, A.; Vaishnava, P.; et al. Enhanced electrochemical performance of graphene modified $LiFePO_4$ cathode material for lithium ion batteries. *Solid State Ion.* **2013**, *253*, 94–100. [CrossRef]

11

Investigation of the Thermodynamic Properties of Surface Ceria and Ceria–Zirconia Solid Solution Films Prepared by Atomic Layer Deposition on Al$_2$O$_3$

Tzia Ming Onn, Xinyu Mao, Chao Lin, Cong Wang and Raymond J. Gorte *

Department of Chemical and Biomolecular Engineering, University of Pennsylvania, 34th Street, Philadelphia, PA 19104, USA; tonn@seas.upenn.edu (T.M.O.); xinyumao@seas.upenn.edu (X.M.); linchao@seas.upenn.edu (C.L.); wangcong@seas.upenn.edu (C.W.)
* Correspondence: gorte@seas.upenn.edu

Abstract: The properties of 20 wt % CeO$_2$ and 21 wt % Ce$_{0.5}$Zr$_{0.5}$O$_2$ films, deposited onto a γ-Al$_2$O$_3$ by Atomic Layer Deposition (ALD), were compared to bulk Ce$_{0.5}$Zr$_{0.5}$O$_2$ and γ-Al$_2$O$_3$-supported samples on which 20 wt % CeO$_2$ or 21 wt % CeO$_2$–ZrO$_2$ were deposited by impregnation. Following calcination to 1073 K, the ALD-prepared catalysts showed much lower XRD peak intensities, implying that these samples existed as thin films, rather than larger crystallites. Following the addition of 1 wt % Pd to each of the supports, the ALD-prepared samples exhibited much higher rates for CO oxidation due to better interfacial contact between the Pd and ceria-containing phases. The redox properties of the ALD samples and bulk Ce$_{0.5}$Zr$_{0.5}$O$_2$ were measured by determining the oxidation state of the ceria as a function of the H$_2$:H$_2$O ratio using flow titration and coulometric titration. The 20 wt % CeO$_2$ ALD film exhibited similar thermodynamics to that measured previously for a sample prepared by impregnation. However, the sample with 21 wt % Ce$_{0.5}$Zr$_{0.5}$O$_2$ on γ-Al$_2$O$_3$ reduced at a much higher P_{O_2} and showed evidence for transition between the Ce$_{0.5}$Zr$_{0.5}$O$_2$ and Ce$_{0.5}$Zr$_{0.5}$O$_{1.75}$ phases.

Keywords: Atomic Layer Deposition; ceria; ceria–zirconia; thermodynamics; improved stability

1. Introduction

Ceria is a component in a number of commercial catalysts, as demonstrated in a recent, comprehensive review [1]. Two important examples include the use of ceria-based materials as oxidation catalysts in Diesel particulate filters [2] and the use of ceria–zirconia solid-solutions as Oxygen-Storage Capacitors (OSC) in automotive three-way catalysts [3–7]. The critical property of ceria that makes it useful in these and other catalytic applications is the relative ease with which it changes oxidation states between Ce^{3+} and Ce^{4+}. This, together with the fact that reduced ceria can be oxidized by steam or CO$_2$ [8] and then transfer oxygen to transition metals with which it is in contact make it a promoter of hydrocarbon oxidation [9,10], water-gas-shift (WGS) reaction [11–14], steam reforming [15–18], and more.

It is known that pure ceria becomes essentially unreducible after harsh redox cycling [19]. Although this loss in reducibility is accompanied by a loss in surface area, it is not simply a kinetic phenomenon, since the ceria–zirconia solid solutions that are used in three-way catalysts also have low surface areas but still maintain their OSC properties [20]. While the loss in surface area does not prevent the application of ceria–zirconia solid solutions for OSC, high surface areas are required in other cases. For example, for CO-oxidation and WGS reactions over Pd/ceria, the sites at the Pd–ceria boundary have been shown to exhibit much higher rates [21]. In these examples, the surface area of the ceria should be as high as possible.

An obvious approach for maintaining a high surface area for the ceria phase is to deposit it onto a high-surface-area support that has good thermal stability, such as γ-Al_2O_3. The ceria phase is usually added by infiltration with metal salts, followed by calcination; however, the use of a support for the ceria phase is often only partially effective because the ceria does not typically cover the entire surface. For example, rates on a catalyst in which Pd is deposited onto a ceria/γ-Al_2O_3 support prepared by infiltration of ceria are often much lower than rates on ceria-supported Pd because much of the Pd is not in contact with ceria in the former case [22]. Finally, it can be difficult to produce a supported, ceria–zirconia solid solution because this requires good mixing of the Ce and Zr cations during the precipitation stage in catalyst preparation.

Work from our laboratory has recently demonstrated that one can use Atomic Layer Deposition (ALD) to prepare thin ceria [22] and ceria–zirconia [23] films that uniformly cover a γ-Al_2O_3 support. The principle behind ALD is that an organometallic precursor is allowed to react with the substrate surface, after which the adsorbed precursor is oxidized in a separate step. Because reaction of the precursor with the substrate is limited to a monolayer at most, ALD allows the formation of uniform, atomic-scale, oxide films. There are comprehensive reviews of ALD topics that are available [24,25]. Mixed oxides are easily formed by alternating between different precursors. Because oxides have low surface energies compared to metals, the uniform films formed by ALD tend to be thermally stable. In the cases of ceria and ceria–zirconia films on γ-Al_2O_3, the γ-Al_2O_3 remains "covered", even after calcination to 1173 K.

As mentioned previously, the reducibility of ceria depends on the surface area and/or crystallite size [19,26] and ceria–zirconia solid solutions are more reducible than pure ceria [20,27]. Because the films prepared by ALD can be sub-nanometer in thickness, it is anticipated that the reducibility of these materials may be very different from even ordinary polycrystalline ceria. Quantification of reducibility can be difficult and kinetic measures, such as Temperature Programmed Reduction peak temperatures, can be dramatically affected by the presence of catalysts and other factors. Therefore, we chose to quantify reducibility by measuring the thermodynamic properties of ALD-prepared films using flow titration and coulometric titration [20,27]. In both of these techniques, the sample is equilibrated at high temperatures in a specified P_{O_2} and the extent of ceria reduction (e.g., x in $CeO_{(2-x)}$) is then determined as a function of that P_{O_2} and temperature. The P_{O_2} is typically established by equilibrium with H_2 oxidation (i.e., $P_{O_2} = K_{equi}^{-1} \frac{P_{H_2O}}{P_{H_2}}$) and is therefore fixed by passing a mixture of H_2 and H_2O over the sample at a specified H_2O:H_2 ratio. Because the activity of solids is one, the equilibrium constant for oxidation of $CeO_{(2-x)}$, and therefore the free energy of oxidation at that value of x, is determine directly from the measured P_{O_2}. Furthermore, the temperature dependence of the equilibrium constant can be used to determine the heat of oxidation through the Clausius-Clapeyron Equation.

What we will show is that 0.4-nm, ALD ceria films on γ-Al_2O_3 have thermodynamic redox properties that are different from that of bulk ceria but essentially identical to that reported previously for polycrystalline ceria formed by impregnation into γ-Al_2O_3. This may be due the film coalescing into particles upon redox cycling. By contrast, the 0.5-nm ceria–zirconia films were stable and very different from either bulk ceria or ceria–zirconia solid solution. The films exhibit evidence for an equilibrium transition between the $Ce_{0.5}Zr_{0.5}O_2$ and $Ce_{0.5}Zr_{0.5}O_{1.75}$ phases.

2. Results

2.1. Sample Characterization

The synthesis and characterization of the $CeO_2(ALD)/\gamma$-Al_2O_3 and the $CZ(ALD)/\gamma$-Al_2O_3 samples have been presented in detail elsewhere [22,23]. Results from Transmission Electron Microscopy (TEM) indicated that the γ-Al_2O_3 was uniformly covered by the ALD films, and there was good agreement between the thicknesses of the films determined from TEM and value of 0.4-nm estimated from the sample weight changes. By contrast, TEM of a $CeO_2(IMP)/\gamma$-Al_2O_3 sample with the same ceria loading as the ALD sample showed the presence of relatively large CeO_2 crystallites that covered only a fraction of alumina surface [22].

The differences between the ALD and impregnated samples are readily apparent from the XRD patterns, shown in Figure 1. All of the samples in this figure were calcined in air to 1073 K for 2 h prior to the measurements. To normalize the intensities, 12.5 wt % carbon was added to each sample and the height of the peak XRD peak at 26.7 degrees 2θ was fixed. The XRD pattern for the unmodified $\gamma\text{-}Al_2O_3$ is shown in Figure 1a for reference. In agreement with the previous report [22], the patterns for the $CeO_2(ALD)/\gamma\text{-}Al_2O_3$ (Figure 1b) and $CeO_2(IMP)/\gamma\text{-}Al_2O_3$ (Figure 1c) samples, both of which had 20 wt % CeO_2, are very different. The diffraction pattern of $CeO_2(ALD)/\gamma\text{-}Al_2O_3$ remained unchanged from that of $\gamma\text{-}Al_2O_3$, either because the CeO_2 film is amorphous or because the thickness of the CeO_2 film is less than the coherence length of the X-rays [22]. By contrast, the $CeO_2(IMP)/\gamma\text{-}Al_2O_3$ shows relatively intense peaks corresponding to the fluorite structure of CeO_2.

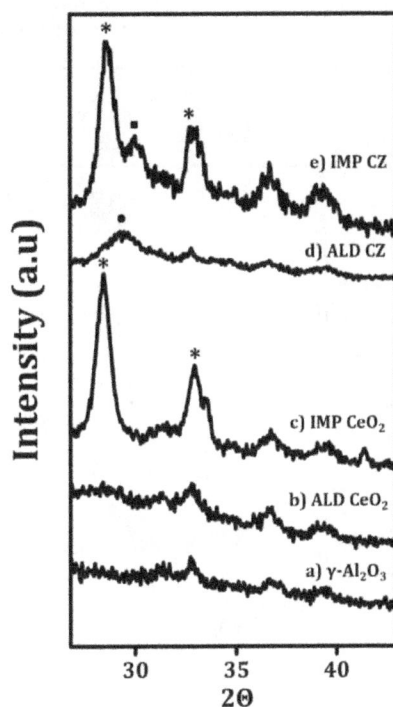

Figure 1. XRD patterns of (**a**) $\gamma\text{-}Al_2O_3$ support after calcination to 1173 K in air, and ALD-modified samples, (**b**) $20CeO_2\text{-}Al_2O_3$ and (**d**) $20Ce_{0.5}Zr_{0.5}O_2\text{-}Al_2O_3$ heated to 1073 K. The patterns in (**c**) and (**e**) were obtained from $CeO_2(IMP)/Al_2O_3$ and $CZ(IMP)/Al_2O_3$ respectively. The loadings of materials are the same for the ALD-modified supports and the infiltrated supports. Peaks were normalized to a distinct graphite peak around $2\theta = 26.7°$. Characteristic peaks for ceria–zirconia solid solution are marked by ●, while peaks for CeO_2 and ZrO_2 are marked by * and ■ respectively.

The XRD patterns for the $CZ(ALD)/\gamma\text{-}Al_2O_3$ (Figure 1d) and $CZ(IMP)/\gamma\text{-}Al_2O_3$ (Figure 1e) samples, both with 21 wt % $Ce_{0.5}Zr_{0.5}O_2$, are similarly different. There is evidence for a fluorite-related feature at 29.4 degrees 2θ on the ALD sample; but the feature is very broad and weak, again suggesting that the oxide remains in the form of amorphous film. The peak position, roughly half way between the expected angles for CeO_2 and ZrO_2, is consistent with formation of a solid, ceria–zirconia solution [28]. Because it is difficult to form a single-phase, ceria–zirconia solid solution by impregnation, the impregnated sample shows two features in this region, at approximately 28 and 30 degrees 2θ, which are due to ceria- and zirconia-rich phases, respectively [28]. Again, the intensities of the peaks on $CZ(IMP)/\gamma\text{-}Al_2O_3$ are much greater, implying the presence of three-dimensional crystallites.

Because redox cycling, alternating between reducing and oxidizing environments at high temperatures, has been shown to affect ceria crystallite size much more strongly than simple, high-temperature calcination [19], we also measured XRD patterns for the $CeO_2(ALD)/\gamma$-Al_2O_3, $CZ(ALD)/\gamma$-Al_2O_3, and $CeO_2(IMP)/\gamma$-Al_2O_3 samples after redox cycling, with results shown in Figure 2. The $CeO_2(IMP)/\gamma$-Al_2O_3 sample underwent three cycles of alternating reduction (90% H_2 and 10% H_2O for 30 min) at 1073 K and oxidation (dry air for 30 min) at 1073 K. The two ALD-prepared samples were used for the thermodynamic measurements that will be discussed later in this paper and were exposed to many oxidation and reduction cycles over a period of several weeks, at temperatures between 873 and 1073 K. Redox cycling significantly narrowed and increased the intensity of the peak at approximately 28 degrees 2θ on the $CeO_2(IMP)/\gamma$-Al_2O_3 sample. Using the width at half height of the (220) diffraction peak, the ceria crystallite size on the $CeO_2(IMP)/\gamma$-Al_2O_3 sample increased from 7 to 18 nm. Redox aging also caused changes in the XRD pattern for the $CeO_2(ALD)/\gamma$-Al_2O_3 in that the fluorite phase now clearly evident, implying that the film had formed at least some crystallites. However, even after this harsh treatment, the CeO_2 crystallite size on the $CeO_2(ALD)/\gamma$-Al_2O_3 sample was only 5 nm. Furthermore, the intensity of the CeO_2 diffraction peak remained much lower than that of even the fresh $CeO_2(IMP)/\gamma$-Al_2O_3 sample. The $CZ(ALD)/\gamma$-Al_2O_3 sample also exhibited a weak peak that can be associated with a $Ce_{0.5}Zr_{0.5}O_2$ phase; but, again, this feature is weak in intensity and very broad.

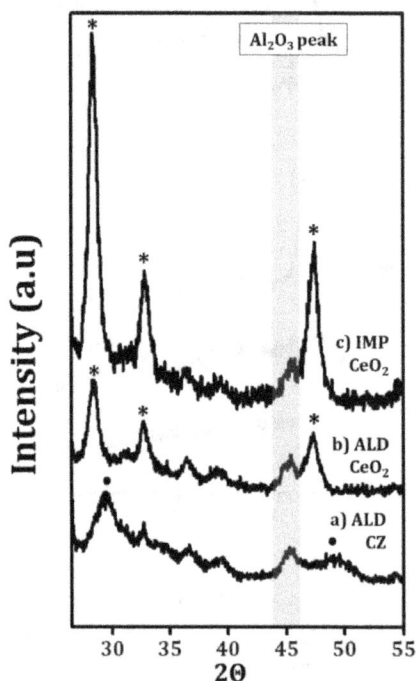

Figure 2. XRD patterns of (**a**) $20Ce_{0.5}Zr_{0.5}O_2$–Al_2O_3, (**b**) $20CeO_2$–Al_2O_3, and (**c**) $CeO_2(IMP)/Al_2O_3$ after redox cycling conditions. The $CeO_2(IMP)/\gamma$-Al_2O_3 sample underwent three cycles of alternating reduction (90% H_2 and 10% H_2O for 30 min) and oxidation (dry air for 30 min) at 1073 K. The ALD-prepared samples, (**a**) and (**b**), were exposed to similar oxidation and reduction cycles over a period of several weeks, at temperatures between 873 and 1073 K. Peaks were normalized to a distinct graphite peak around 2θ = 26.7°. Characteristic peaks for ceria–zirconia solid solution are marked by ●, while peaks for CeO_2 are marked by *.

The sites at the interface between Pd and ceria are known to be considerably more active for CO oxidation than either Pd or ceria sites individually [21,29]. Therefore, another indication of how well CeO_2 or $Ce_{0.5}Zr_{0.5}O_2$ cover the γ-Al_2O_3 in these composite oxides was gained by adding Pd to the oxide supports and then measuring CO oxidation rates. With the exception of the bulk CeO_2, 1 wt % Pd was added to the γ-Al_2O_3, $CeO_2(ALD)/\gamma$-Al_2O_3, $CeO_2(IMP)/\gamma$-Al_2O_3, $CZ(ALD)/\gamma$-Al_2O_3, and $CZ(IMP)/\gamma$-Al_2O_3 after each of the supports had been calcined at 1073 K for 2 h. The bulk CeO_2

was calcined to only 773 K as it may lose significant surface area after high temperature treatment. After adding the Pd, the catalysts were dried and then calcined to 773 K. BET surface areas and the Pd dispersions, measured after reduction at 673 K, are reported in Table 1; and differential reaction rates are shown in Figure 3. All of the catalysts had similar BET surface areas and Pd dispersions but the catalytic activities varied significantly. To obtain rates under differential conditions, it was therefore necessary to measure them over different temperature ranges. Since the activation energies were similar, it was possible to extrapolate the rates to a common temperature and qualitatively estimate the contact between Pd and the ceria phase in each sample.

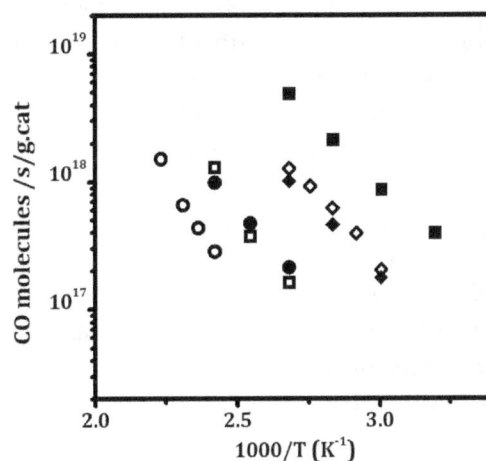

Figure 3. Steady-state, differential reaction rates for CO oxidation reaction with partial pressure of CO and O_2 at 25 Torr and 12.5 Torr, respectively. The supports were heated at 1073 K in air for 2 h before 1 wt % of Pd was introduced. After calcination at 773 K in air, the CO oxidation rates were compared for the following catalysts: ($\bigcirc$)—Pd/Al_2O_3, ($\square$)—Pd/CeO_2(IMP)/Al_2O_3, ($\bullet$)—Pd/CZ(IMP)/Al_2O_3, ($\blacklozenge$) θ—Pd/$20Ce_{0.5}Zr_{0.5}O_2$–Al_2O_3, and ($\blacksquare$) Pd/$20CeO_2$–Al_2O_3. The catalyst, ($\lozenge$)—Pd/CeO_2, heated to 773 K in air, was added for comparison. The CeO_2 support was heated to 773 K.

Table 1. Metal Dispersion and BET Surface Area of Catalysts.

	Pd Dispersion (%)	BET Surface Area (m^2/g)
Pd/Al_2O_3	28	96
Pd/CeO_2(IMP)/Al_2O_3	23	80
Pd/CZ(IMP)/Al_2O_3	26	82
Pd/$20CeO_2$–Al_2O_3	33	77
Pd/$20Ce_{0.5}Zr_{0.5}O_2$/Al_2O_3	34	71
Pd/CeO_2	40	45

First, at 373 K, the Pd/CeO_2(ALD)/γ-Al_2O_3 sample was nearly 100 times more active than Pd/γ-Al_2O_3, demonstrating that the Pd–CeO_2 interfacial sites are indeed more active for this reaction. Although the differential rates on Pd/CeO_2(ALD)/γ-Al_2O_3 were also a factor of three higher than those on Pd/CeO_2, this is likely due to the higher surface area of the ALD sample. In any case, the results on Pd/CeO_2(ALD)/γ-Al_2O_3 suggest that all of the Pd is in contact with CeO_2 ALD film and that the CeO_2 must cover the entire surface of the support. This is the same conclusion reached earlier from WGS measurements on similar samples [22]. The Pd/CZ(ALD)/γ-Al_2O_3 catalyst exhibited rates similar to that found on Pd/CeO_2 and the two impregnated samples showed activities closer to that of Pd/γ-Al_2O_3. These results imply that Pd is not in good contact with the ceria phases on the supports prepared by impregnation. The lower activity of Pd/CZ(ALD)/γ-Al_2O_3 compared to Pd/CeO_2(ALD)/γ-Al_2O_3 may be due to the fact that the presence of ZrO_2 decreases contact area between Pd and CeO_2 in the Pd/CZ(ALD)/γ-Al_2O_3 sample.

2.2. *Thermodynamic Measurements*

Thermodynamic measurements were performed on the bulk $Ce_{0.5}Zr_{0.5}O_2$, the $CeO_2(ALD)/\gamma$-Al_2O_3, and the $CZ(ALD)/\gamma$-Al_2O_3 samples after each sample has been calcined to 1073 K for 2 h. To determine the maximum extents of reduction in the $CeO_2(ALD)/\gamma$-Al_2O_3 and the $CZ(ALD)/\gamma$-Al_2O_3 samples, the samples were initially exposed to dry, flowing H_2 at 973 K in the flow-titration system. Assuming that all the oxygen which could be reversibly removed was due to reduction of Ce^{4+} to Ce^{3+}, the calculated O:Ce ratio for both samples was 1.55 after reduction. This value did not change with repeated oxidation-reduction cycles and differs from the value expected for complete reduction by 10%. The probable reason for the discrepancy is the uncertainty in the ceria loading (e.g., Sample weights can be difficult to measure due to water adsorption upon exposure to laboratory air). The result demonstrates that the ceria in the ALD films is fully reducible.

The thermodynamic, redox properties of the $CeO_2(ALD)/\gamma$-Al_2O_3 sample were measured using both flow-titration (the solid data points) and coulometric-titration (the unfilled points) measurements; and a plot of the cerium oxygen stoichiometry as a function of P_{O_2} is shown in Figure 4. For comparison purposes, the figure also includes data from earlier work on bulk ceria and from an impregnated, 30 wt % ceria on La-treated Al_2O_3 (shown in the inset) [19]. Flow-titration measurements were performed at 873, 973, and 1073 K but coulometric-titration was performed only at 1073 K on this sample. The data at 1073 K showed that there was good agreement between the two techniques in the P_{O_2} range where the measurements overlapped. The redox thermodynamics for the $CeO_2(ALD)/\gamma$-Al_2O_3 sample differ dramatically from that for bulk ceria. As noted in previous publications, bulk ceria is very difficult to reduce [30]. Of equal interest, the thermodynamic data for the $CeO_2(ALD)/\gamma$-Al_2O_3 sample are remarkably similar to the previously published data for impregnated ceria on La-doped alumina in regions where the P_{O_2} values overlap. For example, at 973 K and a P_{O_2} of 10^{-20} atm, the O:Ce ratio was close to 1.75 on both samples. Similarly, at 873 K and a P_{O_2} of 10^{-25} atm, the O:Ce ratio was close to 1.78 on both samples. As shown by the earlier XRD data, the $CeO_2(ALD)/\gamma$-Al_2O_3 showed some formation of CeO_2 crystallites upon redox cycling, but the crystallites on the ALD-prepared sample were clearly much smaller. The fact that the thermodynamic data did not change would suggest that equilibrium data is not a strong function of crystallite size, at least over the range of crystallite sizes investigated here.

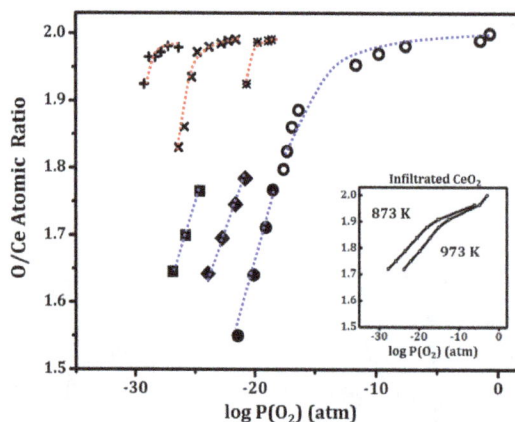

Figure 4. Oxidation isotherms for $20CeO_2$-Al_2O_3 at selected temperatures (873 K, 973 K, 1073 K). The results for pure ceria were obtained from a previous publication from our group [19], and were determined by flow titration at (+)—873 K, (×)—973 K, and (*)—1073 K. The (■), (▲), and (●) symbols show the isotherms for $20CeO_2$-Al_2O_3 determined by flow titration at 873 K, 973 K and 1073 K respectively. The (○) symbol shows the isotherm for $20CeO_2$-Al_2O_3 determined by coulometric titration at 1073 K. For comparison, previous results for infiltrated ceria on alumina-based support obtained from coulometric titration at 873 K and 973 K are shown [19].

The oxidation isotherms for bulk $Ce_{0.5}Zr_{0.5}O_2$ and for $CZ(ALD)/\gamma$-Al_2O_3 are shown in Figure 5. Results for the bulk $Ce_{0.5}Zr_{0.5}O_2$ were only measured at 1073 K but the data again agree well with previously published data for a sample with this composition in regions where measurements were previously performed [27]. Similar to results for bulk CeO_2 and $CeO_2(ALD)/\gamma$-Al_2O_3, the equilibrium oxygen stoichiometry in the bulk $Ce_{0.5}Zr_{0.5}O_2$ sample varied gradually with the P_{O_2}. The data for the $CZ(ALD)/\gamma$-Al_2O_3 sample were significantly different. First, flow titration was not useful in measuring the properties for the $CZ(ALD)/\gamma$-Al_2O_3 sample because the sample was completely reduced at all experimentally accessible H_2O:H_2 ratios between 873 K and 1073 K. It was therefore necessary to measure the equilibrium isotherms at 873 and 1073 K using coulometric titration. Interestingly, the isotherms at both temperatures were nearly vertical lines, which is an indication that there is equilibrium between two well-defined compounds. The transitions for reduction of Ce^{4+} to the Ce^{3+} states occur at P_{O_2} of approximately 10^{-9} atm at 1073 K and 10^{-15} atm at 873 K. These values are much higher than the equilibrium P_{O_2} for the $CeO_2(ALD)/\gamma$-Al_2O_3 sample, showing that the mixed oxide is much easier to reduce.

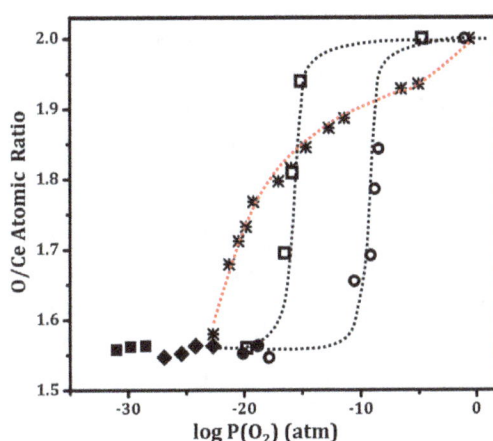

Figure 5. O/Ce ratio as a function of $P(O_2)$ for $20Ce_{0.5}Zr_{0.5}O_2$–Al_2O_3, determined by flow titration, at temperatures (♦)—773K, (■)—873 K, and (●)—1073 K. The (○) symbol shows the isotherm for $20Ce_{0.5}Zr_{0.5}O_2$–Al_2O_3, determined by coulometric titration at 1073 K. Oxidation isotherm result for pure ceria–zirconia ($Ce_{0.5}Zr_{0.5}O_2$) are obtained by flow titration at (*)—1073 K, and the result is consistent with previous literature [26].

Heats of oxidation can be calculated from the temperature dependence of the isotherm using Equation (1):

$$-\Delta H = R\frac{\partial lnP(O_2)}{\partial \frac{1}{T}} \tag{1}$$

The heats, $-\Delta H$, calculated from this equation and the data in Figures 4 and 5 are plotted in Figure 6, together with previously reported numbers for CeO_2 supported on La-Al_2O_3 [19] and for bulk $Ce_{0.5}Zr_{0.5}O_2$ [26]. The heats of oxidation for both the $CeO_2(ALD)/\gamma$-Al_2O_3 and $CZ(ALD)/\gamma$-Al_2O_3 samples were between 550 and 600 kJ/mol O_2, independent of the extent of reduction. This is much lower than the reported heat of oxidation for bulk CeO_2, 760 kJ/mol O_2. In the previous study of impregnated CeO_2 on La-doped Al_2O_3, $-\Delta H$ was reported to vary with stoichiometry, increasing from 500 kJ/mol O_2 at low extents of reduction and to a value close to that of the bulk oxide at high reduction levels. $-\Delta H$ for $Ce_{0.5}Zr_{0.5}O_2$ were reported to be in the range of 500 kJ/mol O_2. The uncertainty in the calculated $-\Delta H$ is relatively high, perhaps ~50 kJ/mol, when the oxide stoichiometry varies with P_{O_2}, as it does for the $CeO_2(ALD)/\gamma$-Al_2O_3 sample. However, when the isotherm is a vertical line, as with the $CZ(ALD)/\gamma$-Al_2O_3 sample, the uncertainty is much less. Therefore, we suggest that the oxidation enthalpies for all of these catalysts may be similar, at least for low extents of reduction.

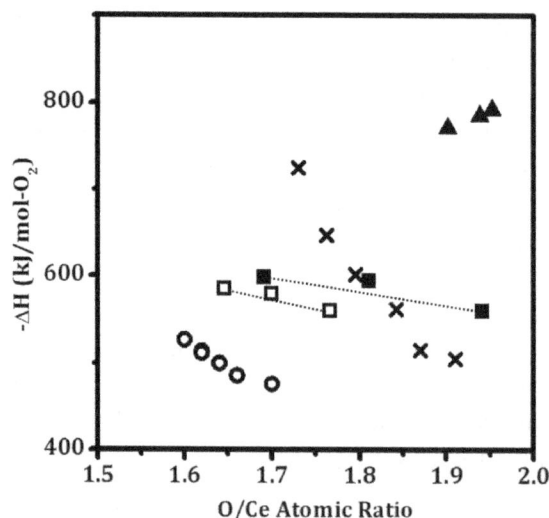

Figure 6. $-\Delta H$ of oxidation as a function of O/Ce ratio for (■)—$20Ce_{0.5}Zr_{0.5}O_2$-Al_2O_3 and (□)—$20CeO_2$–Al_2O_3. Results for (▲)—bulk CeO_2, (○)—bulk $Ce_{0.5}Zr_{0.5}O_2$, and (×)—infiltrated CeO_2 on alumina-based support are obtained from previous work for comparison [19].

3. Discussion

The results in this paper demonstrate that novel ceria and ceria–zirconia supports can be prepared by ALD onto stable supports. In particular, the thin films grown on γ-Al_2O_3 are shown to be stable to high temperature sintering and relatively immune to redox cycling. The ALD-grown materials are structurally very different from materials having the same compositions but prepared by conventional impregnation. Not only do the ALD films form a more uniform coverage over the γ-Al_2O_3 support but one can more easily form mixed oxides like $Ce_{0.5}Zr_{0.5}O_2$ without the use of chelating agents, like citric acid.

Since there was no loss in reducibility upon redox cycling on any of the samples, $CeAlO_3$ formation does not appear to be a serious issue with these materials. However, there was evidence of CeO_2 crystallite growth on the $CeO_2(ALD)/\gamma$-Al_2O_3 sample under redox cycling. This may be partly due to differences in the "wetting" properties of CeO_2 and Ce_2O_3 on γ-Al_2O_3 or it may be due to the expansion and contraction that must occur in the film upon oxidation and reduction. If wetting issues dominate, it may be possible to stabilize the CeO_2 films by using something other than γ-Al_2O_3 as the underlying support; if expansion-contraction issue are responsible, the underlying support will probably not matter. It is interesting that the ceria–zirconia film in the $CZ(ALD)/\gamma$-Al_2O_3 sample appears to be more stable but the reasons for this are still uncertain.

The thin $Ce_{0.5}Zr_{0.5}O_2$ film on the $CZ(ALD)/\gamma$-Al_2O_3 sample appears to exhibit different equilibrium P_{O_2} from that of bulk $Ce_{0.5}Zr_{0.5}O_2$. While this could be due to surface energies, we suggest that it may also due to the confined geometry of the film. In previous work with bulk ceria–zirconia mixed oxides [31], it was argued that the shape of the isotherm results primarily from changes in the entropy of oxidation, rather than changes in the heats of oxidation. This conclusion was based on the fact that $-\Delta H$ was found to be independent of the extent of reduction. To explain how the oxidation entropy would change with the extent of reduction while the enthalpy was constant, it was suggested that each lattice oxygen that is adjacent to a pair of Ce^{4+} ions is energetically equivalent. However, since it should be energetically difficult to remove a lattice oxygen that is next to a vacancy, the number of possible oxygens that can be removed next to each set of adjacent Ce^{4+} ions will decrease as the solid becomes progressively reduced. In a film that is less than one unit cell in thickness, there is no possibility for adjacent vacancies and $-\Delta G$ could be independent of the extent of reduction. According to this hypothesis, the equilibrium properties of ceria–zirconia films should change with thickness, a possibility that would be interesting to test in future work.

Obviously, there is still much to learn about the properties of thin ceria and ceria–zirconia films prepared by ALD. The materials clearly show interesting properties as catalyst supports. Studying their redox properties may help to elucidate how surfaces and interfaces with other oxides affect thermodynamic equilibria.

4. Experimental Methods

4.1. Sample Preparation and Characterization

Samples were prepared by both ALD and conventional impregnation methods, using γ-Al$_2$O$_3$ (Strem Chemicals, Inc., Newburyport, MA, USA) as a support that had been calcined at 1173 K and had a BET surface area of 130 m^2/g. The ALD-prepared samples were synthesized in a home-built apparatus using very similar methods to that described in previous work [32]. Briefly, the ALD cycle consisted of exposing the evacuated γ-Al$_2$O$_3$ substrate to a few Torr of the organometallic precursors, either Ce(TMHD)$_4$ (Strem Chemicals, Inc.) or Zr(TMHD)$_4$ (Strem Chemicals, Inc.), at 503 K for ~300 s, followed by evacuation to ~50 millitorr and precursor oxidation. It should be noted that the substrate was exposed to each gaseous precursor for multiple times to ensure surface saturation. One change in the present study from that in previous work was that precursor oxidation was performed at 503 K for 300 s using an O$_2$ plasma generated by adding ~100 Torr of O$_2$ to the substrate chamber, followed by contacting a tesla coil to a Cu wire that was wrapped around the substrate chamber. Oxidation using a plasma has previously been shown to be very effective at removing difficult-to-oxidize species at low temperatures [33]. To ensure that the ligands were completely oxidized, we removed each sample from the system after every 5 cycles and then heated it to 673 K in a muffle furnace for 5 min. Growth rates for both CeO$_2$ (0.016 nm/cycle) and ZrO$_2$ (0.024 nm/cycle) were determined by measuring the sample mass after every five cycles and were identical to that which was reported in earlier publications from our laboratory [22,23]. The present study focused on just two ALD-prepared samples: a CeO$_2$/γ-Al$_2$O$_3$ made with 20 ALD cycles of ceria (20 wt % CeO$_2$, CeO$_2$(ALD)/γ-Al$_2$O$_3$) and a Ce$_{0.5}$Zr$_{0.5}$O$_2$/γ-Al$_2$O$_3$ made with 12 ALD cycles of ceria and 8 ALD cycles of zirconia (21 wt % Ce$_{0.5}$Zr$_{0.5}$O$_2$, CZ(ALD)/γ-Al$_2$O$_3$). In both samples, the films were estimated to be 0.4-nm thick, based on the sample weight changes, assuming a uniform film over the entire alumina surface.

Conventional alumina-supported catalysts containing 20 wt % CeO$_2$ (CeO$_2$(IMP)/γ-Al$_2$O$_3$) or 21 wt % of a CeO$_2$–ZrO$_2$ mixture (CZ(IMP)/γ-Al$_2$O$_3$) were prepared by infiltration with aqueous solutions of cerium (III) nitrate hexahydrate (Ce(NO$_3$)$_3$·6H$_2$O, Sigma Aldrich, St. Louis, MO, USA) or a mixture of zirconium oxynitrate hydrate (ZrO(NO$_3$)$_2$·xH$_2$O, Sigma Aldrich, degree of hydration of fresh bottle ~6) and Ce(NO$_3$)$_3$·6H$_2$O. The samples were then dried at 333 K overnight, followed by calcination to 673 K for 6 h to remove the remaining nitrates. Bulk CeO$_2$ was prepared by precipitating an aqueous solution of Ce(NO$_3$)$_3$·6H$_2$O with excess ammonium hydroxide (NH$_4$OH, Fisher Scientific, Hampton, NH, USA), then drying the sample overnight at 333 K followed by calcination at 773 K for 6 h [22]. Bulk Ce$_{0.5}$Zr$_{0.5}$O$_2$ was prepared by first dissolving stoichiometric amounts of Ce(NO$_3$)$_3$ and ZrO(NO$_3$)$_2$ in distilled water. The aqueous mixture was titrated at a rate of 5 mL/min into a solution of ammonium hydroxide, while stirring. The precipitate was allowed to dry overnight at 333 K, followed by calcination to 773 K for 6 h [23].

Samples containing 1 wt % Pd were prepared by incipient wetness using aqueous solutions of tetraaminepalladium(II) nitrate (Sigma Aldrich). The materials were dried overnight at 333 K and calcined at 773 K in air for 6 h to remove any organics and nitrates. It is noteworthy that the ceria-containing samples prepared by ALD were relatively hydrophobic, a fact that has been reported previously for flat surfaces modified by CeO$_2$ ALD [34]. However, the aqueous solutions did slowly absorb into the powders and the preparation procedure for adding Pd was the same except for the time required to impregnate.

X-ray Diffraction (XRD) patterns were recorded on a Rigaku Smartlab diffractometer (Toyko, Japan) equipped with a Cu Kα source (λ = 0.15416 nm). Crystalline graphite, physically mixed with the samples in a ratio of 1:8, was used as a reference for the peak positions in XRD. The Pd dispersions were determined by CO chemisorption at room temperature on the reduced catalysts. In this procedure, the samples were first oxidized in 200 Torr O_2 at 673 K and reduced in 200 Torr H_2 at 673 K before measuring CO uptakes. Dispersions were calculated assuming one CO per surface Pd.

Steady-state rates for CO oxidation were measured under differential conditions in a quartz-tube, flow reactor using 25 Torr CO and 12.5 Torr O_2. The total flow rate of He was maintained at 60 mL/min and the mass of catalyst was 0.10 g. Products were analyzed using a gas chromatograph (SRI8610C) (Torrance, CA, USA) equipped with a Hayesep Q column and a TCD detector. All rates in this study were normalized to the mass of the catalyst. Differential conversions were maintained in all cases.

4.2. Redox Equilibrium Measurements

Equilibrium oxidation isotherms were obtained by both flow titration and coulometric titration. In flow titration, the oxidation state of the catalyst is measured by determining the amount of oxygen required to completely re-oxidize the sample after it has been equilibrated at a given P_{O_2}. While the equilibrium stoichiometry for a specified H_2–H_2O ratio was the same whether starting from an oxidized or a reduced sample, we found that equilibrium was achieved more rapidly starting from the reduced form. Therefore, the experiments were performed in the following manner: 0.5 g of sample were placed in a quartz-tube flow reactor, exposed to dry flowing H_2 (30 mL/min) at the temperature of interest for 0.5 h, and then exposed to a flowing H_2–H_2O mixture (30 mL/min) for 1 h. The H_2–H_2O mixture was produced by passing H_2 through a temperature-controlled water bubbler, using the equilibrium vapor pressure to calculate the H_2O partial pressure. After purging the sample with dry He for 0.5 h, flowing air (21% O_2 and 79% N_2, Airgas) was passed over the sample at a rate of 3.1 mL/min and the composition of the effluent gas was determined using a quadrupole mass spectrometer. The amount of oxygen required to re-oxidize the sample was obtained by integrating the difference between the N_2 and O_2 signals. This in turn provides the oxidation state of the ceria that had been in equilibrium with the H_2–H_2O mixture.

Because the H_2–H_2O ratio can only be controlled over a limited range in flow titration, coulometric titration was also used to verify the flow-titration data and to extend the range of P_{O_2} equilibrium measurements [19]. In coulometric titration, a 0.5 g sample was inserted into a YSZ (yttria-stabilized zirconia) tube that had Ag electrodes painted on both inside and outside. The YSZ tube was then placed in a horizontal tube furnace and then heated to the temperature of interest, either 873 K or 1073 K, using a heating rate of 1.0 K/min. During the temperature ramp, a mixture of 5% O_2, 10% H_2O, and 85% Ar was allowed to flow over the sample at a flow rate of 110 mL/min in order to ensure that the sample was completely oxidized at the start of the measurements. After 1 h at the temperature of interest, the flow was stopped and the ends of the YSZ tube were sealed with Cajon fittings. Specified amounts of oxygen were then electrochemically pumped from the inside of the YSZ tube by applying a current across the electrodes using a Gamry instruments potentiostat. After allowing the system to come to equilibrium with the electrodes at open circuit, the equilibrium P_{O_2} was calculated from the Nernst equation and the open circuit potential. To demonstrate equilibrium, oxidation isotherms were also measured starting with the reduced sample.

5. Conclusions

Atomic Layer Deposition (ALD) of CeO_2 and $Ce_{0.5}Zr_{0.5}O_2$ films on γ-Al_2O_3 produces materials in which the alumina surface is coated with either ceria or a ceria–zirconia mixed oxide. The films exhibit good thermal stability upon oxidation and are also reasonably stable to redox cycling. Although the thermodynamic, redox properties of the CeO_2 films prepared by ALD were similar to those of a sample prepared by impregnation, the ALD samples showed better catalytic properties as supports for Pd due to improved contact between the ceria and the Pd. Thermodynamic measurements on the mixed-oxide film indicated that solid solutions were formed and these were much more easily reduced than normal ceria.

Acknowledgments: Tzia Ming Onn and Raymond J. Gorte are grateful to the Department of Energy, Office of Basic Energy Sciences, Chemical Sciences, Geosciences and Biosciences Division, Grant No. DE-FG02-13ER16380 for support of this work.

Author Contributions: Tzia Ming Onn and Raymond J. Gorte conceived and designed the experiments; Xinyu Mao, Chao Lin, and Cong Wang performed the experiments; Tzia Ming Onn and Raymond J. Gorte analyzed the data; Tzia Ming Onn and Raymond J. Gorte wrote the paper.

References

1. Montini, T.; Melchionna, M.; Monai, M.; Fornasiero, P. Fundamentals and catalytic applications of CeO_2-based materials. *Chem. Rev.* **2016**, *116*, 5987–6041. [CrossRef] [PubMed]

2. Fino, D.; Bensaid, S.; Piumetti, M.; Russo, N. A review on the catalytic combustion of soot in diesel particulate filters for automotive applications: from powder catalysts to structured reactors. *Appl. Catal. A* **2016**, *509*, 75–96. [CrossRef]

3. Shelef, M.; Graham, G.W.; McCabe, R.W. Ceria and other oxygen storage components in automotive catalysts. In *Catalysis by Ceria and Related Materials*; Imperial College Press: London, UK, 2002; Volume 2, pp. 343–375.

4. Gandhi, H.; Graham, G.; McCabe, R.W. Automotive exhaust catalysis. *J. Catal.* **2003**, *216*, 433–442. [CrossRef]

5. Sugiura, M.; Ozawa, M.; Suda, A.; Suzuki, T.; Kanazawa, T. Development of innovative three-way catalysts containing ceria–zirconia solid solutions with high oxygen storage/release capacity. *Bull. Chem. Soc. Jpn.* **2005**, *78*, 752–767. [CrossRef]

6. Gorte, R.J. Ceria in catalysis: from automotive applications to the water–gas shift reaction. *AIChE J.* **2010**, *56*, 1126–1135. [CrossRef]

7. Li, J.; Liu, X.; Zhan, W.; Guo, Y.; Guo, Y.; Lu, G. Preparation of high oxygen storage capacity and thermally stable ceria–zirconia solid solution. *Catal. Sci. Technol.* **2016**, *6*, 897–907. [CrossRef]

8. Sharma, S.; Hilaire, S.; Vohs, J.; Gorte, R.J.; Jen, H.-W. Evidence for oxidation of ceria by CO_2. *J. Catal.* **2000**, *190*, 199–204. [CrossRef]

9. Heck, R.M.; Farrauto, R.J. Automobile exhaust catalysts. *Appl. Catal. A* **2001**, *221*, 443–457. [CrossRef]

10. Bueno-Lopez, A.; Krishna, K.; Makkee, M.; Moulijn, J.A. Active oxygen from CeO_2 and its role in catalysed soot oxidation. *Catal. Lett.* **2005**, *99*, 203–205. [CrossRef]

11. Bunluesin, T.; Gorte, R.J.; Graham, G. Studies of the water–gas-shift reaction on ceria-supported Pt, Pd, and Rh: Implications for oxygen-storage properties. *Appl. Catal. B* **1998**, *15*, 107–114. [CrossRef]

12. Hilaire, S.; Wang, X.; Luo, T.; Gorte, R.J.; Wagner, J. A comparative study of water–gas-shift reaction over ceria supported metallic catalysts. *Appl. Catal. A* **2001**, *215*, 271–278. [CrossRef]

13. Fu, Q.; Saltsburg, H.; Flytzani-Stephanopoulos, M. Active nonmetallic Au and Pt species on ceria-based water-gas shift catalysts. *Science* **2003**, *301*, 935–938. [CrossRef] [PubMed]

14. Colussi, S.; Katta, L.; Amoroso, F.; Farrauto, R.J.; Trovarelli, A. Ceria-based palladium zinc catalysts as promising materials for water gas shift reaction. *Catal. Commun.* **2014**, *47*, 63–66. [CrossRef]

15. Wang, X.; Gorte, R.J. A study of steam reforming of hydrocarbon fuels on Pd/ceria. *Appl. Catal. A* **2002**, *224*, 209–218. [CrossRef]

16. Duarte, R.; Safonova, O.; Krumeich, F.; Makosch, M.; Van Bokhoven, J. Oxidation state of Ce in CeO_2-promoted Rh/Al_2O_3 catalysts during methane steam reforming: H_2O activation and alumina stabilization. *ACS Catal.* **2013**, *3*, 1956–1964. [CrossRef]

17. Divins, N.; Casanovas, A.; Xu, W.; Senanayake, S.; Wiater, D.; Trovarelli, A.; Llorca, J. The influence of nano-architectured CeO_x supports in Rh Pd/CeO_2 for the catalytic ethanol steam reforming reaction. *Catal. Today* **2015**, *253*, 99–105. [CrossRef]

18. Moretti, E.; Storaro, L.; Talon, A.; Chitsazan, S.; Garbarino, G.; Busca, G.; Finocchio, E. Ceria–zirconia based catalysts for ethanol steam reforming. *Fuel* **2015**, *153*, 166–175. [CrossRef]

19. Zhou, G.; Shah, P.R.; Montini, T.; Fornasiero, P.; Gorte, R.J. Oxidation enthalpies for reduction of ceria surfaces. *Surf. Sci.* **2007**, *601*, 2512–2519. [CrossRef]

20. He, B.J.-J.; Wang, C.-X.; Zheng, T.-T.; Zhao, Y.-K. Thermally induced deactivation and the corresponding strategies for improving durability in automotive three-way catalysts. *Johnson Matthey Technol. Rev.* **2016**, *60*, 196–203. [CrossRef]

21. Cargnello, M.; Doan-Nguyen, V.V.; Gordon, T.R.; Diaz, R.E.; Stach, E.A.; Gorte, R.J.; Fornasiero, P.; Murray, C.B. Control of metal nanocrystal size reveals metal-support interface role for ceria catalysts. *Science* **2013**, *341*, 771–773. [CrossRef] [PubMed]

22. Onn, T.M.; Zhang, S.; Arroyo-Ramirez, L.; Xia, Y.; Wang, C.; Pan, X.; Graham, G.W.; Gorte, R.J. High-surface-area ceria prepared by ALD on Al_2O_3 support. *Appl. Catal. B* **2017**, *201*, 430–437. [CrossRef]

23. Onn, T.M.; Dai, S.; Chen, J.; Pan, X.; Graham, G.W.; Gorte, R.J. High-surface area ceria–zirconia films prepared by atomic layer deposition. *Catal. Lett.* **2017**, *147*, 1464–1470. [CrossRef]

24. Singh, J.A.; Yang, N.; Bent, S.F. Nanoengineering heterogeneous catalysts by atomic layer deposition. *Annu. Rev. Chem. Biomol. Eng.* **2017**, *8*, 41–62. [CrossRef] [PubMed]

25. Miikkulainen, V.; Leskelä, M.; Ritala, M.; Puurunen, R.L. Crystallinity of inorganic films grown by atomic layer deposition: Overview and general trends. *J. Appl. Phys.* **2013**, *113*, 2. [CrossRef]

26. Bonk, A.; Remhof, A.; Maier, A.C.; Trottmann, M.; Schlupp, M.V.F.; Battaglia, C.; Vogt, U.F. Low-Temperature Reducibility of $M_xCe_{1-x}O_2$ (M = Zr, Hf) under Hydrogen Atmosphere. *J. Phys. Chem. C* **2016**, *120*, 118–125. [CrossRef]

27. Zhou, G.; Shah, P.R.; Kim, T.; Fornasiero, P.; Gorte, R.J. Oxidation entropies and enthalpies of ceria–zirconia solid solutions. *Catal. Today* **2007**, *123*, 86–93. [CrossRef]

28. Kaspar, J.; Fornasiero, P.; Balducci, G.; Di Monte, R.; Hickey, N.; Sergo, V. Effect of ZrO_2 content on textural and structural properties of CeO_2–ZrO_2 solid solutions made by citrate complexation route. *Inorg. Chim. Acta* **2003**, *349*, 217–226. [CrossRef]

29. Bunluesin, T.; Gorte, R.J.; Graham, G. Co oxidation for the characterization of reducibility in oxygen storage components of three-way automotive catalysts. *Appl. Catal. B* **1997**, *14*, 105–115. [CrossRef]

30. Kim, T.; Vohs, J.M.; Gorte, R.J. Thermodynamic investigation of the redox properties of ceria–zirconia solid solutions. *Ind. Eng. Chem. Res.* **2006**, *45*, 5561–5565. [CrossRef]

31. Shah, P.R.; Kim, T.; Zhou, G.; Fornasiero, P.; Gorte, R.J. Evidence for entropy effects in the reduction of ceria−zirconia solutions. *Chem. Mater.* **2006**, *18*, 5363–5369. [CrossRef]

32. Anthony, S.Y.; Küngas, R.; Vohs, J.M.; Gorte, R.J. Modification of SOFC cathodes by atomic layer deposition. *J. Electrochem. Soc.* **2013**, *160*, F1225–F1231.

33. Jia, L.; Farouha, A.; Pinard, L.; Hedan, S.; Comparot, J.-D.; Dufour, A.; Tayeb, K.B.; Vezin, H.; Batiot-Dupeyrat, C. New routes for complete regeneration of coked zeolite. *Appl. Catal. B* **2017**, *219*, 82–91. [CrossRef]

34. Lv, Q.; Zhang, S.; Deng, S.; Xu, Y.; Li, G.; Li, Q.; Jin, Y. Transparent and water repellent ceria film grown by atomic layer deposition. *Surf. Coat. Technol.* **2017**, *320*, 190–195. [CrossRef]

12

Unique Hydrogen Desorption Properties of LiAlH$_4$/h-BN Composites

Yuki Nakagawa *, Shigehito Isobe, Takao Ohki and Naoyuki Hashimoto

Graduate School of Engineering, Hokkaido University, N-13, W-8, Sapporo 060-8278, Japan;
isobe@eng.hokudai.ac.jp (S.I.); takao-ohki@eng.hokudai.ac.jp (T.O.); hasimoto@eng.hokudai.ac.jp (N.H.)

* Correspondence: y-nakagawa@eng.hokudai.ac.jp

Abstract: Hexagonal boron nitride (h-BN) is known as an effective additive to improve the hydrogen de/absorption properties of hydrogen storage materials consisting of light elements. Herein, we report the unique hydrogen desorption properties of LiAlH$_4$/h-BN composites, which were prepared by ball-milling. The desorption profiles of the composite indicated the decrease of melting temperature of LiAlH$_4$, the delay of desorption kinetics in the first step, and the enhancement of the kinetics in the second step, compared with milled LiAlH$_4$. Li$_3$AlH$_6$ was also formed in the composite after desorption in the first step, suggesting h-BN would have a catalytic effect on the desorption kinetics of Li$_3$AlH$_6$. Finally, the role of h-BN on the desorption process of LiAlH$_4$ was discussed by comparison with the desorption properties of LiAlH$_4$/X (X = graphite, LiCl and LiI) composites, suggesting the enhancement of Li ion mobility in the LiAlH$_4$/h-BN composite.

Keywords: alanate; h-BN; hydrogen storage; catalyst; Li ion mobility

---

1. Introduction

Hydrogen storage is a key technology for a future hydrogen energy society [1]. However, it is still challenging to develop high performance hydrogen storage materials with high hydrogen density, fast de/absorption kinetics, and high cycle stability under moderate temperature and pressure conditions [2,3]. LiAlH$_4$ is one of the most promising hydrogen storage materials because of its high hydrogen capacity and relatively low desorption temperature [4]. The hydrogen desorption process of LiAlH$_4$ is described as follows:

Melting:

LiAlH$_4$(s) $\rightarrow$ LiAlH$_4$(l) Endothermic (150–175 °C)

Decomposition in the first step:

3LiAlH$_4$(l) $\rightarrow$ Li$_3$AlH$_6$(s) + 2Al + 3H$_2$ Exothermic (150–200 °C, 5.3 mass % H$_2$)

Decomposition in the second step:

Li$_3$AlH$_6$(s) $\rightarrow$ 3LiH + Al + 3/2H$_2$ Endothermic (200–270 °C, 2.6 mass % H$_2$)

The decomposition in the first step is an exothermic reaction with a ΔH of -10 kJ·mol^{-1} H$_2$, indicating the reversibility of this step is believed to be thermodynamically difficult [4,5]. In the second step, Li$_3$AlH$_6$ decomposes in an endothermic reaction with a ΔH of 25 kJ·mol^{-1} H$_2$ [4]. Thus, the hydrogenation of LiH/Al to Li$_3$AlH$_6$ is thermodynamically possible.

One of the strategies for improving the properties of hydrogen storage materials is the addition of catalysts/dopants [6]. Ti or its compounds are well-known catalysts for the kinetics of alanate [7–9]. Since Bogdanović et al. reported an absence of hysteresis and nearly horizontal pressure plateaus in the TiCl$_3$-doped NaAlH$_4$ [7], many researchers have studied complex hydrides including alanate as potential reversible hydrogen storage materials. In the case of LiAlH$_4$, the improved desorption kinetics was reported by the doping of Ti catalyst using mechanically milling [10–12]. Recent study also reported a single step hydrogen release of LiAlH$_4$, which was induced by the synergetic effects of Ti catalytic coating and nanosizing effects [13]. Although the rehydrogenation of the desorbed material was not achieved in the milled sample, the regeneration of LiAlH$_4$ from LiH and Ti-catalyzed Al was possible through the solution synthesis approach using THF and Me$_2$O [14–16].

Hexagonal boron nitride (h-BN) is known as an effective additive for chemical hydride and complex hydride systems. For instance, NH$_3$BH$_3$/h-BN composite released hydrogen at low temperature with minimum induction time and less exothermicity [17]. The remarkable hydrogen de/absorption properties were also achieved in the milled LiBH$_4$/h-BN composites [18,19]. The 30 mol % h-BN doped LiBH$_4$ composite started to release hydrogen from 180 °C, which was 100 °C lower than the onset hydrogen desorption temperature of ball-milled LiBH$_4$ [18]. For the 75 mol % h-BN doped LiBH$_4$ composite, the on-set desorption temperature of LiBH$_4$ was reduced to 175 °C and the peak desorption temperature was reduced by 80 °C compared with milled LiBH$_4$ [19]. Furthermore, under moderate rehydrogenation conditions of 400 °C and 10 MPa H$_2$ pressure, the dehydrogenation capacity of the composite maintained 3.1 mass % within three cycles, which was very close to its theoretical capacity. It was assumed that the excellent rehydrogenation property of LiBH$_4$ would be related with the enhanced hydrogen and lithium diffusion capability by the nanoscale h-BN, which was synthesized by ball-milling of h-BN at 490 rpm for 20 h [19]. The enhancement of Li$^+$ and/or H$^-$ diffusion by adding h-BN was firstly reported in the LiNH$_2$/LiH system [20]. Hydrogen was fully desorbed from the LiNH$_2$/LiH/h-BN composite in less than 7 h, whereas the LiNH$_2$/LiH composite desorbed hydrogen in several days. They proposed that h-BN is an efficient catalyst that improves Li$^+$ diffusion and hence the kinetics of the reaction between LiNH$_2$ and LiH [20]. The mobility of Li$^+$ ions between LiH and LiNH$_2$ was also enhanced by adding LiTi$_2$O$_4$ catalyst [21].

Thus, h-BN has attracted much attention as an effective additive to improve the hydrogen de/absorption kinetics of hydrogen storage materials, especially for complex hydrides. However, the addition of h-BN to the alanate system has rarely been reported. In the present study, LiAlH$_4$/h-BN composites were synthesized by planetary ball-milling and their hydrogen desorption properties were analyzed. Also, the desorption process of the composite was investigated by using XRD and FT-IR. Finally, the role of h-BN to the desorption properties of LiAlH$_4$ was discussed by comparison with those of LiAlH$_4$/X (X = graphite, LiCl and LiI) composites.

2. Results

Hydrogen desorption properties of LiAlH$_4$/h-BN composites were analyzed by using TG-DTA-MS. Figure 1 shows the DTA and MS (H$_2$, m/z = 2) profiles of the composites. As shown in Figure 1a, ball-milled LiAlH$_4$ (denoted as 0 mass % in Figure 1) started to melt around 150 °C followed by hydrogen desorption in two steps below 250 °C. In the case of LiAlH$_4$/h-BN composites, desorption profiles were clearly changed compared with LiAlH$_4$. First, the melting temperature (T_m) of LiAlH$_4$ was decreased by adding h-BN. For instance, in the 40 mass % h-BN composite, DTA peak value of T_m was 151 °C, which was 11 °C lower than that of milled LiAlH$_4$. Second, the hydrogen desorption temperature (T_d) in the first step was slightly increased by h-BN addition. As shown in Figure 1b, T_d in the first step became high value with the increasing amount of h-BN. Third, the desorption kinetics in the second step was improved by adding h-BN. As shown in Figure 1b, the desorption peak in the second step became sharp as the amount of h-BN increased up to 14 mass %. However, the peak shape became broad in the 40 mass % h-BN composite, suggesting the addition of too much amount of h-BN could have negative effect on improving the kinetics. Figure 2 shows the TG profiles of LiAlH$_4$/h-BN composites. The total mass loss from TG profile of ball-milled LiAlH$_4$ was 7.7 mass %, which was in good agreement with the theoretical hydrogen desorption amount of LiAlH$_4$ (7.9 mass %) [4]. The hydrogen mass loss of 6.9 mass %, 5.8 mass %, and 3.5 mass % were calculated from the profiles of 4 mass %, 14 mass %, and 40 mass % h-BN composites, respectively. It is noted that theoretical hydrogen desorption capacities of these composites were 7.6 mass %, 6.8 mass %, and 4.7 mass %, respectively, when only considering the hydrogen desorption from $LiAlH_4$. Thus, the experimental values of hydrogen desorption amounts were slightly lower than the theoretical values. This result could originate from the hydrogen desorption during ball-milling or the formation of new H-containing solid compound by the reaction between $LiAlH_4$ and h-BN.

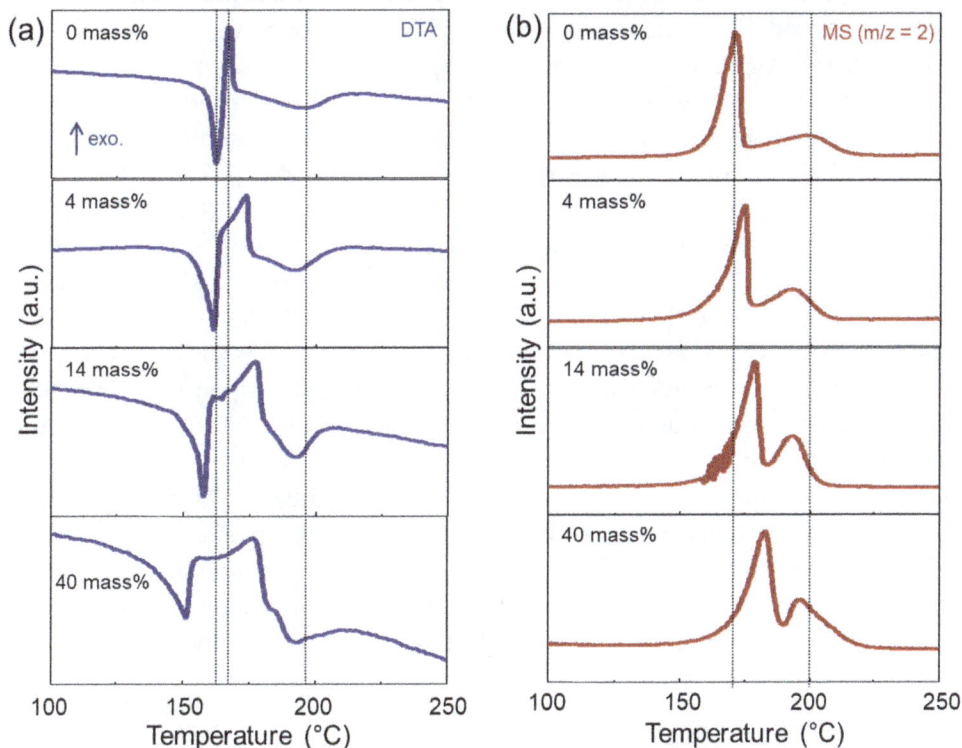

Figure 1. Hydrogen desorption profiles of $LiAlH_4/x$ mass % h-BN (x = 0, 4, 14, 40) composites: (**a**) DTA and (**b**) MS (m/z = 2, H_2) profiles. Heating rate was 5 °C·min^{-1}.

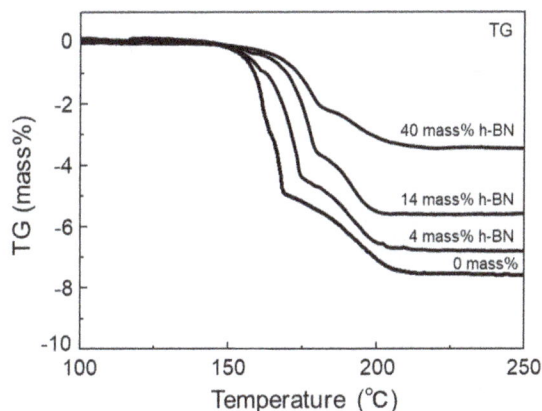

Figure 2. TG profiles of $LiAlH_4/x$ mass % h-BN (x = 0, 4, 14, 40) composites. Heating rate was 5 °C·min^{-1}.

To investigate the interaction between $LiAlH_4$ and h-BN, XRD, and FT-IR measurements were performed for the $LiAlH_4$/h-BN composites. Also, the particle size of composite was observed by using SEM and TEM. Figure 3a shows the XRD profiles of the milled $LiAlH_4$ and $LiAlH_4$/h-BN composites. Only $LiAlH_4$ and h-BN phases were observed in the profiles of the composites. Although the diffraction peaks of LiAlH$_4$ were slightly broadened by h-BN addition, the clear relationship between the broadning and the amount of h-BN was not observed. Figure 3b,c shows the XRD profiles of the milled LiAlH$_4$ and 40 mass % h-BN composite after hydrogen desorption. The 40 mass % h-BN composite formed the similar reaction products compared with LiAlH$_4$. In other words, Li$_3$AlH$_6$ and Al were formed after the hydrogen desorption in the first step, and LiH and Al were formed after the second step. The phase of h-BN was also clearly observed after hydrogen desorption. Broad diffraction peaks around 20° and 27° originate from the polyimide film and grease to prevent the sample oxidation. Figure 4 shows SEM and TEM images of 40 mass % h-BN composite and references. As shown in Figure 4a, the milled LiAlH$_4$ contained a lot of large particles with sizes over 10 μm. On the other hand, the 40 mass % h-BN composite showed the average particle size of a few micrometers, indicating the refinement of LiAlH$_4$ particles occurred in the composite. The submicron particles were also observed in the TEM image of the composite, as shown in Figure 4d. The size of as-received h-BN particle was around 1 μm (Figure 4b). Figure 5 shows the FT-IR spectra of 40 mass % h-BN composite and references. The as-milled composite showed the characteristic Al–H vibrations of LiAlH$_4$ [22] around 1795 cm^{-1} and 1644 cm^{-1}. Also, B–N vibrations of h-BN around 1373 cm^{-1} and 818 cm^{-1} were observed. Although the Al–H vibrations of LiAlH$_4$ disappeared after heating, the B–N vibrations of h-BN still remained. These results were consistent with the results of XRD. However, new IR absorption peak was clearly observed around 2300 cm^{-1} after heating up to 183 °C and 300 °C. Also, another new peak appeared around 1100 cm^{-1} after heating up to 300 °C. Although these peaks were not identified in this work, the peak positions were similar to those of LiBH$_4$ [23], suggesting a such kind of Li–B–H phase exist after heating up to 183 °C and/or 300 °C. The unknown peak around 2300 cm^{-1} was also observed for the IR spectra of the composite consisting of BN^{nano}H$_x$ (ball-milled h-BN under 1.0 MPa H$_2$ for 80 h) and LiH [24]. Thus, the formation of new H-containing solid compound could result in the slightly low hydrogen desorption amount in Figure 2. Considering the results of Figure 5, the possible new compound could be covalently functionalized h-BN species. The details were explained in the next discussion part.

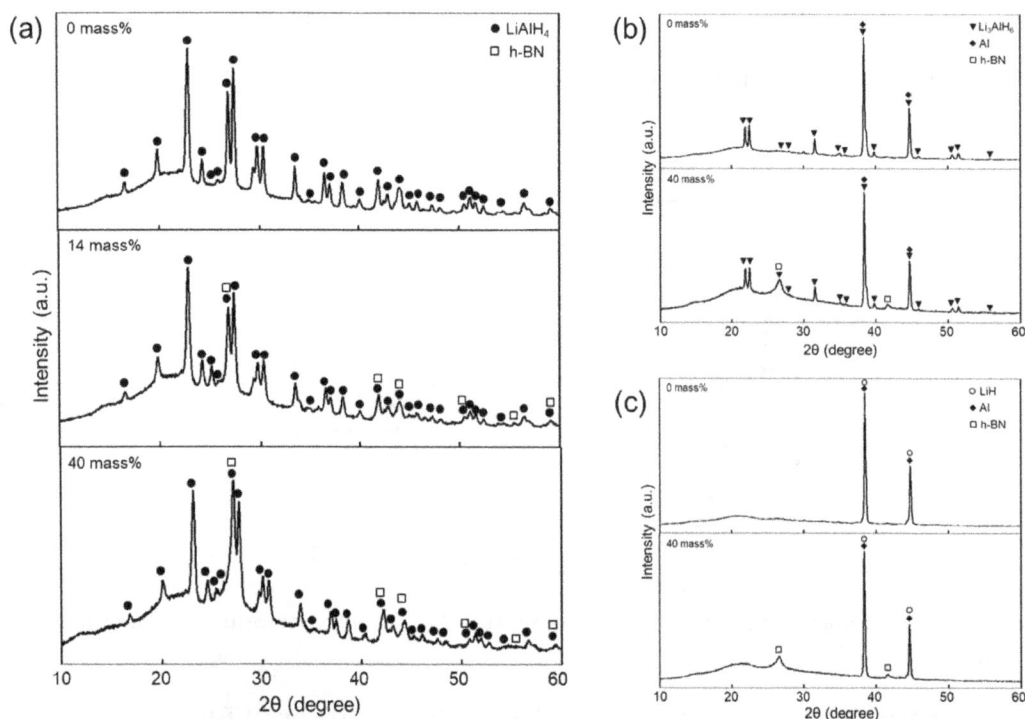

Figure 3. XRD profiles of LiAlH$_4$/x mass % h-BN (x = 0, 14, 40) composites: (**a**) after ball-milling; (**b**) after desorption in the first step; and (**c**) after desorption in the second step. The heating rate was 5 °C·min^{-1}.

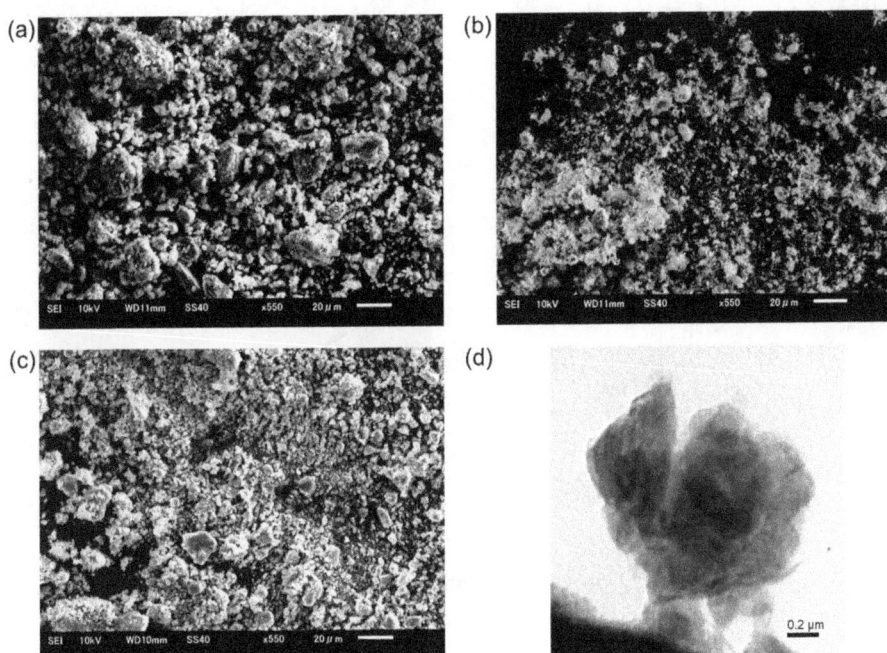

Figure 4. SEM images of (**a**) milled LiAlH$_4$; (**b**) h-BN; (**c**) LiAlH$_4$/40 mass % h-BN composite after milling; and (**d**) TEM image of LiAlH$_4$/40 mass % h-BN composite after milling.

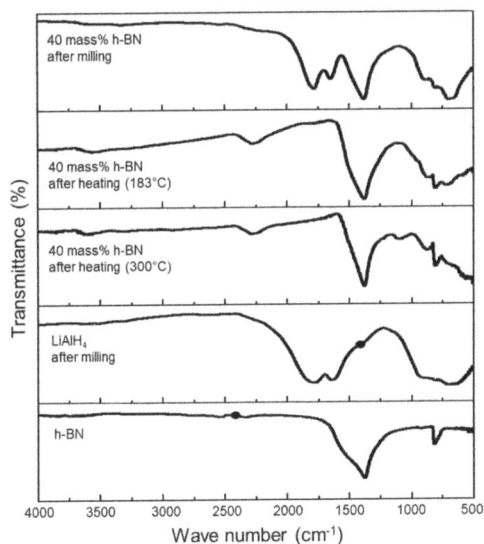

Figure 5. FT-IR spectra of LiAlH$_4$/40 mass % h-BN composite after milling and after hydrogen desorption. LiAlH$_4$ and h-BN spectra are shown as the references. Heating rate was 5 °C·min^{-1}.

As shown in the results of structural characterizaion, LiAlH$_4$/h-BN composites also formed Li$_3$AlH$_6$ as an intermediate product, indicating the similar decomposition pathway with LiAlH$_4$. Thus, h-BN would have the catalytic effect on the hydrogen desorption kinetics of Li$_3$AlH$_6$. The apparent activation energy for hydrogen desorption was calculated by using the Kissinger equation [25],

$$\ln \frac{c}{T_p^2} = -\frac{E_a}{RT_p} + \ln \frac{RA}{E_a}$$

where E_a is the apparent activation energy for hydrogen desorption, c is the heating rate, T_p is the peak temperature, R is gas constant, and A is the frequency factor. Figure 6 shows the Kissinger plots for the hydrogen desorption in the second step of 4 mass % h-BN composite. The obtained apparent activation energy, E_a was 71.5 kJ·mol^{-1}. This value is lower than that reported for Li$_3$AlH$_6$ (92 kJ·mol^{-1}) [26], indicating the desorption kinetics was improved by adding h-BN.

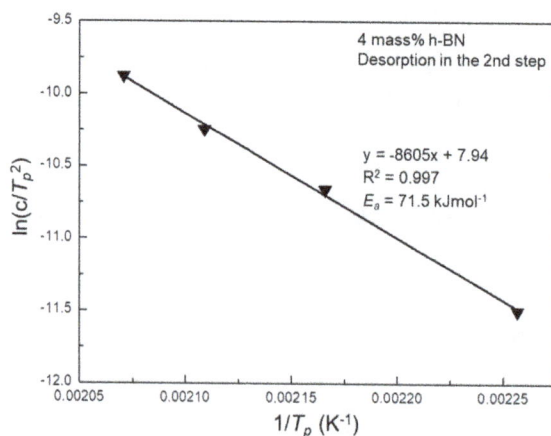

Figure 6. Kissinger plots for the hydrogen desorption in the second step of LiAlH$_4$/4 mass % h-BN composite. Heating rates were 2, 5, 8, and 12 °C·min^{-1}.

In order to understand the role of h-BN, other additives were also ball-milled with LiAlH$_4$ and their desorption properties were analyzed. The detailed results (Figure 7) and the possible role of h-BN on the desorption process were explained in the next discussion part.

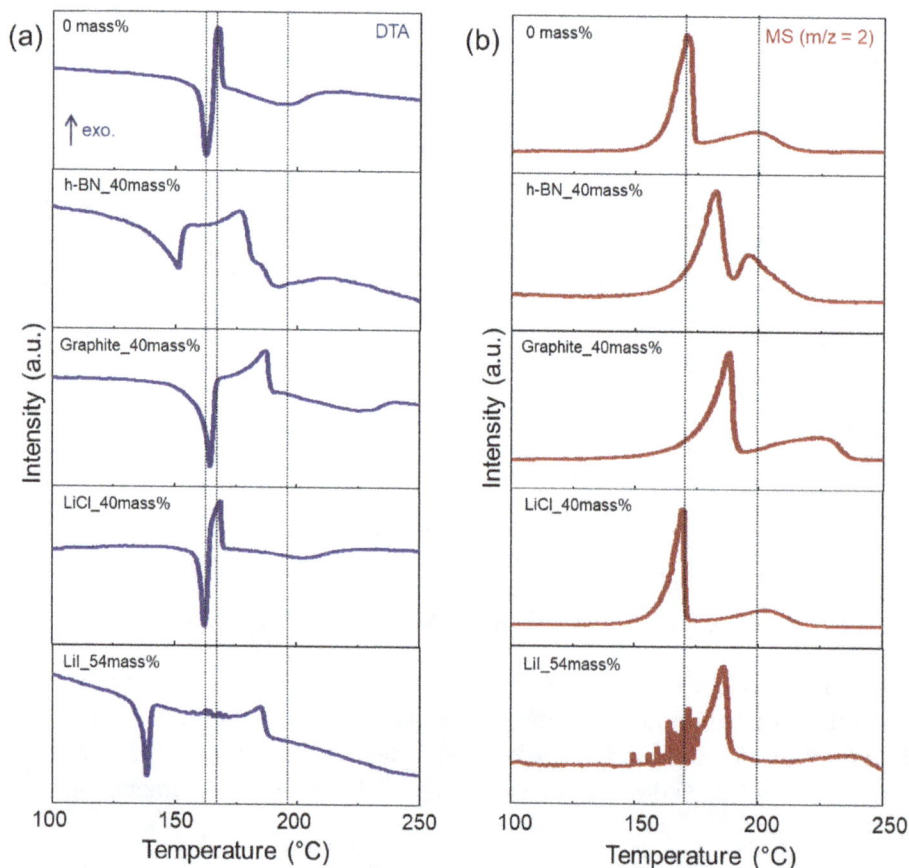

Figure 7. Hydrogen desorption profiles of LiAlH$_4$/X (X = graphite, LiCl, and LiI) composites. The profiles of milled LiAlH$_4$ and LiAlH$_4$/h-BN composites are shown as the references: (a) DTA and (b) MS (*m/z* = 2, H$_2$) profiles. Heating rate was 5 °C·min^{-1}.

3. Discussion

Figure 7 shows the DTA and MS profiles of $LiAlH_4/X$ (X = graphite, LiCl, and LiI) composites. The profiles of $LiAlH_4$ and $LiAlH_4/h$-BN composites are also shown as the references. Graphite was selected as an additive because this compound has a structure similar to h-BN. In spite of its similar structure, hydrogen desorption properties were different from those of h-BN. In the case of $LiAlH_4$/graphite composite, the melting temperature was similar to ball-milled $LiAlH_4$. Also, the desorption kinetics in the second step seemed to be delayed, whereas that of $LiAlH_4/h$-BN was enhanced. Since the both graphite and h-BN are hard materials, the refinement of $LiAlH_4$ particles would occur during the ball-milling process. However, the different desorption profiles were obtained between h-BN and graphite composite, suggesting just the refinement of particles cannot explain this difference. Although the desorption properties of nanoconfined $LiAlH_4$ into graphite with high surface area was reported in the previous study [27], the profiles of graphite composite in this study seemed to be different from those profiles.

Also, LiCl and LiI were selected as additives. Aguey-Zinsou et al. reported that h-BN would enhance the Li ion mobility across the interface of $LiNH_2$ and LiH [20]. Thus, the enhancement of Li ion mobility could be one of the reasons for the unique hydrogen desorption properties of $LiAlH_4/h$-BN composites. Oguchi et al. reported the Li ion conductivity of Li_3AlH_6/LiI composite at 120 °C was much higher than that of Li_3AlH_6, but that of Li_3AlH_6/LiCl at 120 °C was the similar value compared with Li_3AlH_6 [28]. Their results suggest that LiI additive would be effective for increasing the Li ion mobility of $LiAlH_4$, but LiCl would not be effective near the decomposition temperature range of $LiAlH_4$. As shown in Figure 7, the desorption profiles of LiCl composite was similar to those of milled $LiAlH_4$. On the other hand, those of LiI composite showed the decrease of melting temperature and the delay of the first desorption reaction, which was consistent with the results of h-BN composite. This comparative result suggests that Li ion conductivity would increase in the $LiAlH_4/h$-BN composite. The kinetics in the second step wasdelayed in the LiI composite. The possible origin of the high conductivity of $LiAlH_4$ is the anion substitution from complex anion to I^- [28], which partially took place in the case of $LiBH_4$ [29]. Thus, the high decomposition temperature in the second step could originate from the formation of stable solid solution similar to $LiBH_4$–LiI(LiCl) system [29–33]. For clarifying the total desorption process of these composites, the detailed mechanistic study is needed. The analysis of Li ion conductivity of $LiAlH_4/h$-BN composite is currently in progress.

As shown in Figure 1, all the $LiAlH_4/h$-BN composites showed the melting of $LiAlH_4$. This phenomenon was different from the case of $LiAlH_4$ catalyzed by transition metal (Ti, Fe, Co, Nb, etc. [34–36]), which can release hydrogen below the melting temperature. Thus, h-BN would have little interaction with complex anion of $[AlH_4]^-$, whereas transition metal like Ti would destabilize the covalent Al–H bond of complex anion as Sandrock et al. suggested [8]. According to the proposed mechanism by Atakli et al. [37], Li_3AlH_6 is formed by transferring the alkali cation and a hydrogen anion from the two neighboring alanate molecules to central one. In this context, the diffusion distance of Li^+ seems to be very short to form Li_3AlH_6, suggesting Li^+ diffusion would not be the rate-limiting step in the desorption of $LiAlH_4$ in the first step. On the contrary, the desorption properties of h-BN and LiI composite suggested the excess enhancement of Li^+ ion mobility might be related to delaying the formation of Li_3AlH_6. First-principle DFT studies also suggested the formation and migration of $[AlH_4]^-$ vacancy would be the rate-limiting step in the decomposition of $LiAlH_4$ [38]. In the second decomposition step, it was proposed that three LiH are formed from Li_3AlH_6, leaving AlH_3, which spontaneously desorbs hydrogen [37]. Thus, the enhanced mobility of Li^+ may help the destabilization of complex anions to improve the hydrogen desorption kinetics.

As shown in Figure 5, the presence of new bonds similar to those of $LiBH_4$ or ball-milled h-BN with LiH suggests that covalently functionalized h-BN species could be formed after heating the $LiAlH_4/h$-BN composites. It is known that ball-milling of h-BN with a lot of different materials can generate functionalized h-BN nanosheets [39]. For instance, a one-step method for the preparation and functionalization of few-layer BN was developed based on urea-assisted solid exfoliation of commercially available h-BN [40]. Such kind of functionalized BN nanosheets are attractive for a lot of applications such as polymer matrix composites [41], ion conductors [42], and hydrogen storage [43]. Further investigations of $LiAlH_4$/h-BN composites could pave the way for the covalent functionalization of h-BN nanosheets by interaction with $LiAlH_4$.

4. Materials and Methods

4.1. Synthesis of $LiAlH_4$/X Composites

All samples were handled in an Ar-filled glovebox with O_2 and H_2O levels below 2 ppm. $LiAlH_4$ (95%, Sigma-Aldrich, Tokyo, Japan), h-BN (98%, Sigma-Aldrich), graphite (99.99%, Kojundo Chemical Lab., Sakado, Japan), LiCl (99.99%, Sigma-Aldrich) and LiI (99.9%, Sigma-Aldrich) were used as starting materials. The $LiAlH_4$/X (X = h-BN, graphite, LiCl and LiI) composites were synthesized by using planetary ball-milling apparatus (Fritsch Pulverisette 7, Yokohama, Japan) with 20 stainless balls (7 mm in diameter) and 300 mg samples (ball: powder mass ratio = 70: 1). The milling pot was equipped with a quick connector for vacuuming and introducing H_2 gas. The milling was performed under 0.1 MPa H_2 atmosphere with 400 rpm for 2 h with four cycles of 30/15 min operation/interval per each cycle. Also, the milled $LiAlH_4$ was prepared under the same milling conditions for comparison.

4.2. Characterization

Hydrogen desorption properties of the composites were examined by a thermogravimetry and differential thermal analysis equipment (TG-DTA, Bruker, 2000SA, Yokohama, Japan) connected to a mass spectrometer (MS, ULVAC, BGM-102, Chitose, Japan). The desorbed gases were carried from TG-DTA to MS through a capillary by 300 mL·min^{-1} stream of high purity He as a carrier gas. The samples were heated from room temperature to 300 °C with a heating rate of 5 °C·min^{-1}. Structural properties were investigated by powder X-ray diffraction (XRD) measurements (Philips, X'Pert Pro with Cu Kα radiation, Amsterdam, The Netherlands), where all the samples were covered with a polyimide sheet (Kapton, The Nilaco Co., Ltd., Tokyo, Japan) in the glovebox to avoid oxidation during the measurement. Morphology of the composites was observed using scanning electron microscope (SEM, JEOL, JSM-6510LA, Tokyo, Japan) and transmission electron microscope (TEM, JEOL, JEM-2010, Tokyo, Japan). For TEM observations, the samples were dispersed on a molybdenum micro-mesh grid. Fourier transform infrared spectrometer (FT-IR, JASCO, FT/IR 660Plus, Tokyo, Japan) was operated to investigate chemical bonds in the composites. Each sample was put between KBr plates and pressed for measurement.

5. Conclusions

Hydrogen desorption properties of the ball-milled $LiAlH_4$/h-BN composites were investigated. Compared with milled $LiAlH_4$, the composites showed the different desorption profiles, where the decrease of melting temperature (T_m), the delay of desorption kinetics in the first step and the enhancement of the kinetics in the second step were observed. In the 40 mass % h-BN composite, the DTA peak value of T_m was 151 °C, which was 11 °C lower than that of milled $LiAlH_4$. The $LiAlH_4$/h-BN composite formed Li_3AlH_6 after desorption in the first step similar to $LiAlH_4$. Thus, h-BN would have a catalytic effect on the desorption kinetics of Li_3AlH_6. The apparent activation energy in the second step desorption was 71.5 kJ·mol^{-1} for the 4 mass % h-BN composite. From SEM and TEM observations, the refinement of $LiAlH_4$ particle was confirmed in the 40 mass % h-BN composite. The particle size of the composite was around a few micrometers. The hydrogen mass loss of the composite was slightly lower than the theoretical value. The new chemical bond similar to Li–B–H species was observed in the FT-IR spectra of the 40 mass % h-BN composite after the desorption. This result suggested covalently functionalized h-BN nanosheets could be formed in the composite. Finally, the desorption properties of $LiAlH_4$/h-BN composite were compared with those of

LiAlH$_4$/X (X = graphite, LiCl, and LiI) composites, suggesting the enhancement of Li ion mobility in the LiAlH$_4$/h-BN composite compared with LiAlH$_4$. The present work first demonstrates the effect of h-BN addition on the hydrogen desorption properties of alanate.

Acknowledgments: A part of this study was conducted at "Joint-use facilities: Laboratory of Nano-Micro Material Analysis, Laboratory of XPS Analysis, and the Open Facility" at Hokkaido University, supported by the "Material Analysis and Structure Analysis Open Unit (MASAOU)" and "Nanotechnology Platform" program of the Ministry of Education, Culture, Sports, Science and Technology (MEXT), Japan.

Author Contributions: Yuki Nakagawa was involved in all stages of the work, including designing the work, conducting experiments, and analyzing the data; Shigehito Isobe acted as co-supervisor and was involved in the discussion of the results and work planning; Takao Ohki performed the TG-DTA-MS and XRD experiments; Naoyuki Hashimoto acted as the main supervisor and was involved in the discussion of the results; Yuki Nakagawa wrote the paper; and all the authors contributed to the revision of the paper.

References

1. Züttel, A. Materials for hydrogen storage. *Mater. Today* **2003**, *6*, 24–33. [CrossRef]
2. Lai, Q.; Paskevicius, M.; Sheppard, D.A.; Buckley, C.E.; Thornton, A.W.; Hill, M.R.; Gu, Q.; Mao, J.; Huang, Z.; Liu, H.K.; et al. Hydrogen storage materials for mobile and stationary applications: Current state of the art. *ChemSusChem* **2015**, *8*, 2789–2825. [CrossRef] [PubMed]
3. Ley, M.B.; Jensen, L.H.; Lee, Y.S.; Cho, Y.W.; Bellosta von Colbe, J.M.; Dornheim, M.; Rokni, M.; Jensen, J.O.; Sloth, M.; Filinchuk, Y.; et al. Complex hydrides for hydrogen storage—New perspectives. *Mater. Today* **2014**, *17*, 122–128. [CrossRef]
4. Orimo, S.; Nakamori, Y.; Eliseo, J.R.; Züttel, A.; Jensen, C.M. Complex hydrides for hydrogen storage. *Chem. Rev.* **2007**, *107*, 4111–4132. [CrossRef] [PubMed]
5. Jang, J.W.; Shim, J.H.; Cho, Y.W.; Lee, B.J. Thermodynamic calculation of LiH $\leftrightarrow$ Li$_3$AlH$_6$ $\leftrightarrow$ LiAlH$_4$ reactions. *J. Alloys Compd.* **2006**, *420*, 286–290. [CrossRef]
6. Wu, H. Strategies for the improvement of the hydrogen storage properties of metal hydride materials. *ChemPhysChem* **2008**, *9*, 2157–2162. [CrossRef] [PubMed]
7. Bogdanović, B.; Schwickardi, M. Ti-doped alkali metal aluminum hydrides as potential novel reversible hydrogen storage materials. *J. Alloys Compd.* **1997**, *253–254*, 1–9. [CrossRef]
8. Sandrock, G.; Gross, K.; Thomas, G. Effect of Ti-catalyst content on the reversible hydrogen storage properties of the sodium alanates. *J. Alloys Compd.* **2002**, *339*, 299–308. [CrossRef]
9. Gremaud, R.; Borgschulte, A.; Lohstroh, W.; Schreuders, H.; Züttel, A.; Dam, B.; Griessen, R. Ti-catalyzed Mg(AlH$_4$)$_2$: A reversible hydrogen storage material. *J. Alloys Compd.* **2005**, *404–406*, 775–778. [CrossRef]
10. Easton, D.S.; Schneibel, J.H.; Speakman, S.A. Factors affecting hydrogen release from lithium alanate (LiAlH$_4$). *J. Alloys Compd.* **2005**, *398*, 245–248. [CrossRef]
11. Amama, P.B.; Grant, J.T.; Shamberger, P.J.; Voevodin, A.A.; Fisher, T.S. Improved dehydrogenation properties of Ti-doped LiAlH$_4$: Role of Ti precursors. *J. Phys. Chem. C* **2012**, *116*, 21886–21894. [CrossRef]
12. Isobe, S.; Ikarashi, Y.; Yao, H.; Hino, S.; Wang, Y.; Hashimoto, N.; Ohnuki, S. Additive effects of TiCl$_3$ on dehydrogenation reaction of LiAlH$_4$. *Mater. Trans.* **2014**, *55*, 1138–1140. [CrossRef]
13. Wang, L.; Aguey-Zinsou, K.F. Synthesis of LiAlH$_4$ nanoparticles leading to a single hydrogen release step upon Ti coating. *Inorganics* **2017**, *5*, 38. [CrossRef]
14. Liu, X.; McGrady, G.S.; Langmi, H.W.; Jensen, C.M. Facile cycling of Ti-doped LiAlH$_4$ for high performance hydrogen storage. *J. Am. Chem. Soc.* **2009**, *131*, 5032–5033. [CrossRef]
15. Graetz, J.; Wegrzyn, J.; Reilly, J.J. Regeneration of lithium aluminum hydride. *J. Am. Chem. Soc.* **2008**, *130*, 17790–17794. [CrossRef] [PubMed]
16. Wang, J.; Ebner, A.D.; Ritter, J.A. Physiochemical pathway for cyclic dehydrogenation and rehydrogenation of LiAlH$_4$. *J. Am. Chem. Soc.* **2006**, *128*, 5949–5954. [CrossRef] [PubMed]
17. Neiner, D.; Karkamkar, A.; Linehan, J.C.; Arey, B.; Autrey, T.; Kauzlarich, S.M. Promotion of hydrogen release from ammonia borane with mechanically activated hexagonal boron nitride. *J. Phys. Chem. C* **2009**, *113*, 1098–1103. [CrossRef]

18. Tu, G.; Xiao, X.; Qin, T.; Jiang, Y.; Li, S.; Ge, H.; Chen, L. Significantly improved de/rehydrogenation properties of lithium borohydride modified with hexagonal boron nitride. *RSC Adv.* **2015**, *5*, 51110–51115. [CrossRef]

19. Zhu, J.; Wang, H.; Cai, W.; Liu, J.; Ouyang, L.; Zhu, M. The milled LiBH$_4$/h-BN composites exhibiting unexpected hydrogen storage kinetics and reversibility. *Int. J. Hydrog. Energy* **2017**, *42*, 15790–15798. [CrossRef]

20. Aguey-Zinsou, K.F.; Yao, J.; Guo, Z.X. Reaction paths between LiNH$_2$ and LiH with effects of nitrides. *J. Phys. Chem. B* **2007**, *111*, 12531–12536. [CrossRef] [PubMed]

21. Zhang, T.; Isobe, S.; Matsuo, M.; Orimo, S.; Wang, Y.; Hashimoto, N.; Ohnuki, S. Effect of lithium ion conduction on hydrogen desorption of LiNH$_2$–LiH composite. *ACS Catal.* **2015**, *5*, 1552–1555. [CrossRef]

22. Ares, J.R.; Aguey-Zinou, K.F.; Porcu, M.; Sykes, J.M.; Dornheim, M.; Klassen, T.; Bormann, R. Thermal and mechanically activated decomposition of LiAlH$_4$. *Mater. Res. Bull.* **2008**, *43*, 1263–1275. [CrossRef]

23. D'Anna, V.; Spyratou, A.; Sharma, M.; Hagemann, H. FT-IR spectra of inorganic borohydrides. Spectrochim. *Acta Part A* **2014**, *128*, 902–906. [CrossRef] [PubMed]

24. Miyaoka, H.; Ichikawa, T.; Fujii, H.; Kojima, Y. Hydrogen desorption reaction between hydrogen-containing functional groups and lithium hydride. *J. Phys. Chem. C* **2010**, *114*, 8668–8674. [CrossRef]

25. Kissinger, H.E. Reaction kinetics in differential thermal analysis. *Anal. Chem.* **1957**, *29*, 1702–1706. [CrossRef]

26. Andreasen, A.; Vegge, T.; Pedersen, A.S. Dehydrogenation kinetics of as-received and ball-milled LiAlH$_4$. *J. Solid State Chem.* **2005**, *178*, 3672–3678. [CrossRef]

27. Wang, L.; Rawal, A.; Quadir, M.Z.; Aguey-Zinsou, K.F. Nanoconfined lithium aluminum hydride (LiAlH$_4$) and hydrogen reversibility. *Int. J. Hydrog. Energy* **2017**, *42*, 14144–14153. [CrossRef]

28. Oguchi, H.; Matsuo, M.; Sato, T.; Takaumura, H.; Maekawa, H.; Kuwano, H.; Orimo, S. Lithium-ion conduction in complex hydrides LiAlH$_4$ and Li$_3$AlH$_6$. *J. Appl. Phys.* **2010**, *107*, 096104. [CrossRef]

29. Maekawa, H.; Matsuo, M.; Takamura, H.; Ando, M.; Noda, Y.; Karahashi, T.; Orimo, S. Halide-stabilized LiBH$_4$, a room-temperature lithium fast-ion conductor. *J. Am. Chem. Soc.* **2009**, *131*, 894–895. [CrossRef] [PubMed]

30. Matsuo, M.; Takamura, H.; Maekawa, H.; Li, H.W.; Orimo, S. Stabilization of lithium superionic conduction phase and enhancement of conductivity of LiBH$_4$ by LiCl addition. *Appl. Phys. Lett.* **2009**, *94*, 084103. [CrossRef]

31. Oguchi, H.; Matsuo, M.; Hummelshøj, J.S.; Vegge, T.; Nørskov, J.K.; Sato, T.; Miura, Y.; Takamura, H.; Maekawa, H.; Orimo, S. Experimental and computational studies on structural transitions in the LiBH$_4$–LiI pseudobinary system. *Appl. Phys. Lett.* **2009**, *94*, 141912. [CrossRef]

32. Mosegaard, L.; Møller, B.; Jørgensen, J.E.; Filinchuk, Y.; Cerenius, Y.; Hanson, J.C.; Dimasi, E.; Besenbacher, F.; Jensen, T.R. Reactivity of LiBH$_4$: In situ synchrotron radiation power X-ray diffraction study. *J. Phys. Chem. C* **2008**, *112*, 1299–1303. [CrossRef]

33. Rude, L.H.; Groppo, E.; Arnbjerg, L.M.; Ravnsbæk, D.B.; Malmkjær, R.A.; Filinchuk, Y.; Baricco, M.; Besenbacher, F.; Jensen, T.R. Iodide substitution in lithium borohydride, LiBH$_4$–LiI. *J. Alloys Compd.* **2011**, *509*, 8299–8305. [CrossRef]

34. Langmi, H.W.; McGrady, G.S.; Liu, X.; Jensen, C.M. Modification of the H$_2$ desorption properties of LiAlH$_4$ through doping with Ti. *J. Phys. Chem. C* **2010**, *114*, 10666–10669. [CrossRef]

35. Li, Z.; Li, P.; Wan, Q.; Zhai, F.; Liu, Z.; Zhao, K.; Wang, L.; Lü, S.; Zou, L.; Qu, X.; et al. Dehydrogenation improvement of LiAlH$_4$ catalyzed by Fe$_2$O$_3$ and Co$_2$O$_3$ nanoparticles. *J. Phys. Chem. C* **2013**, *117*, 18343–18352. [CrossRef]

36. Ismail, M.; Zhao, Y.; Yu, X.B.; Dou, S.X. Effects of NbF$_5$ addition on the hydrogen storage properties of LiAlH$_4$. *Int. J. Hydrog. Energy* **2010**, *35*, 2361–2367. [CrossRef]

37. Atakli, Z.Ö.K.; Callini, E.; Kato, S.; Mauron, P.; Orimo, S.; Züttel, A. The catalyzed hydrogen sorption mechanism in alkali alanates. *Phys. Chem. Chem. Phys.* **2015**, *17*, 20932. [CrossRef] [PubMed]

38. Hoang, K.; Janotti, A; Van de Walle, C.G. Decomposition mechanism and the effects of metal additives on the kinetics of lithium alanate. *Phys. Chem. Chem. Phys.* **2012**, *14*, 2840–2848. [CrossRef] [PubMed]

39. Weng, Q.; Wang, X.; Wang, X.; Bando, Y.; Golberg, D. Functional hexagonal boron nitride nanomaterials: Emerging properties and applications. *Chem. Soc. Rev.* **2016**, *45*, 3989–4012. [CrossRef] [PubMed]

40. Lei, W.; Mochalin, V.N.; Liu, D.; Qin, S.; Gogotsi, Y.; Chen, Y. Boron nitride colloidal solutions, ultralight aerogels and freestanding membranes through one-step exfoliation and functionalization. *Nat. Commun.* **2015**, *6*, 8849. [CrossRef] [PubMed]

41. Liu, D.; He, L.; Lei, W.; Klika, K.D.; Kong, L.; Chen, Y. Multifunctional polymer/porous boron nitride nanosheet membranes for superior trapping emulsified oils and organic molecules. *Adv. Mater. Interfaces* **2015**, *2*, 1500228. [CrossRef]

42. Hu, S.; Lozada-Hidalgo, M.; Wang, F.C.; Mishchenko, A.; Schedin, F.; Nair, R.R.; Hill, E.W.; Boukhvalov, D.W.; Katsnelson, M.I.; Dryfe, R.A.W.; et al. Proton transport through one-atom-thick crystals. *Nature* **2014**, *516*, 227–230. [CrossRef] [PubMed]

43. Lei, W.; Zhang, H.; Wu, Y.; Zhang, B.; Liu, D.; Qin, S.; Liu, Z.; Liu, L.; Ma, Y.; Chen, Y. Oxygen-doped boron nitride nanosheets with excellent performance in hydrogen storage. *Nano Energy* **2014**, *6*, 219–224. [CrossRef]

13

Si–H Bond Activation of a Primary Silane with a Pt(0) Complex: Synthesis and Structures of Mononuclear (Hydrido)(dihydrosilyl) Platinum(II) Complexes

Norio Nakata * (iD), Nanami Kato, Noriko Sekizawa and Akihiko Ishii *

Department of Chemistry, Graduate School of Science and Engineering, Saitama University, 255 Shimo-okubo, Sakura-ku, Saitama 338-8570, Japan; rdhns681@yahoo.co.jp (N.K.); jjhs61ns@yahoo.co.jp (N.S.)
* Correspondence: nakata@chem.saitama-u.ac.jp (N.N.); ishiiaki@chem.saitama-u.ac.jp (A.I.)

Abstract: A hydrido platinum(II) complex with a dihydrosilyl ligand, [*cis*-PtH(SiH$_2$Trip)(PPh$_3$)$_2$] (**2**) was prepared by oxidative addition of an overcrowded primary silane, TripSiH$_3$ (**1**, Trip = 9-triptycyl) with [Pt(η^2-C$_2$H$_4$)(PPh$_3$)$_2$] in toluene. The ligand-exchange reactions of complex **2** with free phosphine ligands resulted in the formation of a series of (hydrido)(dihydrosilyl) complexes (**3–5**). Thus, the replacement of two PPh$_3$ ligands in **2** with a bidentate bis(phosphine) ligand such as DPPF [1,2-bis(diphenylphosphino)ferrocene] or DCPE [1,2-bis(dicyclohexylphosphino)ethane] gave the corresponding complexes [PtH(SiH$_2$Trip)(L-L)] (**3**: L-L = dppf, **4**: L-L = dcpe). In contrast, the ligand-exchange reaction of **2** with an excess amount of PMe$_3$ in toluene quantitatively produced [PtH(SiH$_2$Trip)(PMe$_3$)(PPh$_3$)] (**5**), where the PMe$_3$ ligand is adopting *trans* to the hydrido ligand. The structures of complexes **2–5** were fully determined on the basis of their NMR and IR spectra, and elemental analyses. Moreover, the low-temperature X-ray crystallography of **2**, **3**, and **5** revealed that the platinum center has a distorted square planar environment, which is probably due to the steric requirement of the *cis*-coordinated phosphine ligands and the bulky 9-triptycyl group on the silicon atom.

Keywords: platinum; primary silane; hydrido complex; oxidative addition; ligand-exchange reaction; X-ray crystallography

---

1. Introduction

The transition metal catalyzed synthesis of functionalized organosilicon compounds gained substantial momentum during the past few decades [1]. Among these catalytic conversions, the oxidative addition of hydrosilanes with platinum(0) complexes is an efficient method for the generation of the platinum(II) hydride species, which has been proposed as a key intermediate in platinum-catalyzed hydrosilylations [2–6] and bis-silylations [7,8], as well as the dehydrogenative couplings of hydrosilanes [9–13]. While a number of reactions of hydrosilanes with platinum(0) complexes affording mononuclear bis(silyl) [14–18] and silyl-bridged multinuclear complexes [19–29] have been described so far, the isolation of mononuclear hydrido(silyl) complexes has been less well studied due to the high reactivity of a Pt–H bond [30–34]. In particular, the synthesis of hydrido(dihydrosilyl) platinum(II) complexes, which are anticipated as the initial products in the Si–H bond activation reactions of primary silanes with platinum(0) complexes, is quite rare. Indeed, only two publications have previously reported the characterization of hydrido(dihydrosilyl) platinum(II) complexes. In 2000, Tessier et al. reported that the reaction of a primary silane with a bulky *m*-terphenyl group with [Pt(PPr$_3$)$_3$] produced the first example of a stable hydrido(dihydrosilyl) complex [*cis*-PtH(SiH$_2$Ar)(PPr$_3$)$_2$] (Ar = 2,6-MesC$_6$H$_3$) [35]. Quite recently, Lai et al. also described the synthesis of a bis(phosphine) hydrido(dihydrosilyl) complex [PtH(SiH$_2$Si^{t}Bu$_2$Me)(dcpe)] (dcpe = 1,2-bis(dicyclohexylphosphino)ethane) containing a Si–Si bond [36]. Meanwhile, we succeeded in the first isolation of a series of hydrido(dihydrogermyl) platinum(II) complexes [PtH(GeH$_2$Trip)(L)] (Trip = 9-triptycyl) using a bulky substituent, 9-triptycyl group [37]. In addition, we reported the first syntheses and structural characterizations of hydrido palladium(II) complexes with a dihydrosilyl- or dihydrogermyl ligand, [PdH(EH$_2$Trip)(dcpe)] (E = Si, Ge) [38]. Very recently, we also found that hydride-abstraction reactions of [MH(EH$_2$Trip)(dcpe)] (M = Pt, Pd, E = Si, Ge) with B(C$_6$F$_5$)$_3$ led to the formations of new cationic dinuclear complexes with bridging hydrogermylene and hydrido ligands, [{M(dcpe)}$_2$(μ-GeHTrip)(μ-H)]$^+$ [39]. As an extension of our previous work and taking into account the interest devoted to hydrido platinum(II) complexes, we present here the synthesis and characterization of a series of mononuclear (hydrido)(dihydrosilyl) complexes [PtH(SiH$_2$Trip)(L)$_2$].

2. Results

2.1. Synthesis and Characterization of [cis-PtH(SiH$_2$Trip)(PPh$_3$)$_2$] (2)

The reaction of TripSiH$_3$ **1** with [Pt(η^2-C$_2$H$_4$)(PPh$_3$)$_2$] in toluene proceeded efficiently at room temperature under inert atmosphere to form the corresponding complex [cis-PtH(SiH$_2$Trip)(PPh$_3$)$_2$] (**2**) in 91% yield as colorless crystals (Scheme 1). In the ^{1}H NMR spectrum of **2**, the characteristic signals of the platinum hydride were observed at δ = −2.15, which were split by 19 and 157 Hz of ^{31}P–^{1}H couplings accompanying 958 Hz of satellite signals from the ^{195}Pt isotope. This chemical shift is comparable with those of the related (hydrido)(dihydrosilyl) complexes, [cis-PtH(SiH$_2$Ar)(PPr$_3$)$_2$] (Ar = 2,6-MesC$_6$H$_3$) (δ = −3.40) [35] and [cis-PtH(SiH$_2$Si^{t}Bu$_2$Me)(dcpe)] (δ = −0.89) [36]. The SiH$_2$ resonance appeared as a multiplet at δ = 4.68 ppm. The ^{31}P{^{1}H} NMR spectrum of **2** exhibited two doublets ($^2J_{P–P}$ = 15 Hz) at δ = 33.8 and 34.5 with ^{195}Pt–^{31}P coupling constants, 2183 and 1963 Hz, which were assigned to the phosphorus atoms lying *trans* to the hydrido and dihydrosilyl ligands, respectively, in agreement with the NMR data for reported germanium congener [cis-PtH(GeH$_2$Trip)(PPh$_3$)$_2$] [δ = 31.2 ($^1J_{Pt–P}$ = 2317 Hz) and 31.6 ($^1J_{Pt–P}$ = 2252 Hz)] [37]. The silicon atom of **2** gave rise to a resonance around δ = −40.6 with splitting due to ^{31}P–^{29}Si couplings ($^2J_{P(trans)–Si}$ = 161, $^2J_{P(cis)–Si}$ = 15 Hz) and ^{195}Pt satellites ($^1J_{Pt–Si}$ = 1220 Hz) in the ^{29}Si{^{1}H} NMR spectrum. In the solid state IR spectrum for **2**, Pt–H and Si–H stretching vibrations were observed at 2041 and 2080 cm^{-1}, respectively. Complex **2** is thermally and air stable in the solid state (melting point: 123 °C (dec.)) or in solution, and no dimerization or dissociation of phosphine ligands was observed.

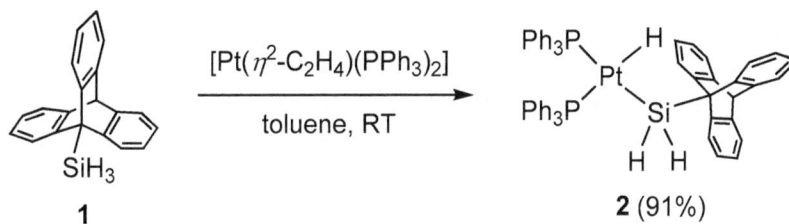

Scheme 1. Synthesis of [cis-PtH(SiH$_2$Trip)(PPh$_3$)$_2$] **2**.

The molecular structure of **2** was determined unambiguously by X-ray crystallographic analysis, as depicted in Figure 1. The X-ray crystallographic analysis of **2** revealed that the platinum center attains a distorted square-planar environment, which was probably due to the steric requirement of the *cis*-coordinated PPh$_3$ ligands and the bulky 9-triptycyl group on the silicon atom. The P1–Pt1–P2 angle of 101.63(3)° and P1–Pt1–Si1 angle of 96.17(3)° deviated considerably from the ideal 90° of square-planar geometry. The Pt–Si bond length is 2.3458(9) Å, which is comparable to those ranging from 2.321 to 2.406 Å observed in the related mononuclear platinum(II) complexes bearing silyl ligands [1]. The hydrogen atom on the platinum atom was located in the electron density map and has a Pt–H distance of 1.59(4) Å. The Pt1–P1 bond length [2.2945(8) Å] is slightly shorter than the Pt1–P2 bond length [2.3401(8) Å], which indicates the stronger *trans* influence of the silicon atom compared with that of the hydride in this complex. This result is consistent with the ^{195}Pt–^{31}P coupling constants (2183 and 1963 Hz) observed in the $^{31}P\{^1H\}$ NMR spectrum.

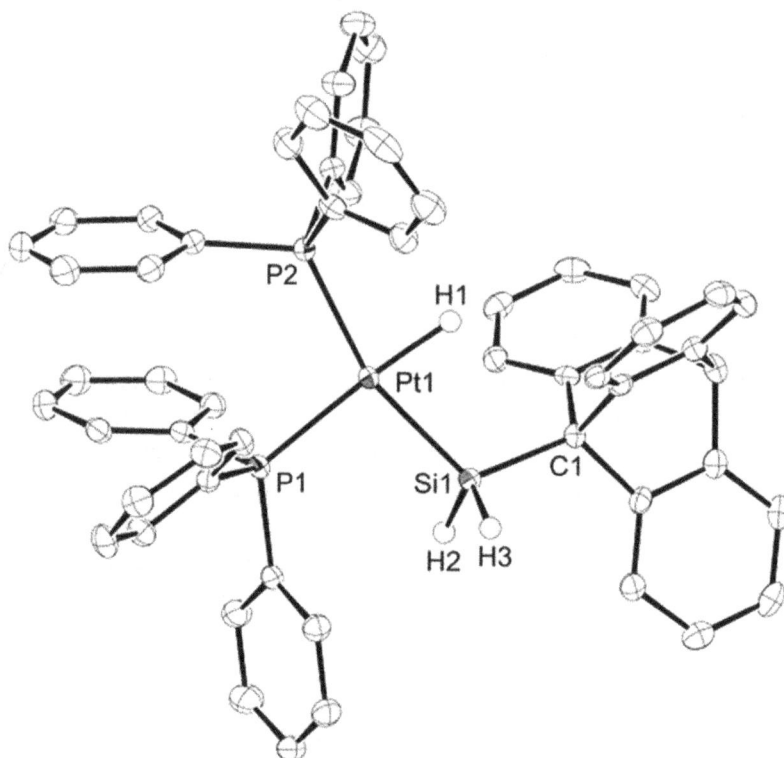

Figure 1. ORTEP of [*cis*-PtH(SiH$_2$Trip)(PPh$_3$)$_2$] **2** (50% thermal ellipsoids, a solvation toluene molecule, and hydrogen atoms, except H1, H2, and H3 were omitted for clarity). Selected bond lengths (Å) and bond angles (°): Pt1–Si1 = 2.3458(9), Pt1–P1 = 2.2945(8), Pt1–P2 = 2.3401(8), Pt1–H1 = 1.59(4), Si1–C1 = 1.918(3), P1–Pt1–P2 = 101.63(3), Si1–Pt1–P1 = 96.17(3), Si1–Pt1–H1 = 79.7(16), P2–Pt1–H1 = 82.5(16), Si1–Pt1–P2 = 162.13(3), P1–Pt1–H1 = 175.8(16).

2.2. *Ligand Exchange Reactions of* **2** *with Free Phosphine Ligands*

We next examined the ligand-exchange reactions of complex **2** with free phosphine ligands. The replacement of two PPh$_3$ ligands in **2** with a bidentate bis(phosphine) ligand such as DPPF (1,2-bis(diphenylphosphino)ferrocene) or DCPE gave the corresponding complexes [PtH(SiH$_2$Trip)(L-L)] (**3**: L-L = dppf, **4**: L-L = dcpe) in 87% and 80% yields, respectively (Scheme 2). In the ^1H NMR spectra of **3** and **4**, the hydride resonated as a doublet of doublets at δ = –1.62 ($^2J_{P-H}$ = 20, 164, $^1J_{Pt-H}$ = 995 Hz) for **3** and –0.46 ($^2J_{P-H}$ = 13, 166, $^1J_{Pt-H}$ = 1004 Hz) for **4**. These chemical shifts are shifted downfield in comparison with that of the starting complex **2** (δ = −2.15), which is probably due to the stronger electron-donating ability of chelating phosphines compared with PPh$_3$. The spectrum for **3** also displayed a multiplet signal centering at δ = 4.61 corresponding to the SiH$_2$ protons, which is shifted upfield by 0.89 ppm in comparison with that of **4**. The $^{31}P\{^1H\}$ NMR spectrum of **3** showed two doublets ($^2J_{P-P}$ = 21 Hz) with ^{195}Pt satellites at δ = 30.5 ($^1J_{Pt-P}$ = 2247 Hz) and 34.5 ($^1J_{Pt-P}$ = 1837 Hz), which are close to those of **2** [δ = 33.8 ($^1J_{Pt-P}$ = 2183 Hz) and 34.5 ($^1J_{Pt-P}$ = 1963 Hz)]. The observation of P-P coupling indicates a large deviation of the P–Pt–P angle from 90° of the ideal square planar geometry around the Pt(II) center (vide infra). In contrast, the $^{31}P\{^1H\}$ resonances for **4** were observed as two singlets at δ = 69.2 ($^1J_{Pt-P}$ = 1809 Hz) and 85.3 ($^1J_{Pt-P}$ = 1678 Hz), respectively, which are relatively shifted downfield relative to those of **2** and **3**. The larger $^1J_{Pt-P}$ values

(2183 Hz for **3**, 1809 Hz for **4**) are assigned to the phosphorus atom *trans* to the hydrido ligand, as in the case of **2**. Furthermore, the $^{29}Si\{^1H\}$ NMR spectra of **3** and **4** showed a doublet of doublets signals at $\delta = -39.0$ ($^2J_{Si-P} = 167, 12$ Hz) for **3**, and -44.6 ($^2J_{Si-P} = 173, 11$ Hz) for **4**, which were accompanied by ^{195}Pt satellites of 1207 Hz for **3** and 1253 Hz for **4**, respectively.

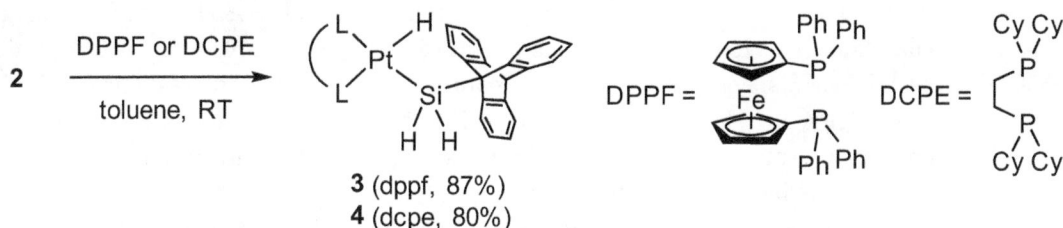

Scheme 2. Ligand-exchange reaction of [*cis*-PtH(SiH$_2$Trip)(PPh$_3$)$_2$] **2** with chelating bis(phosphine)s.

The molecular structure of DPPF-derivative **3** in the crystalline state was confirmed by X-ray crystallography (Figure 2). The platinum atom lies in a distorted square-planar geometry; the sum of the bond angles around the platinum atom is 360.48°. The P1–Pt1–P2 angle is 102.29(9)°, and other angles around the platinum atom are less than 90°, except the P1–Pt1–Si1 angle [95.19(9)°]. The Pt–Si [2.331(3) Å] and two Pt–P bond lengths [2.286(2), 2.319(2) Å] are comparable to those of the corresponding DPPF-ligated hydrido complex [PtH(SiHPh$_2$)(dppf)] [2.3366(4), 2.2830(4), and 2.3192(4) Å, respectively] [40].

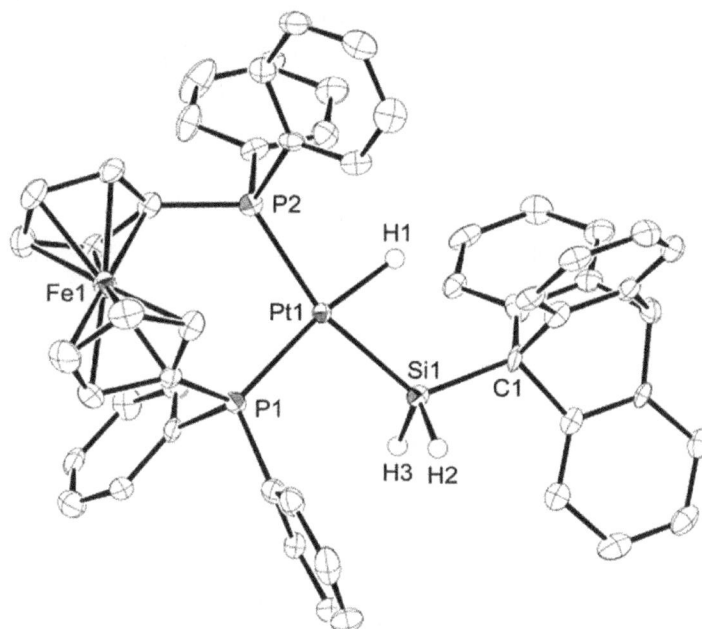

Figure 2. ORTEP of [PtH(SiH$_2$Trip)(dppf)] **3** 50% thermal ellipsoids, a solvation CH$_2$Cl$_2$ molecule, and hydrogen atoms, except H1, H2, and H3 were omitted for clarity. Selected bond lengths (Å) and bond angles (°): Pt1–Si1 = 2.331(3), Pt1–P1 = 2.286(2), Pt1–P2 = 2.319(2), Pt1–H1 = 1.687(10), Si1–C1 = 1.920(10), P1–Pt1–P2 = 102.29(9), Si1–Pt1–P1 = 95.19(9), Si1–Pt1–H1 = 76(5), P2–Pt1–H1 = 87(5), Si1–Pt1–P2 = 162.10(9), P1–Pt1–H1 = 171(5).

It is well known that trimethylphosphine (PMe$_3$) is a strong σ-donating ligand for a wide variety of transition-metal complexes. Therefore, one can reasonably expect the formation of [PtH(SiH$_2$Trip)(PMe$_3$)$_2$] in a similar ligand-exchange reaction of **2** with PMe$_3$. However, we found that the reaction of **2** with 2.2 equivalents of PMe$_3$ in toluene at room temperature did not proceed completely, which resulted in the formation of [PtH(SiH$_2$Trip)(PMe$_3$)(PPh$_3$)] (**5**), where the PPh$_3$ ligand *trans* to hydrido in **2** is exchanged with a PMe$_3$ ligand (Scheme 3). Complex **5** was isolated as colorless crystals in quantitative yield after workup. In the ^{1}H NMR spectrum of **5** at 298 K, the characteristic broad doublet signal due to the platinum hydride was observed centering at $\delta = -1.91$ with splitting by ^{31}P–^{1}H ($^2J_{P(trans)-H} = 157$ Hz) and ^{195}Pt–^{1}H ($^1J_{Pt-H} = 899$ Hz) couplings. The SiH$_2$ protons also appeared as a broad multiplet at $\delta = 5.95$, which is shifted downfield relative to those of the above complexes **2–4** ($\delta = 4.68$–5.50). The ^{31}P{^{1}H} NMR spectrum of **5** at 298 K exhibited two nonequivalent broad singlet signals at $\delta = -22.9$ and 37.4, with two sets of ^{195}Pt satellites of 2115 and 1833 Hz. The former signal was assigned to the PMe$_3$ *trans* to the hydrido ligand using a non ^{1}H-decoupled ^{31}P NMR technique at 253 K. The ^{29}Si{^{1}H} NMR spectrum of **5** at 223 K featured one ^{29}Si resonance as a broad doublet signal at $\delta = -43.9$ ($^2J_{Si-P} = 151$ Hz). The broadening of NMR signals possibly implies the existence of the Si–H σ-complex intermediate **6** in the NMR time scale [38,41,42]. Unfortunately, attempts to probe further the fluxional behavior of **5** by VT (variable temperature)-NMR in solution revealed no appreciable changes in spectroscopic features by ^{1}H and ^{31}P NMR spectroscopy. The molecular structure of **5** is also determined by X-ray analysis, as shown in Figure 3. Distortions from square planar geometry at the platinum center were observed, similar to the cases of **2** and **4**. The Pt–Si bond length of **5** [2.3414(13) Å] is almost equal to those of **2** [2.3458(9) Å] and **4** [2.331(3) Å]. As expected, the Pt1–P1 bond length for the PMe$_3$ ligand of **5** [2.2978(12) Å] is shortened compared with the Pt1–P2 bond length for the PPh$_3$ ligand of **5** [2.3203(11) Å] due to the different ligands in the *trans* positions of the phosphorus atoms. While only a few cationic platinum complexes containing different phosphine ligands have been reported [43–45], complex **5** is the first example of a neutral platinum complex bearing a weakly electron-donating PPh$_3$ and strongly electron-donating PMe$_3$.

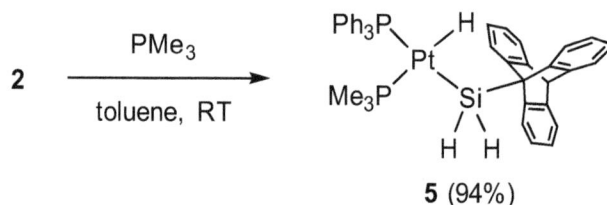

Scheme 3. Ligand-exchange reaction of [*cis*-PtH(SiH$_2$Trip)(PPh$_3$)$_2$] **2** with trimethylphosphine (PMe$_3$).

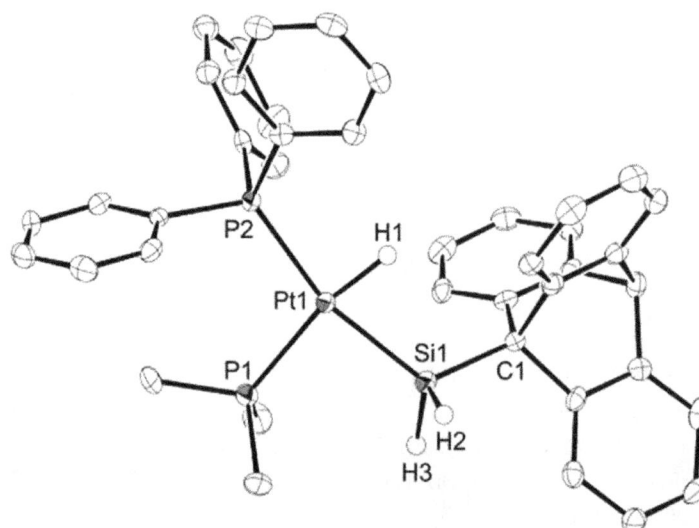

Figure 3. ORTEP of [PtH(SiH$_2$Trip)(PMe$_3$)(PPh$_3$)] **5** 50% thermal ellipsoids, a solvation CH$_2$Cl$_2$ molecule, and hydrogen atoms, except H1, H2, and H3 were omitted for clarity. Selected bond lengths (Å) and bond angles (°): Pt1–Si1 = 2.3414(13), Pt1–P1 = 2.2978(12), Pt1–P2 = 2.3203(11), Pt1–H1 = 1.44(6), Si1–C1 = 1.930(5), P1–Pt1–P2 = 103.47(4), Si1–Pt1–P1 = 92.34(4), Si1–Pt1–H1 = 75(2), P2–Pt1–H1 = 89(2), Si1–Pt1–P2 = 164.19(4), P1–Pt1–H1 = 167(2).

A plausible formation mechanism for **5** is shown in Scheme 4. According to the stronger *trans* influence of the silyl ligand than that of the hydrido ligand, the ligand-exchange of a PPh$_3$ *trans* to silyl ligand takes place to yield the intermediate **5'** in the first step, while **5** might be formed directly from the corresponding coordinatively unsaturated intermediate (3-coordinated 14-electron complexes) [46,47]. Then, the intramolecular interchange of coordination environments between the silyl and hydrido ligands through the Si–H σ-complex intermediate **6** would occur, probably due to the steric repulsion between the bulky 9-triptycyl group on the silicon atom and the *cis*-PPh$_3$ ligand in **5'**. Finally, the corresponding complex **5** was obtained as the thermodynamic product. As another pathway, it is likely that the direct formation of **5** is caused by an initial dissociation of the PPh$_3$ ligand at the *trans* position of the hydrido ligand in **2** due to steric reason.

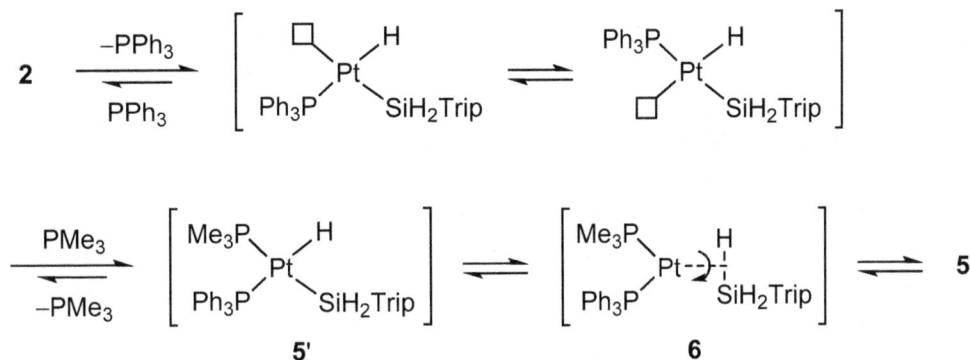

Scheme 4. Plausible reaction pathway for the formation of [PtH(SiH$_2$Trip)(PMe$_3$)(PPh$_3$)] **5**.

3. Materials and Methods

3.1. General Procedures

All of the experiments were performed under an argon atmosphere unless otherwise noted. Solvents were dried by standard methods and freshly distilled prior to use. ^{1}H, ^{13}C, and ^{31}P NMR spectra were recorded on Bruker DPX-400 or DRX-400 (400, 101 and 162 MHz, respectively), Avance-500 (500, 126 and 202 MHz, respectively) (Karlsruhe, Germany) spectrometers using CDCl$_3$ or C$_6$D$_6$ as the solvent at room temperature. ^{29}Si NMR spectra were recorded on Bruker Avance-500 (Karlsruhe, Germany) or JEOL EX-400 (Tokyo, Japan) (99.4 and 79.3 MHz, respectively) spectrometers using CDCl$_3$, CD$_2$Cl$_2$, C$_6$D$_6$, or THF-d_8 as the solvent at room temperature, unless otherwise noted. IR spectra were obtained on a Perkin-Elmer System 2000 FT-IR spectrometer (Walham, MA, USA). Elemental analyses were carried out at the Molecular Analysis and Life Science Center of Saitama University. All of the melting points were determined on a Mel-Temp capillary tube apparatus (Stafford, UK) and are uncorrected. 9-Triptycylsilane (TripSiH$_3$, **1**) [48] and [Pt(η^2-C$_2$H$_4$)(PPh$_3$)$_2$] [49] were prepared according to the reported procedures.

3.1.1. [cis-PtH(SiH$_2$Trip)(PPh$_3$)$_2$] (**2**)

A solution of TripSiH$_3$ **1** (50.9 mg, 0.179 mmol) and [Pt(η^2-C$_2$H$_4$)(PPh$_3$)$_2$] (149.3 mg, 0.199 mmol) in toluene (3 mL) was stirred at room temperature for 1 h to form a pale yellow solution. After the removal of the solvent in vacuo, the residual colorless solid was purified by washing with hexane to give [cis-PtH(SiH$_2$Trip)(PPh$_3$)$_2$] (**2**) (164.3 mg, 91%) as colorless crystals.

^{1}H NMR (400 MHz, CDCl$_3$): δ = −2.15 (dd, $^2J_{H–P(trans)}$ = 157, $^2J_{H–P(cis)}$ = 19, $^1J_{H–Pt}$ = 958 Hz, 1H, PtH), 4.68 (m, 2H, SiH$_2$), 5.28 (s, 1H, TripCH), 6.83–6.89 (m, 6H, Ar), 7.02–7.06 (m, 6H, Ar), 7.15–7.31 (m, 21H, Ar), 7.49 (t, J = 7 Hz, 6H, Ar), 7.81 (d, J = 7 Hz, 3H, Ar). ^{13}C{^{1}H} NMR (101 MHz, CDCl$_3$): δ = 46.6 (TripC), 55.3 (TripCH), 122.7 (Ar(CH)), 123.7 (Ar(CH) × 2), 126.2 (Ar(CH)), 127.8 (Ar(CH)), 127.9 (Ar(CH)), 129.4 (Ar(CH)), 129.6 (Ar(CH)), 134.1 (Ar(CH)), 134.2 (Ar(CH)),

133.7–134.8 (m, Ar(C)), 135.5 (d, $^1J_{C-P}$ = 38 Hz, Ar(C)), 148.4 (Ar(C)), 149.5 (Ar(C)). $^{31}P\{^1H\}$ NMR (162.0 MHz, CDCl$_3$): δ = 33.8 (d, $^2J_{P-P}$ = 15, $^1J_{P-Pt}$ = 2183 Hz), 34.5 (d, $^2J_{P-P}$ = 15, $^1J_{P-Pt}$ = 1963 Hz). $^{29}Si\{^1H\}$ NMR (79.3 MHz, CD$_2$Cl$_2$): δ = −40.6 (dd, $^2J_{Si-P(trans)}$ = 161, $^2J_{Si-P(cis)}$ = 15, $^1J_{Si-Pt}$ = 1220 Hz). IR (KBr, cm^{-1}): ν = 2041 (Pt–H), 2080 (Si–H). Anal. Calcd. for C$_{56}$H$_{46}$P$_2$PtSi: C, 66.99; H, 4.62. Found: C, 66.55; H, 4.61. Melting point: 123 °C (dec.).

3.1.2. [PtH(SiH$_2$Trip)(dppf)] (3)

A solution of **2** (40.5 mg, 0.040 mmol) and DPPF (28.2 mg, 0.048 mmol) in toluene (3 mL) was stirred at room temperature for 5 h. After the removal of the solvent in vacuo, the residual colorless solid was purified by washing with Et$_2$O and hexane to give [PtH(SiH$_2$Trip)(dppf)] (**3**) (37.1 mg, 0.035 mmol, 87%) as yellow crystals.

^{1}H NMR (400 MHz, CDCl$_3$): δ = −1.62 (dd, $^2J_{H-P(trans)}$ = 164, $^2J_{H-P(cis)}$ = 20, $^1J_{H-Pt}$ = 995 Hz, 1H, PtH), 3.86 (s, 2H, Cp), 4.20 (s, 2H, Cp), 4.37 (s, 2H, Cp), 4.58–4.64 (m, 4H, Cp and SiH$_2$), 5.30 (s, 1H, TripCH), 6.87–6.89 (m, 6H, Ar), 7.26 (m, 6H, Ar), 7.41 (br, 6 H, Ar), 7.66–7.81 (m, 14H, Ar). $^{13}C\{^1H\}$ NMR (101 Hz, CDCl$_3$): δ = 46.5 (m, TripC), 55.2 (TripCH), 71.3 (d, $^3J_{C-P}$ = 5 Hz, Cp(CH)), 72.1 (d, $^3J_{C-P}$ = 6 Hz, Cp(CH)), 74.5 (d, $^2J_{C-P}$ = 7 Hz, Cp(CH)), 75.6 (d, $^2J_{C-P}$ = 8 Hz, Cp(CH)), 79.3 (dd, $^1J_{C-P}$ = 42, $^3J_{C-P}$ = 5 Hz, Cp(C)), 80.7 (dd, $^1J_{C-P}$ = 47, $^3J_{C-P}$ = 6 Hz, Cp(C)), 122.7 (Ar(CH)), 123.71 (Ar(CH)), 123.68 (Ar(CH)), 126.4 (Ar(CH)), 127.7 (d, $^3J_{C-P}$ = 11 Hz, Ar(CH)), 128.0 (d, $^3J_{C-P}$ = 10 Hz, Ar(CH)), 129.9 (Ar(CH)), 130.1 (Ar(CH)), 134.3 (d, $^2J_{C-P}$ = 13 Hz, Ar(CH)), 134.6 (d, $^2J_{C-P}$ = 14 Hz, Ar(CH)), 134.8–135.5 (m, Ar(C)), 136.2 (d, $^1J_{C-P}$ = 43 Hz, Ar(C)), 148.4 (Ar(C)), 149.7 (Ar(C)). $^{31}P\{^1H\}$ NMR (202 MHz, CDCl$_3$): δ = 30.5 (d, $^2J_{P-P}$ = 21 Hz, $^1J_{Pt-P}$ = 2247 Hz), 34.5 (d, $^2J_{P-P}$ = 21, $^1J_{Pt-P}$ = 1837 Hz). $^{29}Si\{^1H\}$ NMR (79.3 MHz, CDCl$_3$): δ = −39.0 (dd, $^2J_{Si-P(trans)}$ = 167, $^2J_{Si-P(cis)}$ = 12, $^1J_{Si-Pt}$ = 1207 Hz). IR (KBr, cm^{-1}): ν = 2055 (Pt–H), 2081 (Si–H). Anal. Calcd. for C$_{54}$H$_{44}$FeP$_2$PtSi: C, 62.73; H, 4.29. Found: C, 62.70; H, 4.27. Melting point: 183 °C (dec.).

3.1.3. [PtH(SiH$_2$Trip)(dcpe)] (4)

A solution of **2** (35.6 mg, 0.035 mmol) and DCPE (18.8 mg, 0.044 mmol) in toluene (3 mL) was stirred at room temperature for 5 h. After the removal of the solvent in vacuo, the residual colorless solid was purified by washing with Et$_2$O and hexane to give [PtH(SiH$_2$Trip)(dcpe)] (**4**) (25.5 mg, 0.028 mmol, 80%) as colorless crystals.

^{1}H NMR (500 MHz, CDCl$_3$): δ = −0.45 (dd, $^2J_{H-P(trans)}$ = 165, $^2J_{H-P(cis)}$ = 13, $^1J_{Pt-H}$ = 1004 Hz, 1H, PtH), 1.16–1.51 (m, 20H, Cy), 1.65–1.86 (m, 24H, Cy), 2.16–2.19 (m, 2H, Cy), 2.30–2.37 (m, 2H, Cy), 5.32 (s, 1H, TripCH), 5.50 (dd, $^2J_{H-H}$ = 15, $^3J_{H-P(trans)}$ = 6, $^2J_{Pt-H}$ = 31 Hz, 2H, SiH$_2$), 6.84–6.90 (m, 6H, Ar), 7.30–7.32 (d, J = 7 Hz, 3H, Ar), 7.94–7.96 (d, J = 7 Hz, 3H, Ar). $^{13}C\{^1H\}$ NMR (101 Hz, CDCl$_3$): δ = 23.2 (dd, $^3J_{C-P}$ = 21, 16 Hz, PCH$_2$), 26.2 (dd, $^3J_{C-P}$ = 23, 21 Hz, PCH$_2$), 26.9 (d, $^3J_{C-P}$ = 14 Hz, PCy(CH$_2$)), 26.4 (d, $^3J_{C-P}$ = 13 Hz, PCy(CH$_2$)), 26.8 (d, $^3J_{C-P}$ = 10 Hz, PCy(CH$_2$)), 27.0 (d, $^3J_{C-P}$ = 12 Hz, PCy(CH$_2$)), 28.9 (m, PCy(CH$_2$) × 2), 29.7 (m, PCy(CH$_2$)), 35.3–35.8 (m, PCy(CH) × 2), 46.8 (TripC), 55.4 (TripCH), 122.7 (Ar(CH)), 123.6 (Ar(CH)), 126.9 (Ar(CH)), 148.6 (Ar(C)), 159.2 (Ar(C)). $^{31}P\{^1H\}$ NMR (162 Hz, CDCl$_3$): δ = 69.2 (s, $^1J_{Pt-P}$ = 1809 Hz), 85.3 (s, $^1J_{Pt-P}$ = 1678 Hz). $^{29}Si\{^1H\}$ NMR (79.3 MHz, CD$_2$Cl$_2$): δ = −44.6 (dd, $^2J_{Si-P(trans)}$ = 173, $^2J_{Si-P(cis)}$ = 11, $^1J_{Si-Pt}$ = 1253 Hz). IR (KBr, cm^{-1}): ν = 2057 (Pt–H), 2081 (Si–H). Anal. Calcd for C$_{46}$H$_{64}$P$_2$PtSi: C, 61.24; H, 7.15. Found: C, 61.10; H, 7.10. Melting point: 134 °C (dec.).

3.1.4. [PtH(SiH$_2$Trip)(PMe$_3$)(PPh$_3$)] (5)

A toluene solution of PMe$_3$ (1.0 M, 0.2 mL, 0.200 mmol) was added to a solution of **2** (98.0 mg, 0.098 mmol) in toluene (3.5 mL) at room temperature. The reaction mixture was stirred at room temperature for 30 min. After the removal of the solvent in vacuo, the residual colorless solid was purified by washing with Et$_2$O and hexane to give [PtH(SiH$_2$Trip)(PMe$_3$)(PPh$_3$)] **5** (75.6 mg, 94%) as colorless crystals.

^{1}H NMR (400 MHz, C_6D_6): $\delta = -1.91$ (d, $^2J_{H-P(trans)} = 157$, $^1J_{H-Pt} = 899$ Hz, 1H, PtH), 1.03–1.11 (m, 9H, PMe), 5.34 (s, 1H, TripCH), 5.95–5.97 (m, 2H, SiH_2), 6.82–6.97 (m, 15H, Ar), 7.31 (d, $J = 7$ Hz, 3H, Ar), 7.57 (br, 6H, Ar), 8.50 (d, $J = 7$ Hz, 3H, Ar). ^{13}C{^{1}H}-NMR (101 Hz, $CDCl_3$): $\delta = 16.4$–17.1 (m, PMe), 53.6 (TripC), 55.4 (TripCH), 123.0 (Ar(CH)), 123.9 (Ar(CH)), 124.1 (Ar(CH)), 126.5 (Ar(CH)), 128.5 (d, $^3J_{C-P} = 10$ Hz, Ar(CH)), 130.0 (Ar(CH)), 134.4 (d, $^2J_{C-P} = 13$ Hz, Ar(CH)), 135.6 (d, $^1J_{C-P} = 36$ Hz, Ar(CH), 148.5 (Ar(C)), 149.7 (Ar(C)). ^{31}P{^{1}H}-NMR (162 Hz, C_6D_6): $\delta = -22.9$ (s, $^1J_{Pt-P} = 2115$ Hz), 37.4 (br, $^1J_{Pt-P} = 1833$ Hz). ^{29}Si{^{1}H}-NMR (79.3 MHz, THF-d_8, 223 K) δ -43.9 (d, $^2J_{Si-P(trans)} = 151$ Hz). IR (KBr, cm^{-1}) $\nu = 2029$ (Pt–H), 2054 (Si–H). Anal. Calcd for $C_{41}H_{40}P_2PtSi$: C, 60.21; H, 4.93. Found: C, 60.57; H, 5.00. Melting point: 115 °C (dec.).

3.2. X-ray Crystallographic Studies of 2, 3, and 5

Colorless single crystals of **2** were grown by the slow evaporation of its saturated toluene solution, and single crystals of **3** and **5** were grown by the slow evaporation of its saturated CH_2Cl_2 and hexane solution. The intensity data were collected at 103 K on a Bruker AXS SMART diffractometer (Karlsruhe, Germany) employing graphite-monochromatized Mo Kα radiation ($\lambda = 0.71073$ Å). The structures were solved by direct methods and refined by full-matrix least-squares procedures on F^2 for all reflections (*SHELX*-97) [50]. Hydrogen atoms, except for the PtH and SiH hydrogens of **2**, **3**, and **5**, were located by assuming ideal geometry, and were included in the structure calculations without further refinement of the parameters. Full details of the crystallographic analysis and accompanying cif files (see Supplementary Materials) may be obtained free of charge from the Cambridge Crystallographic Data Centre (CCDC numbers 1577277, 1577278, and 1577279) via http://www.ccdc.cam.ac.uk/conts/retrieving.html (or from the CCDC, 12 Union Road, Cambridge CB2 1EZ, UK; Fax: +44-1223-336033; E-mail: deposit@ccdc.cam.ac.uk).

3.2.1. [cis-PtH(SiH$_2$Trip)(PPh$_3$)$_2$] (2)

$C_{56}H_{46}P_2PtSi$, C_7H_8, $M_W = 1096.18$, triclinic, space group P-1, $a = 12.7971(6)$ Å, $b = 13.4847(6)$ Å, $c = 14.8861(7)$ Å, $\alpha = 99.2791(10)°$, $\beta = 99.2791(10)°$, $\gamma = 90.3020(10)°$, $V = 2472.4(2)$ Å^3, $Z = 2$, $D_{calc.} = 1.472$ g·cm^{-3}, R_1 ($I > 2\sigma I$) = 0.0310, wR_2 (all data) = 0.0718 for 11639 reflections and 617 parameters, $GOF = 1.025$.

3.2.2. [PtH(SiH$_2$Trip)(dppf)] (3)

$C_{54}H_{44}FeP_2PtSi$, CH_2Cl_2, MW = 1118.79, monoclinic, space group $P2_1$, $a = 12.2585(6)$ Å, $b = 15.5003(7)$ Å, $c = 12.9367(6)$ Å, $\beta = 110.8060(10)°$, $V = 2297.81(19)$ Å^3, $Z = 2$, $D_{calc} = 1.617$ g·cm^{-3}, R_1 ($I > 2\sigma I$) = 0.0484, wR_2 (all data) = 0.1171 for 8350 reflections, 571 parameters, and 2 restraints, $GOF = 1.018$.

3.2.3. [PtH(SiH$_2$Trip)(PMe$_3$)(PPh$_3$)] (5)

$C_{41}H_{40}P_2PtSi$, CH_2Cl_2, $M_W = 902.78$, orthorhombic, space group $Pbca$, $a = 15.9074(6)$ Å, $b = 20.9224(8)$ Å, $c = 22.6113(9)$ Å, $V = 7525.5(5)$ Å^3, $Z = 8$, $D_{calc} = 1.594$ g·cm^{-3}, R_1 ($I > 2\sigma I$) = 0.0325, wR_2 (all data) = 0.0706 for 7011 reflections and 448 parameters, $GOF = 1.026$.

4. Conclusions

We have demonstrated that the oxidative addition of the sterically bulky primary silane, TripSiH$_3$ **1** with [Pt(η^2-C_2H_4)(PPh$_3$)$_2$] in toluene, resulted in the formation of the mononuclear (hydrido)(dihydrosilyl) complex [cis-PtH(SiH$_2$Trip)(PPh$_3$)$_2$] **2**. The ligand-exchange reactions of **2** with free chelating bis(phosphine)s such as DPPF or DCPE resulted in the formations of a series of (hydrido)(dihydrosilyl) complexes [PtH(SiH$_2$Trip)(L)] (**3**: L = dppf, **4**: L = dcpe). In contrast, the reaction of **2** with an excess amount of PMe$_3$ in toluene quantitatively produced [PtH(SiH$_2$Trip)(PMe$_3$)(PPh$_3$)] **5**. The latter is of particular interest, as it represents the first platinum complex having different simple phosphine ligands such as a weakly electron-donating PPh$_3$ and a strongly electron-donating PMe$_3$. Further investigations on the reactivity of these complexes are currently in progress.

Acknowledgments: This work was partially supported by JSPS KAKENHI Grant Number T17K05771 (to Norio Nakata). We are grateful to Kohtaro Osakada and Makoto Tanabe (Tokyo Institute of Technology, Japan) for their assistance in measurement of ^{29}Si NMR spectroscopy.

Author Contributions: Norio Nakata, Nanami Kato, and Norio Sekizawa contributed to the synthesis and data analysis; Norio Nakata performed X-ray crystallography and wrote the manuscript; Norio Nakata and Akihiko Ishii proposed idea on the design of the experiment, reviewed and approved the final manuscript.

References

1. Corey, J.Y. Reactions of hydrosilanes with transition metal complexes. *Chem. Rev.* **2016**, *116*, 11291–11435. [CrossRef] [PubMed]

2. Marciniec, B. *Comprehensive Handbook on Hydrosilylation*; Pergamon Press: Oxford, UK, 1992; ISBN 0-08-040272-0.

3. Ojima, I.; Li, Z.; Zhu, J. Recent Advances in Hydrosilylation and Related Reactions. In *The Chemistry of Organic Silicon Compounds*; Patai, S., Rappoport, Z., Eds.; John Wiley and Sons: New York, NY, USA, 1998; pp. 1687–1792.

4. Marciniec, B.; Maciejewski, H.; Pietraszuk, C.; Pawluc, P. *Hydrosilylation: A Comprehensive Review on Recent Advances*; Marciniec, B., Ed.; Advances in Silicon Science; Springer: Berlin, Germany, 2009; Volume 1, ISBN 978-1-4020-8172-9.

5. Chalk, A.J.; Harrod, J.F. Homogeneous catalysis. II. The mechanism of the hydrosilation of olefins catalyzed by group VIII metal complexes. *J. Am. Chem. Soc.* **1965**, *87*, 16–21. [CrossRef]

6. Ozawa, F. The chemistry of organo(silyl)platinum(II) complexes relevant to catalysis. *J. Organomet. Chem.* **2000**, *611*, 332–342. [CrossRef]

7. Sharma, H.; Pannell, K.H. Activation of the Si–Si bond by transition metal complexes. *Chem. Rev.* **1995**, *95*, 1351–1374. [CrossRef]

8. Suginome, M.; Ito, Y. Transition-metal-catalyzed additions of silicon—silicon and silicon—heteroatom bonds to unsaturated organic molecules. *Chem. Rev.* **2000**, *100*, 3221–3256. [CrossRef] [PubMed]

9. Ojima, I.; Inaba, S.-I.; Kogure, T.; Nagai, Y. The action of tris(triphenylphosphine)chlororhodium on polyhydromonosilanes. *J. Organomet. Chem.* **1973**, *55*, C7–C8. [CrossRef]

10. Chang, L.S.; Johnson, M.P.; Fink, J. Polysilyl complexes of platinum—Synthesis and thermochemistry. *Organometallics* **1989**, *8*, 1369–1371. [CrossRef]

11. Harrod, J.F.; Mu, Y.; Samuel, E. Catalytic dehydrocoupling: A general method for the formation of element-element bonds. *Polyhedron* **1991**, *10*, 1239–1245. [CrossRef]

12. Woo, H.G.; Walzer, J.F.; Tilley, T.D. A σ-bond metathesis mechanism for dehydropolymerization of silanes to polysilanes by d^0 metal catalysts. *J. Am. Chem. Soc.* **1992**, *114*, 7047–7055. [CrossRef]

13. Grimmond, B.J.; Corey, J.Y. Amino-functionalized zirconocene (C$_5$H$_4$CH(Me)NMe$_2$)$_2$ZrCl$_2$, a catalyst for the dehydropolymerization of PhSiH$_3$: Probing of peripheral substituent effects on catalyst dehydropolymerization activity. *Inorg. Chim. Acta* **2002**, *330*, 89–94. [CrossRef]

14. Kim, Y.-J.; Park, J.-I.; Lee, S.-C.; Osakada, K.; Tanabe, M.; Choi, J.-C.; Koizumi, T.; Yamamoto, T. *Cis* and *trans* isomers of Pt(SiHAr$_2$)$_2$(PR$_3$)$_2$ (R = Me, Et) in the solid state and in solutions. *Organometallics* **1999**, *18*, 1349–1352. [CrossRef]

15. Braddock-Wilking, J.; Corey, J.Y.; Trankler, K.A.; Xu, H.; French, L.M.; Praingam, N.; White, C.; Rath, N.P. Spectroscopic and reactivity studies of platinum—silicon monomers and dimers. *Organometallics* **2006**, *25*, 2859–2871. [CrossRef]

16. Arii, H.; Takahashi, M.; Noda, A.; Nanjo, M.; Mochida, K. Spectroscopic and structural studies of thermally unstable intermediates generated in the reaction of [Pt(PPh$_3$)$_2$(η^2-C$_2$H$_4$)] with dihydrodisilanes. *Organometallics* **2008**, *27*, 1929–1935. [CrossRef]

17. Ahrens, T.; Braun, T.; Braun, B. Activation of 1,2-dihydrodisilanes at platinum(0) phosphine complexes: Syntheses and structures of bissilyl platinum(II) complexes. *Z. Anorg. Allg. Chem.* **2014**, *640*, 93–99. [CrossRef]

18. Li, Y.-H.; Huang, Z.-F.; Li, X.-A.; Lai, W.-Y.; Wang, L.-H.; Ye, S.-H.; Cui, L.-F.; Wang, S. Synthesis and structural characterization of a novel bis(silyl) platinum(II) complex bearing SiH_3 ligand. *J. Orgnomet. Chem.* **2014**, *749*, 246–250. [CrossRef]

19. Auburn, M.; Ciriano, M.; Howard, J.A.K.; Murray, M.; Pugh, N.J.; Spencer, J.L.; Stone, F.G.A.; Woodward, P. Synthesis of bis-μ-diorganosilanediyl-*af*-dihydridobis(triorganophosphine)diplatinum complexes: Crystal and molecular structure of [{PtH(μ-SiMe2)[P(C6H11)3]}2]. *J. Chem. Soc. Dalton Trans.* **1980**, 659–666. [CrossRef]

20. Zarate, E.A.; Tessier-Youngs, C.A.; Young, W.J. Synthesis and structural characterization of platinum–silicon dimers with unusually short cross-ring silicon–silicon interactions. *J. Am. Chem. Soc.* **1988**, *110*, 4068–4070. [CrossRef]

21. Levchinsky, Y.; Rath, N.P.; Braddock-Wilking, J. Reaction of a symmetrical diplatinum complex containing bridging μ-η^2-H–SiH(IMP) ligands (IMP = 2-isopropyl-6-methylphenyl) with PMe_2Ph. Formation and characterization of {(PhMe2P)2Pt[μ-SiH(IMP)]}2. *Organometallics* **1999**, *18*, 2583–2586. [CrossRef]

22. Sanow, L.M.; Chai, M.; McConnville, D.B.; Galat, K.J.; Simons, R.S.; Rinaldi, P.L.; Young, W.; Tessier, C.A. Platinum–silicon four-membered rings of two different structural types. *Organometallics* **2000**, *19*, 192–205. [CrossRef]

23. Braddock-Wilking, J.; Levchinsky, Y.; Rath, N.P. Synthesis and characterization of diplatinum complexes containing bridging μ-η^2-H–SiHAr ligands. X-ray crystal structure determination of {(Ph3P)Pt[μ-η^2-H–SiHAr]}2 (Ar = 2,4,6-(CF3)3C6H2, C6Ph5). *Organometallics* **2000**, *19*, 5500–5510. [CrossRef]

24. Osakada, K.; Tanabe, M.; Tanase, T. A triangular triplatinum complex with electron-releasing SiPh2 and PMe3 Ligands: [{Pt(μ-SiPh2)(PMe3)}3]. *Angew. Chem. Int. Ed.* **2000**, *39*, 4053–4054. [CrossRef]

25. Braddock-Wilking, J.; Corey, J.Y.; Trankler, K.A.; Dill, K.M.; French, L.M.; Rath, N.P. Reaction of silafluorenes with (Ph3P)2Pt(η2-C2H4): Generation and characterization of Pt–Si monomers, dimers, and trimers. *Organometallics* **2004**, *23*, 4576–4584. [CrossRef]

26. Braddock-Wilking, J.; Corey, J.Y.; French, L.M.; Choi, E.; Speedie, V.J.; Rutherford, M.F.; Yao, S.; Xu, H.; Rath, N.P. Si–H bond activation by (Ph3P)2Pt(η2-C2H4) in dihydrosilicon tricycles that also contain O and N heteroatoms. *Organometallics* **2006**, *25*, 3974–3988. [CrossRef]

27. Tanabe, M.; Ito, D.; Osakada, K. Diplatinum complexes with bridging silyl ligands. Si–H bond activation of μ-silyl ligand leading to a new platinum complex with bridging silylene and silane ligands. *Organometallics* **2007**, *26*, 459–462. [CrossRef]

28. Tanabe, M.; Ito, D.; Osakada, K. Ligand exchange of diplatinum complexes with bridging silyl ligands involving Si–H bond cleavage and formation. *Organometallics* **2008**, *27*, 2258–2267. [CrossRef]

29. Tanabe, M.; Kamono, M.; Tanaka, K.; Osakada, K. Triangular triplatinum complex with four bridging Si ligands: Dynamic behavior of the molecule and catalysis. *Organometallics* **2017**, *36*, 1929–1935. [CrossRef]

30. Mullica, D.F.; Sappenfield, E.L. Synthesis and X-ray structural investigation of *cis*-hydrido(triphenylsilyl)-1,4-butanediyl-*bis*-(dicyclohexylphosphine)platinum(II). *Polyhedron* **1991**, *10*, 867–872. [CrossRef]

31. Latif, L.A.; Eaborn, C.; Pidcock, A.P. Square planar platinum(II) complexes. Crystal structures of *cis*- bis(triphenylphosphine)hydro (triphenylstannyl)platinum(II) and *cis*-bis(triphenylphosphine)-hydro(triphenylsilyl)platinum(II). *J. Organomet. Chem.* **1994**, *474*, 217–221. [CrossRef]

32. Koizumi, T.; Osakada, K.; Yamamoto, T. Intermolecular transfer of triarylsilane from RhCl(H)(SiAr3)[P(*i*-Pr)3]2 to a platinum(0) complex, giving *cis*-PtH(SiAr3)(PEt3)2 (Ar = C6H5, C6H4F-*p*, C6H4Cl-*p*). *Organometallics* **1997**, *16*, 6014–6016. [CrossRef]

33. Chan, D.; Duckett, S.B.; Heath, S.L.; Khazal, I.G.; Perutz, R.N.; Sabo-Etienne, S.; Timmins, P.L. Platinum bis(tricyclohexylphosphine) silyl hydride complexes. *Organometallics* **2004**, *23*, 5744–5756. [CrossRef]

34. Arii, H.; Takahashi, M.; Nanjo, M.; Mochida, K. Syntheses of mono- and dinuclear silylplatinum complexes bearing a diphosphino ligand via stepwise bond activation of unsymmetric disilanes. *Dalton Trans.* **2010**, *39*, 6434–6440. [CrossRef] [PubMed]

35. Simons, R.S.; Sanow, L.M.; Galat, K.J.; Tessier, C.A.; Young, W.J. Synthesis and Structural Characterization of the Mono(silyl)platinum(II) Complex *cis*-(2,6-Mes$_2$C$_6$H$_3$(H)$_2$Si)Pt(H)(PPr$_3$)$_2$. *Organometallics* **2000**, *19*, 3994–3996. [CrossRef]

36. Wang, S.; Yu, X.-F.; Li, N.; Yang, T.; Lai, W.-Y.; Mi, B.-X.; Li, J.-F.; Li, Y.-H.; Wang, L.-H.; Huang, W. Synthesis and structural studies of a rare bis(phosphine) (hydrido) (silyl) platinum(II) complex containing a Si–Si single bond. *J. Orgnomet. Chem.* **2015**, *776*, 113–116. [CrossRef]

37. Nakata, N.; Fukazawa, S.; Ishii, A. Synthesis and crystal structures of the first stable mononuclear dihydrogermyl(hydrido) platinum(II) complexes. *Organometallics* **2009**, *28*, 534–538. [CrossRef]

38. Nakata, N.; Fukazawa, S.; Kato, N.; Ishii, A. Palladium(II) hydrido complexes having a primary silyl or germyl ligand: Synthesis, crystal structures, and dynamic behavior. *Organometallics* **2011**, *30*, 4490–4493. [CrossRef]

39. Nakata, T.; Sekizawa, N.; Ishii, A. Cationic dinuclear platinum and palladium complexes with bridging hydrogermylene and hydrido ligands. *Chem. Commun.* **2015**, *51*, 10111–10114. [CrossRef] [PubMed]

40. Kalläne, S.; Laubenstein, R.; Braun, T.; Dietrich, M. Activation of Si–Si and Si–H bonds at a platinum bis(diphenylphosphanyl)ferrocene (dppf) complex: Key steps for the catalytic hydrogenolysis of disilanes. *Eur. J. Inorg. Chem.* **2016**, 530–537. [CrossRef]

41. Boyle, R.C.; Mague, J.T.; Fink, M.J. The first stable mononuclear silyl palladium hydrides. *J. Am. Chem. Soc.* **2003**, *125*, 3228–3229. [CrossRef] [PubMed]

42. Boyle, R.C.; Pool, D.; Jacobsen, H.; Fink, M.J. Dynamic processes in silyl palladium complexes: Evidence for intermediate Si−H and Si−Si σ-complexes. *J. Am. Chem. Soc.* **2006**, *128*, 9054–9055. [CrossRef] [PubMed]

43. Feducia, J.A.; Campbell, A.N.; Doherty, M.Q.; Gagné, M.R. Modular catalysts for diene cycloisomerization: Rapid and enantioselective variants for bicyclopropane synthesis. *J. Am. Chem. Soc.* **2006**, *128*, 13290–13297. [CrossRef] [PubMed]

44. Campbell, A.N.; Gagné, M.R. Room-temperature β-H elimination in (P$_2$P)Pt(OR) cations: Convenient synthesis of a platinum hydride. *Organometallics* **2007**, *26*, 2788–2790. [CrossRef]

45. Pan, B.; Xu, Z.; Bezpalko, M.W.; Foxman, B.M.; Thomas, C.M. *N*-Heterocyclic phosphenium ligands as sterically and electronically-tunable isolobal analogues of nitrosyls. *Inorg. Chem.* **2012**, *51*, 4170–4179. [CrossRef] [PubMed]

46. Frey, U.; Helm, L.; Merbach, A.E.; Romeo, R. High-pressure NMR kinetics. Part 41. Dissociative substitution in four-coordinate planar platinum(II) complexes as evidenced by variable-pressure high-resolution proton NMR magnetization transfer experiments. *J. Am. Chem. Soc.* **1989**, *111*, 8161–8165. [CrossRef]

47. Romeo, R.; Grassi, A.; Scolaro, L.M. Factors affecting reaction pathways in nucleophilic substitution reactions on platinum(II) complexes: A comparative kinetic and theoretical study. *Inorg. Chem.* **1992**, *31*, 4383–4390. [CrossRef]

48. Brynda, M.; Bernardinelli, G.; Dutan, C.; Geoffroy, M. Kinetic stabilization of primary hydrides of main group elements. The synthesis of an air-stable, crystalline arsine and silane. *Inorg. Chem.* **2003**, *42*, 6586–6588. [CrossRef] [PubMed]

49. Cook, C.D.; Jauhal, G.S. Chemistry of low-valent complexes. II. Cyclic azo derivatives of platinum. *J. Am. Chem. Soc.* **1968**, *90*, 1464–1467. [CrossRef]

50. Sheldrick, G.M. A short history of *SHELX*. *Acta Cryst. A* **2008**, *64*, 112–122. [CrossRef] [PubMed]

14

Synthesis of 1,3-Diols from Isobutene and HCHO via Prins Condensation-Hydrolysis Using CeO$_2$ Catalysts: Effects of Crystal Plane and Oxygen Vacancy

Zhixin Zhang [1] [ID], Yehong Wang [1], Jianmin Lu [1], Min Wang [1], Jian Zhang [1], Xuebin Liu [2] and Feng Wang [1,*]

[1] State Key Laboratory of Catalysis, Dalian National Laboratory for Clean Energy, Dalian Institute of Chemical Physics, Chinese Academy of Sciences, Dalian 116023, China; zhangzhixin@dicp.ac.cn (Z.Z.); wangyehong@dicp.ac.cn (Y.W.); lujianmin@dicp.ac.cn (J.L.); wangmin@dicp.ac.cn (M.W.); zjian@dicp.ac.cn (J.Z.)

[2] Energy Innovation Laboratory, BP Office (Dalian Institute of Chemical Physics), Dalian 116023, China; xuebin.liu@se1.bp.com

* Correspondence: wangfeng@dicp.ac.cn

Abstract: We herein report the synthesis of 3-methyl-1,3-butanediol from isobutene and HCHO in water via a Prins condensation-hydrolysis reaction over CeO$_2$, which is a water-tolerant Lewis acid catalyst. The CeO$_2$ exhibits significant catalytic activity for the reaction, giving 95% HCHO conversion and 84% 3-methyl-1,3-butanediol selectivity at 150 °C for 4 h. The crystal planes of CeO$_2$ have a significant effect on the catalytic activity for the Prins reaction. The (110) plane shows the highest catalytic activity among the crystal planes investigated (the (100), (110), and (111) planes), due to its higher concentration of Lewis acid sites, which is in line with the concentration of oxygen vacancies. Detailed characterizations, including NH$_3$-TPD, pyridine-adsorbed FT-IR spectroscopy, and Raman spectroscopy, revealed that the concentration of Lewis acid sites is proportional to the concentration of oxygen vacancies. This study indicates that the Lewis acidity induced by oxygen vacancy can be modulated by selective synthesis of CeO$_2$ with different morphologies, and that the Lewis acidity and oxygen vacancy play an important role in Prins condensation and hydrolysis reaction.

Keywords: CeO$_2$; crystal plane effect; Prins condensation; hydrolysis; oxygen vacancy; Lewis acidity

1. Introduction

The Prins condensation of olefins with aldehydes is one of the most important organic reactions, allowing one to obtain alkyl-*m*-dioxanes, 1,3-diols, conjugated diolefins, and other valuable compounds [1,2]. Among these chemicals, 1,3-diols is a commodity chemical that is mainly used as a building block in polymerization and as a surfactant. For example, dehydration of 3-methyl-1,3-butanediol produces isoprene [3], which is mainly used as a monomer for manufacture of polyisoprene rubber and butyl rubber [4,5]. 3-methyl-1,3-butanediol can be synthesized by Prins condensation of isobutene with formaldehyde to 4,4-dimethyl-1,3-dioxane followed by its hydrolysis [6], which is a promising route because the feedstocks (isobutene [7–9] and formaldehyde [10,11]) can be found in bio-refineries, based on recent research and industrial progress [12–14]. In the synthesis route, both the Prins condensation and hydrolysis reaction require acid catalysts [3,15,16]. Various solid acid catalysts, such as metal oxide catalysts [17], zeolite catalysts [18], phosphate catalysts [15,19,20], and heteropolyacids catalysts [21,22], have been developed for the two-step reaction (Prins condensation and hydrolysis reaction). Alternatively, a single-stage 3-methyl-1,3-butanediol synthesis from isobutene and HCHO in water is one interesting route to be investigated, and a water-tolerant acid catalyst is critical.

Our previous work indicates that CeO_2 is a water-tolerant Lewis acid catalyst [23,24]. It has been proved that the coordinatively unsaturated Ce cations on CeO_2 surface act as the Lewis acid sites [25–28]. According to experimental data and theory calculations [24,29–32], the concentration of surface coordinatively unsaturated Ce cations is associated with the population of oxygen vacancies over CeO_2. Additionally, the formation energy of oxygen vacancies varies greatly among different CeO_2 surfaces. For the three low-index surfaces of CeO_2 ((111), (110), and (100)), the formation energies of oxygen vacancy follow the order of (110) < (100) < (111) [33], which implies that the order of oxygen vacancy formation is (110) > (100) > (111). Therefore, selective synthesis of CeO_2 with different crystal planes would modulate the concentration of oxygen vacancies. Furthermore, the Lewis acidity of CeO_2 catalyst may be manipulated by morphology control.

In this study, 3-methyl-1,3-butanediol will be synthesized from isobutene and formaldehyde using CeO_2 catalysts via Prins condensation-hydrolysis reaction in water. CeO_2 with different morphologies (rod, octahedron, and cube) will be tested for the Prins condensation-hydrolysis reaction, and the effect of crystal planes ((111), (110), and (100)) and oxygen vacancy will also be investigated. NH_3-TPD, pyridine adsorption IR, and Raman spectroscopy were used to characterize the acid properties and defect structures of these CeO_2 catalysts. Based on these experiments, the structure-property-activity relationship will be established.

2. Results and Discussion

CeO_2 has been proved a highly active and water-tolerant catalyst for the synthesis of 1,3-butanediol from propylene and HCHO via Prins condensation-hydrolysis in water [24]. This work encouraged us to apply the CeO_2 catalyst for isobutene-formaldehyde condensation in water to synthesize 3-methyl-1,3-butanediol, which is an important higher-molecule alkanediol for the polymer industries; its synthesis is obviously difficult, but scientifically interesting. Our synthetic route includes the Prins condensation of isobutene with HCHO to 4,4-dimethyl-1,3-dioxane (**3**), and then hydrolysis of **3** to 3-methyl-1,3-butanediol (**4**) (Scheme 1).

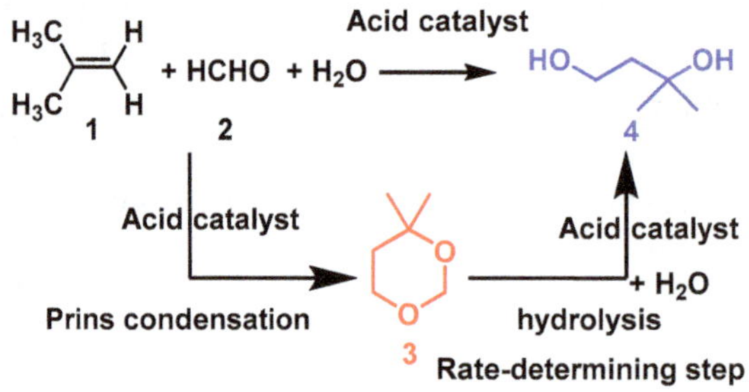

Scheme 1. Prins condensation-hydrolysis of isobutene with HCHO in water.

2.1. Effect of Reaction Temperature

The Prins condensation-hydrolysis of isobutene with formaldehyde (FA) in water over pristine CeO_2 was conducted at different temperatures ranging from 60 to 150 °C for 4 h (Figure 1). It can be seen from Figure 1 that the FA conversion increases slowly from 1% at 60 °C to 13% at 100 °C and then rapidly increases to 95% at 150 °C. The total selectivities of 4,4-dimethyl-1,3-dioxane (**3**) and 3-methyl-1,3-butanediol (**4**) are above 98% throughout the temperature range (60–150 °C) investigated. No dehydration products (such as 3-methyl-butenol isomers and isoprene) [22,34] derived from 3-methyl-1,3-butanediol, which are usually obtained via vapor-phase dehydration at high temperatures (≥300 °C), were observed. In our catalytic system, the reactions were conducted at liquid phase (Reaction temperature ≤150 °C) and in water, which did not favor the dehydration of 3-methyl-1,3-butanediol. The selectivity of 4,4-dimethyl-1,3-dioxane (**3**) increases with increasing temperature below 130 °C, and then decreases with further increase in temperature; meanwhile, the selectivity of the target product, 3-methyl-1,3-butanediol (**4**), has the opposite behavior.

The hydrolysis of **3** to **4** seems to be the rate-determining step in the Prins condensation-hydrolysis reaction under our reaction conditions, and the high temperature (>130 °C) favors the hydrolysis of 4,4-dimethyl-1,3-dioxane (**3**) to target product (**4**) (Scheme 1). These results show that the best yield (80%) of target product **4** can be achieved at 150 °C among the reaction temperature range investigated.

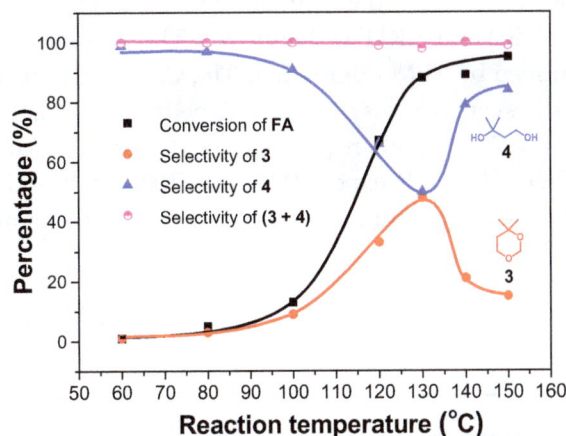

Figure 1. Effect of reaction temperature for the Prins condensation-hydrolysis of isobutene with FA in water over pristine CeO$_2$. Reaction conditions: 50 mg CeO$_2$, 1.5 mL H$_2$O, 0.21 mL HCHO (38 wt %), 3.0 g isobutene, 4 h.

2.2. Time-On-Stream Profile

Prins condensation-hydrolysis of isobutene with formaldehyde (FA) in water was conducted at 150 °C over pristine CeO$_2$ to explore the optimal reaction time (Figure 2). The FA conversion rapidly increases to 95% after 4 h under the reaction conditions, and then levels off at ca. 95% with reaction times increased to 8 h. The highest selectivity of the target product 3-methyl-1,3-butanediol (**4**) is 84% during the time-on-stream investigated. Further increasing the reaction time, the selectivity of **4** seems to decrease slightly; simultaneously, the selectivities of **3** and others increase slightly. This behavior indicates the Prins condensation-hydrolysis reaction reaches equilibrium. The total selectivities of **3** and **4** are over 95% throughout the course of the reaction, and slightly decrease along with the increase in reaction time, indicating that side reactions may occur after long reaction times at such high temperature (150 °C). These results indicate that pristine CeO$_2$ is a highly effective catalyst for the Prins condensation-hydrolysis reaction, and the best catalytic activity (95% FA conversion and 84% 3-methyl-1,3-butanediol selectivity) can be obtained at 150 °C for 4 h.

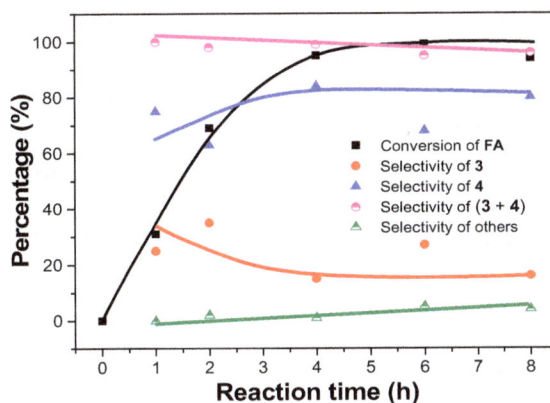

Figure 2. Reaction profiles for the Prins condensation-hydrolysis of isobutene with FA in water over pristine CeO$_2$. Reaction conditions: 50 mg CeO$_2$, 1.5 mL H$_2$O, 0.21 mL HCHO (38 wt %), 3.0 g isobutene, 150 °C.

2.3. Effect of Crystalline Plane

It has been reported that the exposed crystalline planes of CeO_2 (the (100), (110) and (111) crystalline planes) possess different surface concentrations of oxygen vacancies because of the difference in the formation energy of oxygen vacancies [29,35]. We successfully prepared CeO_2 with different morphologies via the previously reported methods [23,24,36,37], and confirmed their morphology by XRD (Figure S1) and HRTEM (Figure S2). The average sizes of CeO_2 with different morphologies were determined by TEM (Figure S2). The CeO_2-rods are about 7.4 nm in diameter and 69 nm in length, and the average sizes of CeO_2 octahedron and CeO_2-cube are 118 and 61 nm, respectively. Our studies [24,38,39] and others [33,40] have also proved that the nanorods, nanocubes, and nanooctahedrons of CeO_2 selectively expose (110) and (100), (100), and (111) planes, respectively (Table 1). The theoretical exposed crystalline planes and their ratio are shown in Table 1, as obtained from the literature [37].

Table 1. Prins condensation-hydrolysis of isobutene with FA in water over CeO_2 with different morphologies [1].

Entry	Catalyst	Exposed Crystalline Planes	Co A_{595}/A_{462} [2]	S [4]	Conv. (%)	Sel. (%)	
						3	4
1	CeO_2-rod	(110)/(100) = 2/1	0.077	86	32	13	87
2	CeO_2-cube	(100)	0.001	21	8	14	86
3	CeO_2-octahedron	(111)	0.003	9	1	12	88
4	Pristine CeO_2	(111), (110), (100)	0.009 [3]	67	13	9	91

[1] Reaction conditions: 50 mg catalyst, 1.5 mL H_2O, 0.21 mL HCHO (38 wt %), 3.0 g isobutene, 100 °C, 4 h. [2] A_{595}/A_{462} means the relative concentration of intrinsic oxygen vacancy concentration (Co) of CeO_2 samples determined by Raman spectra in Figure S3. [3] The A_{595}/A_{462} value of pristine CeO_2 are referenced from literature [25]. [4] Specific surface area S_{BET} ($m^2 \cdot g^{-1}$).

We tested the catalytic activities of CeO_2 with the three different morphologies in the Prins condensation-hydrolysis reaction, and the results are shown in Table 1. In order to distinguish the difference in catalytic performance of these three CeO_2 catalysts, relatively low reaction temperature (100 °C) was selected due to the similar catalytic activity of CeO_2 under harsh reaction conditions (such as 150 °C for 2 h) (Table S1). Under 100 °C for 4 h, 32% FA conversion and 87% 3-methyl-1,3-butanediol (**4**) selectivity were obtained for CeO_2-rod, meanwhile only 8% and 1% FA conversion were achieved for CeO_2-cube and CeO_2-octahedron with similar product distributions, respectively. For comparison, the catalytic activities of pristine CeO_2 (13% FA conversion and 91% 3-methyl-1,3-butanediol selectivity) were also listed in entry 4, Table 1. The order of catalytic activity is in line with the sequence of specific surface area (S_{BET}), which indicates that the S_{BET} is another key parameter for the catalytic performance over CeO_2. These results also indicate the CeO_2-rod is the best catalyst for the reaction among these catalysts. According to the exposed crystalline planes, we can conclude the (110) plane is the most active crystalline plane for this reaction.

As the complex crystal planes over the pristine CeO_2 surface, the well-defined CeO_2 with different morphologies were selected as models in this study to make clear the correlation between the catalytic performance and their physicochemical properties. NH_3-TPD, pyridine adsorption IR, and Raman spectroscopy were used to characterize the acid properties and oxygen-defect structures, respectively.

NH_3-TPD technique was used to investigate the acid strength of CeO_2 with different morphologies (Figure 3). All three CeO_2 samples have two broader NH_3-desorption peaks in the range from 50 to 600 °C in the desorption profiles. The high desorption peak around 110–156 °C can be ascribed to the weak acid sites, and the weak desorption peak centered at 340–420 °C is related to the medium acid sites on the CeO_2 surface. It can be found that these three samples mainly possess weak acid sites, along with a small number of medium ones. The two strongest peaks, at around 145 °C and 340 °C, which are ascribed to NH_3 adsorbed on weak acid sites and medium acid sites, respectively, were observed for CeO_2-rod. Two types of acid sites at low temperature around 110 °C and 156 °C and a weak peak around 420 °C were observed for CeO_2-octahedron. Additionally, two weak peaks around 122 °C and 360 °C were detected for CeO_2-cube. Furthermore, the peak areas, which reflect the acidity, vary with the CeO_2 morphologies, and the CeO_2-rod catalyst shows the highest acidity among the three morphologies. These results also imply that the predominant factor responded for the different catalytic performance between the three catalysts may be the concentration of the acid sites.

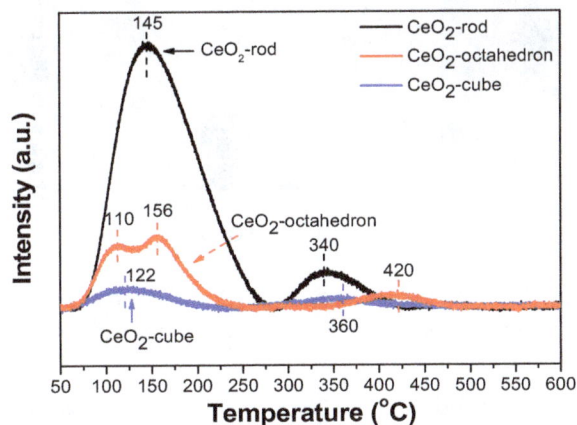

Figure 3. NH_3-TPD of CeO_2 with different morphologies.

Pyridine-adsorption IR is an effective characterization for measuring the concentration of the acid sites and distinguishing the acid type of solid catalysts [41–43]. From the pyridine adsorption IR spectra of CeO_2 with different morphologies (Figure 4), it can be seen that no characteristic band attributed to Brönsted acid sites, which usually appear around 1540 cm^{-1} [42], are observed on these samples. Strong bands around 1440 and 1595 cm^{-1} are observed, which are the characteristic bands of the coordinatively bound pyridine on Lewis acid sites [43]. The surface concentration of Lewis acid sites can be calculated based on the integrated band area at 1440 cm^{-1} and Formula (1) [44], and the CeO_2-rod presents the highest concentration of Lewis acid sites (188 $\mu mol \cdot g^{-1}$), which is about threefold higher than that of CeO_2-octahedron (64 $\mu mol \cdot g^{-1}$), and fivefold higher than that of CeO_2-cube (39 $\mu mol \cdot g^{-1}$). Additionally, the concentration of Lewis acid sites over CeO_2-rod is 3.5 times higher than that of pristine CeO_2 (54 $\mu mol \cdot g^{-1}$) [24,25]. We plot the concentration of Lewis acid sites and the catalytic performance versus the catalyst morphologies showed in Figure 5. It can be found that the CeO_2-rod obtains the best catalytic activity because of its higher Lewis acidity.

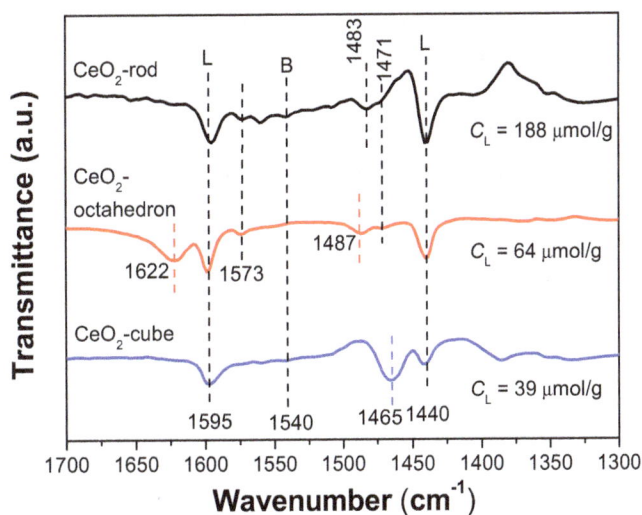

Figure 4. Pyridine adsorption IR spectra of CeO_2 with different morphologies.

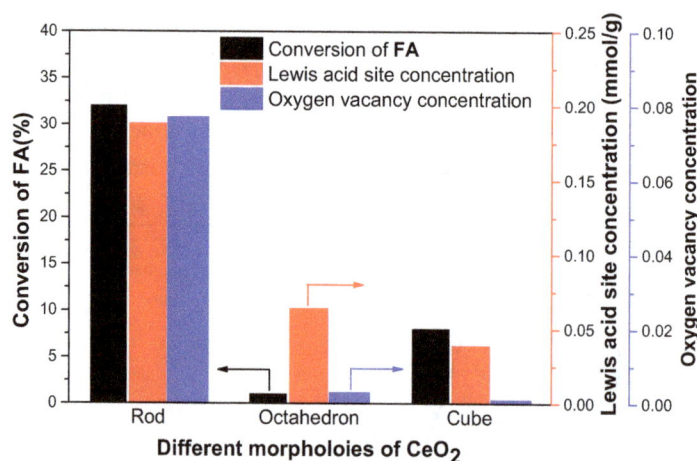

Figure 5. The relationship between the conversion of FA and the concentration of Lewis acid sites (unit: mmol/g) or the concentration of oxygen vacancies versus the CeO_2 with different morphologies.

It should be noted that some bands around 1622, 1573, 1485, and 1465 cm^{-1} were observed, which can be ascribed to the pyridine adsorbed on Lewis acid sites; these results indicate that the CeO_2 surface is predominately covered by Lewis acid sites. This result is in agreement with the previous study that only Lewis acid sites exist for CeO_2 [45,46] and metal-doped CeO_2 surfaces [25]. However, these complex bands imply the different Lewis acid sites for CeO_2 with different morphologies. The bands around 1573 and 1485 cm^{-1} for CeO_2-rod and CeO_2-octahedron surfaces are weak, and the influence on the concentration of Lewis acid sites may be neglected. However, the effect of 1622 and 1465 cm^{-1} on the concentration of Lewis acid sites should be considered, because these bands are obvious on the octahedron and cube, respectively. Furthermore, the peak area of 1465 cm^{-1} is larger than that of the 1622 cm^{-1} peak; this result may indicate that the concentration of Lewis acid sites over CeO_2-cube is larger than that of CeO_2-octahedron. This may be another reason for the opposite trend of HCHO conversion over CeO_2-cube and octahedron when the conversion is correlated with the Lewis acid site concentration. Unfortunately, no consensus has been reached on the assignments of these bands, and the calculated formula has not yet been established. In the literature, the band at 1623 cm^{-1} could be assigned to the 8a mode for pyridine adsorbed on more acidic sites [47], but others have proposed that it is more likely to be due to the (1 + 6a) combination mode of pyridine [48]. The assignment of the 1465 cm^{-1} band is more complex. There are four assignments based on the literature [49]. Thus, more work is needed to assign and quantify these bands.

It is well known that the structure determines the performance. The above results spur us on to in-depth investigations of the CeO_2 structures with different morphologies. The surface oxygen of a perfect CeO_2-rod, which exposes the (110) and (100) crystalline planes, is 9.5×10^{19} atom per gram, which is three-times (2.9×10^{19}) and seven-times (1.4×10^{19}) the values for a perfect cube and octahedron, respectively [23], implying the probability that oxygen vacancy formation for CeO_2-rod is higher than that for cube and octahedron. DFT calculations have proven that the formation energies of oxygen vacancies for different CeO_2 surfaces follow the order of (110) < (100) < (111) [33], meaning that the order of intrinsic oxygen vacancy formation follows (110) > (100) > (111). Raman spectroscopy is a powerful technique for detecting the defect structures over metal oxides; here, we used it to determine the practical concentration of observed oxygen vacancies over the three CeO_2 catalysts. Figure S3 shows the visible (532 nm) Raman spectra of the CeO_2 samples with different morphologies, providing the information of several surface layers over CeO_2. A strong band around 462 cm^{-1} and a weak band around 595 cm^{-1} were observed in all three samples. The former can be ascribed to the F_{2g} vibrational mode of CeO_2, which possesses a fluorite-type structure [50]. The CeO_2-rod gives a much broader 462 cm^{-1} peak than that of CeO_2-octahedron and CeO_2-cube, which is an inhomogeneous strain broadening caused by the differences in particle size between these three samples. This is a size-dependent phenomenon observed on CeO_2 nanoparticles [51]. This is consistent with TEM measurement, where CeO_2-rod has the smallest size among these three morphologies (7.4 nm $\times$ 69 nm for CeO_2-rod, 118 nm for CeO_2-octahedron, and 61 nm for CeO_2-cube) (Figure S2). Interestingly, the HCHO conversion trend over these three morphologies is line with the particle size of these catalysts, indicating that the particle size may also affect the catalytic performance for the Prins reaction. The weak band around 595 cm^{-1} is related to intrinsic oxygen vacancy sites [52]. The oxygen vacancies detected by Raman spectroscopy are primarily those on several surface layers of the CeO_2. The ratio of the integrated peak areas (A_{595}/A_{462}) was used to quantify the relative concentrations of oxygen vacancy [53]. The relative oxygen vacancy concentration of CeO_2 samples are shown in Table 1 and Figure S3. A plot of the oxygen vacancy concentration (measured by Raman spectra) and Lewis acid site concentration (measured by pyridine adsorption IR) against the catalyst morphologies was drawn (Figure 5). It can be found that the more oxygen vacancies, the more Lewis acid sites, and the CeO_2-rod shows the highest concentration of oxygen vacancies and Lewis acid sites among the three morphologies.

Combined with the catalytic performances, Lewis acid site concentration, and oxygen vacancy concentration (Figure 5), we found an increase in FA conversion when changing the shape of CeO_2. The CeO_2-rod shows the best FA conversion because of its having the highest concentration of Lewis acid sites and oxygen vacancies among the three morphologies investigated.

It must be noted that the conversion trend does not seem to correlate with the concentration of Lewis acid sites and the number of oxygen vacancies for octahedrons and cubes. This may be caused by the following two factors: (1) the different specific surface area (S_A) between the CeO_2-cube and octahedron (the S_A of CeO_2-cube is 21 m$^2 \cdot$g^{-1}, which is 2.3-times higher than that of CeO_2-octahedron, see Table 1); (2) the different particle sizes between CeO_2-cube and octahedron (the particle size of CeO_2-cube is about 61 nm, which is about 2-times larger than that of CeO_2-octahedron (~118 nm), see Figure S2). The smaller and higher S_A of CeO_2-cube could be attributed to the higher catalytic performance than that of CeO_2-octahedron.

3. Materials and Methods

3.1. Materials

All chemicals and reagents were handled in air and used without further purification. $Ce(NO_3)_3 \cdot 6H_2O$ was of analytical grade (AR), obtained from Aladdin Chemicals. NaOH (AR), $NH_3 \cdot H_2O$ (28–30%), $Na_3PO_4 \cdot 12H_2O$ (99.5%), and HCHO (38 wt %) were obtained from Sinopharm Chemical Reagent Co., Ltd. (Shanghai, China).

3.2. Preparation of the CeO_2 Catalysts

Pristine CeO_2 were prepared by a traditional precipitation process described in the literature [23,24]. In a typical experiment, $Ce(NO_3)_3 \cdot 6H_2O$ (5.0 g) was dissolved in Millipore-purified water (18 m$\Omega \cdot$cm, 100 mL), and then the solution was adjusted to a pH of 11.0 by the addition of $NH_3 \cdot H_2O$ under stirring at room temperature. The obtained precipitate was filtered, washed with deionized water, and then dried at 120 °C for 12 h. The obtained solid was calcined at 500 °C under the flow of air (50 mL$\cdot$min^{-1}) for 4 h.

CeO_2 samples with different morphologies were prepared by hydrothermal methodology reported by our and other groups previously [23,24,36,37].

Preparation of CeO_2-rod and cube. Typically, 5.209 g of $Ce(NO_3)_3 \cdot 6H_2O$ and a desired amount of NaOH (57.6–60 g) were dissolved in 20 mL and 210 mL of Millipore-purified water in a Teflon bottle, respectively. Then, the $Ce(NO_3)_3$ solution was added into the solution of NaOH under stirring at room temperature for 30 min, and a homogeneous milky slurry was formed. After the bottle was placed into a stainless steel vessel autoclave and sealed, the autoclave was heated to the desired temperature (100–180 °C) in an oven for 24 h. After the hydrothermal treatment, the solution above the precipitate was removed by centrifugation, and washed with deionized water and ethanol. Finally, the obtained sample was dried at 60 °C in an oven for 12 h.

Preparation of CeO_2-octahedron. Typically, 0.023 g of $Na_3PO_4 \cdot 12H_2O$ was added into the solution of $Ce(NO_3)_3$ (2.618 g in 240 mL of Millipore-purified water) under stirring at room temperature for 30 min in a Teflon bottle. Subsequently, the Teflon bottle was transferred into a stainless steel autoclave, and then placed in an oven at a temperature range of 170 °C for 12 h. After the hydrothermal treatment, the solution above the precipitate was removed by centrifugation, and washed with deionized water and ethanol. Finally, the obtained sample was dried at 60 °C in an oven for 12 h.

3.3. Prins Condensation-Hydrolysis Reaction

Aqueous formaldehyde (FA) (38 wt % HCHO, 0.21 mL, 3.0 mmol of HCHO), CeO_2 (50 mg), H_2O (1.5 mL), and a magnetic stir bar were loaded into a 15 ml Teflon-lined autoclave reactor. Quantified isobutene (99.99%) was charged into the reactor from a cylinder via weighting the reactor before and after the charging and then sealed. The sealed reactor was heated to the desired reaction temperature via a mantle. The 3-methyl-1,3-butanediol, 4,4-dimethyl-1,3-dioxane, etc., were analyzed by gas chromatography–mass spectrometry (GC–MS) using an Agilent 7890A/5975C (Santa Clara, CA, USA) instrument equipped with an HP-5MS column. The formaldehyde solution was analyzed by a GC (Tianmei, Shanghai, China) equipped with a packed column (GDX-401) and a TCD detector. An external standard was used to quantify the formaldehyde conversion.

3.4. Acidity Characterization by NH_3 Temperature-Programmed Desorption (NH_3-TPD)

The NH_3-TPD profiles of CeO_2 with different morphologies were recorded in a U-type quartz tube combined with an on-line mass spectrometer (MS, GSD320 Thermostar, Shanghai, China). In a typical experiment, about 40 mg of CeO_2 sample was placed in the tube and pretreated at 210 °C for 60 min under an argon flow (30 mL·min^{-1}), then cooling to 30 °C; several pulses of NH_3 were injected until no change in the NH_3 concentration was detected by the on-line MS. After that, the CeO_2 sample was flushed with argon for 60 min. Finally, the NH_3 desorption was conducted via increasing the oven temperature from 30 to 600 °C in the rate of 10 °C·min^{-1}. On-line MS recorded the gas effluents during the NH_3 desorption process.

3.5. Acidity Characterization by Pyridine Adsorption IR Spectroscopy

The acid type and concentration were determined by pyridine-adsorption IR spectra, which was conducted on a Bruker 70 IR spectrometer. A self-supporting sample disk of about 13 mm diameter (around 30 mg) was made by pressing in a mould, then the sample disk was placed into a homemade IR cell attached to a closed glass-circulation system. The disk was pretreated under a flow of argon (10 mL·min^{-1}) at 350 °C for 30 min, and then vacuumed for 30 min. After the temperature had cooled to 30 °C and the sample chamber had been evacuated to <10^{-3} mbar, a reference spectrum was recorded. After that, pyridine adsorption over sample disk was conducted via exposure to pyridine vapor. When the sample disk adsorbed pyridine was vacuumed at 150 °C for 30 min and cooled to 30 °C, the IR spectrum was recorded. The concentration of Lewis acid sites were calculated based on the integrated absorbance of the L band (1440 cm^{-1}) and the following formula [44]:

$$C = 1.42 \times IA \times R^2 / W \tag{1}$$

where C is the concentration of Lewis acid sites (mmol (g of catalyst)$^{-1}$), IA is the integrated absorbance of the L band (cm^{-1}), R is the radius of the catalyst disk (cm), and W is the mass of the sample disk (mg).

4. Conclusions

We demonstrate that CeO_2 is a highly effective catalyst for the synthesis of 1,3-diols from isobutene and HCHO via Prins condensation-hydrolysis reaction in water. 95% HCHO conversion and 84% 3-methyl-1,3-butanediol selectivity were obtained over the pristine CeO_2 catalyst under 150 °C for 4 h. The crystal planes of CeO_2 have a significant effect on the catalytic activities for the condensation-hydrolysis reaction. The (110) plane over CeO_2-rod surface is the most active crystalline plane for the reaction because of its having the richest intrinsic oxygen vacancies and highest Lewis acidity among the three low-index crystal planes investigated (the (100), (110), and (111) planes). Furthermore, the concentration of Lewis acid sites is proportional to the concentration of relative oxygen vacancies, indicating the strong contact between oxygen vacancies and Lewis acid sites. This study also implies that the Lewis acidity induced by oxygen vacancy can be modulate by the selective synthesis of CeO_2 with different crystal planes.

Acknowledgments: This work was supported by the National Natural Science Foundation of China (21422308, 21403216 and 21690084), Strategic Priority Research Program of Chinese Academy of Sciences (XDB17020300), and by Dalian Institute of Chemical Physics (DICP DMTO201406).

Author Contributions: Zhixin Zhang and Feng Wang conceived and designed the experiments; Zhixin Zhang, Yehong Wang and Jian Zhang performed the experiments; Jianmin Lu and Min Wang analyzed the data; Xuebin Liu contributed reagents/materials/analysis tools; Zhixin Zhang, Xuebin Liu and Feng Wang wrote and revised the paper.

References

1. Isagulyants, V.I.; Khaimova, T.G.; Melikyan, V.R.; Pokrovskaya, S.V. Condensation of unsaturated compounds with formaldehyde (the Prins reaction). *Russ. Chem. Rev.* **1968**, *37*, 17–25. [CrossRef]

2. Arundale, E.; Mikeska, L.A. The olefin-aldehyde condensation. The Prins reaction. *Chem. Rev.* **1952**, *51*, 505–555. [CrossRef]

3. Ivanova, I.; Sushkevich, V.L.; Kolyagin, Y.G.; Ordomsky, V.V. Catalysis by coke deposits: Synthesis of isoprene over solid catalysts. *Angew. Chem. Int. Ed.* **2013**, *52*, 12961–12964. [CrossRef] [PubMed]

4. Chattopadhyay, D.K.; Raju, K.V.S.N. Structural engineering of polyurethane coatings for high performance applications. *Prog. Polym. Sci.* **2007**, *32*, 352–418. [CrossRef]

5. Fenouillot, F.; Rousseau, A.; Colomines, G.; Saint-Loup, R.; Pascault, J.P. Polymers from renewable 1,4:3,6-dianhydrohexitols (isosorbide, isomannide and isoidide): A review. *Prog. Polym. Sci.* **2010**, *35*, 578–622. [CrossRef]

6. Yashima, T.; Katoh, Y.; Komatsu, T. Synthesis of 3-methyl-3-butene-1-ol from isobutene and formaldehyde on FeMCM-22 zeolites. In *Studies in Surface Science and Catalysis*; Kiricsi, I., Pál-Borbély, G., Nagy, J.B., Karge, H.G., Eds.; Elsevier: Amsterdam, The Netherlands, 1999; Volume 125, pp. 507–514.

7. Sun, J.; Zhu, K.; Gao, F.; Wang, C.; Liu, J.; Peden, C.H.; Wang, Y. Direct conversion of bio-ethanol to isobutene on nanosized $Zn_xZr_yO_z$ mixed oxides with balanced acid-base sites. *J. Am. Chem. Soc.* **2011**, *133*, 11096–11099. [CrossRef] [PubMed]

8. Liu, C.; Sun, J.; Smith, C.; Wang, Y. A study of $Zn_xZr_yO_z$ mixed oxides for direct conversion of ethanol to isobutene. *Appl. Catal. A* **2013**, *467*, 91–97. [CrossRef]

9. Sun, J.; Wang, Y. Recent advances in catalytic conversion of ethanol to chemicals. *ACS Catal.* **2014**, *4*, 1078–1090. [CrossRef]

10. Deo, G.; Wachs, I.E. Reactivity of supported vanadium-oxide catalysts—The partial oxidation of methanol. *J. Catal.* **1994**, *146*, 323–334. [CrossRef]

11. Routray, K.; Zhou, W.; Kiely, C.J.; Wachs, I.E. Catalysis science of methanol oxidation over iron vanadate catalysts: Nature of the catalytic active sites. *ACS Catal.* **2011**, *1*, 54–66. [CrossRef]

12. Martín, M.; Grossmann, I.E. Optimal simultaneous production of *i*-butene and ethanol from switchgrass. *Biomass Bioenergy* **2014**, *61*, 93–103. [CrossRef]

13. Crisci, A.J.; Dou, H.; Prasomsri, T.; Román-Leshkov, Y. Cascade reactions for the continuous and selective production of isobutene from bioderived acetic acid over zinc-zirconia catalysts. *ACS Catal.* **2014**, *4*, 4196–4200. [CrossRef]

14. De la Cruz, V.; Hernández, S.; Martín, M.; Grossmann, I.E. Integrated synthesis of biodiesel, bioethanol, isobutene, and glycerol ethers from algae. *Ind. Eng. Chem. Res.* **2014**, *53*, 14397–14407. [CrossRef]

15. Sushkevich, V.L.; Ordomsky, V.V.; Ivanova, I.I. Synthesis of isoprene from formaldehyde and isobutene over phosphate catalysts. *Appl. Catal. A* **2012**, *441–442*, 21–29. [CrossRef]

16. Sreevardhan Reddy, S.; David Raju, B.; Siva Kumar, V.; Padmasri, A.H.; Narayanan, S.; Rama Rao, K.S. Sulfonic acid functionalized mesoporous SBA-15 for selective synthesis of 4-phenyl-1,3-dioxane. *Catal. Commun.* **2007**, *8*, 261–266. [CrossRef]

17. Yamaguchi, T.; Nishimichi, C. Olefin-aldehyde condensation reaction on solid acids. *Catal. Today* **1993**, *16*, 555–562. [CrossRef]

18. Dumitriu, E.; Trong On, D.; Kaliaguine, S. Isoprene by Prins condensation over acidic molecular sieves. *J. Catal.* **1997**, *170*, 150–160. [CrossRef]

19. Krzywicki, A.; Wilanowicz, T.; Malinowski, S. Catalytic and physicochemical properties of the Al_2O_3–H_3PO_4 system, I. Vapor phase condensation of isobutylene and formaldehyde—The Prins reaction. *React. Kinet. Catal. Lett.* **1979**, *11*, 399–403. [CrossRef]

20. Ai, M. The formation of isoprene by means of a vapor-phase prins reaction between formaldehyde and isobutene. *J. Catal.* **1987**, *106*, 280–286. [CrossRef]

21. Li, G.; Gu, Y.; Ding, Y.; Zhang, H.; Wang, J.; Gao, Q.; Yan, L.; Suo, J. Wells–Dawson type molybdovanadophosphoric heteropolyacids catalyzed Prins cyclization of alkenes with paraformaldehyde under mild conditions—a facile and efficient method to 1,3-dioxane derivatives. *J. Mol. Catal. A Chem.* **2004**, *218*, 147–152. [CrossRef]

22. Songsiri, N.; Rempel, G.L.; Prasassarakich, P. Liquid-phase synthesis of isoprene from methyltert-butyl ether and formalin using Keggin-type heteropolyacids. *Ind. Eng. Chem. Res.* **2016**, *55*, 8933–8940. [CrossRef]

23. Wang, Y.H.; Wang, F.; Zhang, C.F.; Zhang, J.; Li, M.R.; Xu, J. Transformylating amine with DMF to formamide over CeO_2 catalyst. *Chem. Commun.* **2014**, *50*, 2438–2441. [CrossRef] [PubMed]

24. Wang, Y.H.; Wang, F.; Song, Q.; Xin, Q.; Xu, S.T.; Xu, J. Heterogeneous ceria catalyst with water-tolerant Lewis acidic sites for one-pot synthesis of 1,3-diols via Prins condensation and hydrolysis reactions. *J. Am. Chem. Soc.* **2013**, *135*, 1506–1515. [CrossRef] [PubMed]

25. Zhang, Z.; Wang, Y.; Lu, J.; Zhang, C.; Wang, M.; Li, M.; Liu, X.; Wang, F. Conversion of isobutene and formaldehyde to diol using praseodymium-doped CeO_2 catalyst. *ACS Catal.* **2016**, *6*, 8248–8254. [CrossRef]

26. Tamura, M.; Siddiki, S.; Shimizu, K. CeO_2 as a versatile and reusable catalyst for transesterification of esters with alcohols under solvent-free conditions. *Green Chem.* **2013**, *15*, 1641–1646. [CrossRef]

27. Tamura, M.; Sawabe, K.; Tomishige, K.; Satsuma, A.; Shimizu, K.-I. Substrate-specific heterogeneous catalysis of CeO_2 by entropic effects via multiple interactions. *ACS Catal.* **2015**, *5*, 20–26. [CrossRef]

28. Tamura, M.; Noro, K.; Honda, M.; Nakagawa, Y.; Tomishige, K. Highly efficient synthesis of cyclic ureas from CO_2 and diamines by a pure CeO_2 catalyst using a 2-propanol solvent. *Green Chem.* **2013**, *15*, 1567–1577. [CrossRef]

29. Paier, J.; Penschke, C.; Sauer, J. Oxygen defects and surface chemistry of ceria: Quantum chemical studies compared to experiment. *Chem. Rev.* **2013**, *113*, 3949–3985. [CrossRef] [PubMed]

30. Zhang, S.; Huang, Z.Q.; Ma, Y.; Gao, W.; Li, J.; Cao, F.; Li, L.; Chang, C.R.; Qu, Y. Solid frustrated-Lewis-pair catalysts constructed by regulations on surface defects of porous nanorods of CeO_2. *Nat. Commun.* **2017**, *8*, 15266. [CrossRef] [PubMed]

31. Montini, T.; Melchionna, M.; Monai, M.; Fornasiero, P. Fundamentals and catalytic applications of CeO_2-based materials. *Chem. Rev.* **2016**, *116*, 5987–6041. [CrossRef] [PubMed]

32. Sun, C.W.; Li, H.; Chen, L.Q. Nanostructured ceria-based materials: Synthesis, properties, and applications. *Energy Environ. Sci.* **2012**, *5*, 8475–8505. [CrossRef]

33. Si, R.; Flytzani-Stephanopoulos, M. Shape and crystal-plane effects of nanoscale ceria on the activity of Au-CeO_2 catalysts for the water-gas shift reaction. *Angew. Chem. Int. Ed.* **2008**, *47*, 2884–2887. [CrossRef] [PubMed]

34. Dumitriu, E.; Hulea, V.; Fechete, I.; Catrinescu, C.; Auroux, A.; Lacaze, J.-F.; Guimon, C. Prins condensation of isobutylene and formaldehyde over Fe-silicates of MFI structure. *Appl. Catal. A* **1999**, *181*, 15–28. [CrossRef]

35. Qiao, Z.A.; Wu, Z.L.; Dai, S. Shape-controlled ceria-based nanostructures for catalysis applications. *ChemSusChem* **2013**, *6*, 1821–1833. [CrossRef] [PubMed]

36. Wang, M.; Wang, F.; Ma, J.P.; Li, M.R.; Zhang, Z.; Wang, Y.H.; Zhang, X.C.; Xu, J. Investigations on the crystal plane effect of ceria on gold catalysis in the oxidative dehydrogenation of alcohols and amines in the liquid phase. *Chem. Commun.* **2014**, *50*, 292–294. [CrossRef] [PubMed]

37. Mai, H.X.; Sun, L.D.; Zhang, Y.W.; Si, R.; Feng, W.; Zhang, H.P.; Liu, H.C.; Yan, C.H. Shape-selective synthesis and oxygen storage behavior of ceria nanopolyhedra, nanorods, and nanocubes. *J. Phys. Chem. B* **2005**, *109*, 24380–24385. [CrossRef] [PubMed]

38. Zhang, Z.; Wang, Y.; Wang, M.; Lu, J.; Zhang, C.; Li, L.; Jiang, J.; Wang, F. The cascade synthesis of α,β-unsaturated ketones via oxidative C–C coupling of ketones and primary alcohols over a ceria catalyst. *Catal. Sci. Technol.* **2016**, *6*, 1693–1700. [CrossRef]

39. Zhang, Z.; Wang, Y.; Wang, M.; Lü, J.; Li, L.; Zhang, Z.; Li, M.; Jiang, J.; Wang, F. An investigation of the effects of CeO_2 crystal planes on the aerobic oxidative synthesis of imines from alcohols and amines. *Chin. J. Catal.* **2015**, *36*, 1623–1630. [CrossRef]

40. Huang, X.S.; Sun, H.; Wang, L.C.; Liu, Y.M.; Fan, K.N.; Cao, Y. Morphology effects of nanoscale ceria on the activity of Au/CeO_2 catalysts for low-temperature CO oxidation. *Appl. Catal. B* **2009**, *90*, 224–232. [CrossRef]

41. Parry, E.P. An infrared study of pyridine adsorbed on acidic solids. Characterization of surface acidity. *J. Catal.* **1963**, *2*, 371–379. [CrossRef]

42. Chakraborty, B.; Viswanathan, B. Surface acidity of MCM-41 by in situ IR studies of pyridine adsorption. *Catal. Today* **1999**, *49*, 253–260. [CrossRef]

43. Tamura, M.; Shimizu, K.-I.; Satsuma, A. Comprehensive IR study on acid/base properties of metal oxides. *Appl. Catal. A* **2012**, *433–434*, 135–145. [CrossRef]

44. Emeis, C.A. Determination of integrated molar extinction coefficients for infrared absorption bands of pyridine adsorbed on solid acid catalysts. *J. Catal.* **1993**, *141*, 347–354. [CrossRef]

45. Tamura, M.; Wakasugi, H.; Shimizu, K.; Satsuma, A. Efficient and substrate-specific hydration of nitriles to amides in water by using a CeO_2 catalyst. *Chem. Eur. J.* **2011**, *17*, 11428–11431. [CrossRef] [PubMed]

46. Wu, Z.; Mann, A.K.P.; Li, M.; Overbury, S.H. Spectroscopic investigation of surface-dependent acid–base property of ceria nanoshapes. *J. Phys. Chem. C* **2015**, *119*, 7340–7350. [CrossRef]

47. Zaki, M.I.; Hussein, G.A.M.; Mansour, S.A.A.; El-Ammawy, H.A. Adsorption and surface reactions of pyridine on pure and doped ceria catalysts as studied by infrared spectroscopy. *J. Mol. Catal.* **1989**, *51*, 209–220. [CrossRef]

48. Binet, C.; Daturi, M.; Lavalley, J.-C. IR study of polycrystalline ceria properties in oxidised and reduced states. *Catal. Today* **1999**, *50*, 207–225. [CrossRef]

49. Flego, C.; Kiricsi, I.; Perego, C.; Bellussi, G. The Origin of the Band at 1462 cm^{-1} Generally Appearing Upon Desorption of Pyridine from Acidic Solids-Steps Towards a More Comprehensive Understanding. *Catal. Lett.* **1995**, *35*, 125–133. [CrossRef]

50. Shyu, J.Z.; Weber, W.H.; Gandhi, H.S. Surface characterization of alumina-supported ceria. *J. Phys. Chem.* **1988**, *92*, 4964–4970. [CrossRef]

51. Wu, Z.; Li, M.; Howe, J.; Meyer, H.M.; Overbury, S.H. Probing defect sites on CeO_2 nanocrystals with well-defined surface planes by Raman spectroscopy and O_2 adsorption. *Langmuir* **2010**, *26*, 16595–16606. [CrossRef] [PubMed]

52. Popovic, Z.V.; Dohcevic-Mitrovic, Z.; Konstantinovic, M.J.; Scepanovic, M. Raman scattering characterization of nanopowders and nanowires (rods). *J. Raman Spectrosc.* **2007**, *38*, 750–755. [CrossRef]

53. Li, L.; Chen, F.; Lu, J.Q.; Luo, M.F. Study of defect sites in $Ce_{1-x}M_xO_{2-\delta}$ ($x = 0.2$) solid solutions using Raman spectroscopy. *J. Phys. Chem. A* **2011**, *115*, 7972–7977. [CrossRef] [PubMed]

15

Mechanistic Implications for the Ni(I)-Catalyzed Kumada Cross-Coupling Reaction

Linda Iffland [1], Anette Petuker [1], Maurice van Gastel [2] and Ulf-Peter Apfel [1],* ⓘ

[1] Anorganische Chemie I, Ruhr-Universität Bochum, Universitätsstraße 150, 44801 Bochum, Germany; Linda.Iffland@rub.de (L.I.); Anette.Petuker@rub.de (A.P.)

[2] Max-Planck-Institut für Chemische Energiekonversion, Stiftstraße 34-36, 45470 Mülheim, Germany; maurice.van-gastel@cec.mpg.de

* Correspondence: ulf.apfel@rub.de

Abstract: Herein we report on the cross-coupling reaction of phenylmagnesium bromide with aryl halides using the well-defined tetrahedral Ni(I) complex, [(Triphos)Ni^{I}Cl] (Triphos = 1,1,1-tris(diphenylphosphinomethyl)ethane). In the presence of 0.5 mol % [(Triphos)Ni^{I}Cl], good to excellent yields (75–97%) of the respective coupling products within a reaction time of only 2.5 h at room temperature were achieved. Likewise, the tripodal Ni(II)complexes [(κ^2-Triphos)Ni^{II}Cl$_2$] and [(κ^3-Triphos)Ni^{II}Cl](X) (X = ClO$_4$, BF$_4$) were tested as potential pre-catalysts for the Kumada cross-coupling reaction. While the Ni(II) complexes also afford the coupling products in comparable yields, mechanistic investigations by UV/Vis and electron paramagnetic resonance (EPR) spectroscopy indicate a Ni(I) intermediate as the catalytically active species in the Kumada cross-coupling reaction. Based on experimental findings and density functional theory (DFT) calculations, a plausible Ni(I)-catalyzed reaction mechanism for the Kumada cross-coupling reaction is presented.

Keywords: nickel; tripodal ligands; cross-coupling; EPR; Grignard; DFT

---

1. Introduction

Metal-catalyzed cross-coupling reactions are an essential tool for the formation of carbon–carbon bonds [1]. Coupling reactions of organometallic reagents with organic electrophiles are industrially utilized for the synthesis of pharmaceuticals, agrochemicals and polymers. Although the first cross-coupling reactions of organomagnesium bromides described by Kumada or Corriu utilized nickel catalysts, for example, [Ni(dppe)Cl$_2$] and Ni(acac)$_2$ [2,3], the most efficient and commonly employed catalysts for such C–C coupling reactions nowadays are based on noble metals [4]. In particular, palladium and its complexes are exhaustively reported for C–C bond formation reactions and mechanistic details of Pd-based catalysts are well understood [5]. Due to the high cost and low abundance of noble metals, the development of alternative catalytic systems based on earth-abundant and less expensive metals, such as nickel and iron, recently attracted much interest. Likewise, Ni(0), Ni(I), Ni(II) and Ni(III) complexes have been proposed as potential intermediates for various C–C coupling reactions including Heck, Hiyama, Kumada, Negishi, Suzuki–Miyaura, Sonogashira and Stille coupling techniques [6–8]. Along this line, Ni(I) as well as Ni(III) species were proposed as reactive intermediates in the nickel-catalyzed coupling of aryl halides by *trans*-ArNiBr(PEt$_3$)$_2$ [9]. Notably, nickel is substantially more nucleophilic as compared to Pd and Pt due to its smaller atomic size. Because of their higher nucleophilicity, nickel catalysts allow reactions under milder conditions and more challenging electrophiles than can be used in palladium catalysis.

While numerous examples of low-valent nickel(I) complexes were reported, well-defined stable Ni(I) precursors as catalysts in cross-coupling reactions are rare [10]. Such Ni(I) complexes are frequently used as potent pre-catalysts in the Negishi coupling of alkyliodides and alkylzinc bromides [11–13], in the Suzuki–Miyaura reaction for the formation of C–C bonds from arylhalides and phenylboronic acid [14–16], as well as in Ni(I)-catalyzed Kumada coupling reactions [17,18]. Here, most commonly, bulky N-heterocyclic carbenes (NHCs) are utilized as a ligand platform for the stabilization of the monovalent nickel centers. Along this line, Whittlesey and coworkers reported a series of extremely air-sensitive [Ni^I(PPh$_3$)(NHC)X] complexes (X = Br, Cl) (Scheme 1) comprising bulky NHC ligands [17]. Herein, the sterically demanding substitution at the central heteroaromatic ring is required to stabilize the low-coordinate and low-valent Ni(I)-complex and to avoid intermolecular reactions. The Ni(I) complexes were subsequently studied for the Kumada coupling of aryl chlorides, for example, chlorobenzene and 4-chlorotoluene, with phenylmagnesium chloride leading to 1,1'-biphenyl or 4-methyl-1,1'-biphenyl.

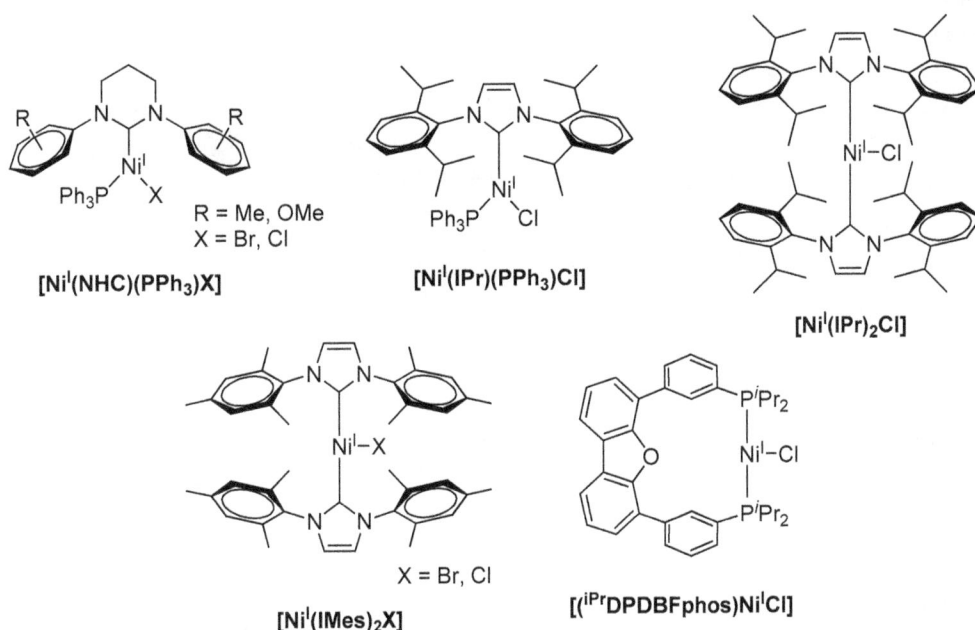

Scheme 1. Literature-known monovalent nickel catalysts for the Kumada coupling [10].

Notably, an increase in both the carbene ring size and steric demand of the N-substituents led to a decrease in catalytic activity. Likewise, bulkier cross-coupling partners like mesitylmagnesium bromide or more challenging substrates (e.g., aryl fluorides) led to a decrease in the yield of the bi-aryl products to less than 30%. Among all Ni(I) species used for these C–C bond formation studies, [Ni^I(6-Mes)(PPh$_3$)Br] (6-Mes = 1,3-bis(2,4,6-trimethylphenyl)hexahydropyrimidine-2-ylidene) showed highest activity for the coupling of chlorobenzene and phenylmagnesium chloride with yields up to 83% of 1,1'-biphenyl [5]. The efficiency of the coupling reaction could further be increased when complexes [Ni^I(IPr)(PPh$_3$)Cl] and [Ni^I(IPr)$_2$Cl] (IPr = 1,3-bis(2,6-diisopropylphenyl)imidazolin-2-ylidene) were applied with catalyst loadings of 1 mol %, affording yields of up to 98% for the desired coupling product within reaction times of 3 and 18 h [18,19]. In analogy, [Ni^I(IMes)$_2$X] (IMes = 1,3-bis(2,4,6-trimethylphenyl)imidazole-2-ylidene, X = Br, Cl) acts as a potent catalyst for the Kumada cross-coupling reaction (Scheme 1) [20]. When compared to [Ni^0(IMes)$_2$] and [Ni^{II}(IMes)$_2$X$_2$], all three complexes revealed similar catalytic activity for the reaction of chlorobenzene and mesitylmagnesium bromide under the same experimental conditions and afforded 2,4,6-trimethyl-1,1'-biphenyl in 73–79% yield when 3 mol % catalyst was applied. It is worth mentioning that biscarbene Ni(I) complexes were shown to dimerize under the applied reaction conditions, forming dinuclear Ni(I)(μ-Cl)$_2$Ni(I) intermediates that were suggested as the active catalyst, and the involvement of both [Ni(I)–Ni(I)] and [Ni(II)–Ni(II)] intermediates was subsequently suggested [21]. [(^{iPr}DPDBFphos)Ni^{I}Cl] (^{iPr}DPDBFphos = 4,6-bis(diisopropylphosphinophenyl)dibenzofuran) even allowed for the coupling of vinyl chloride and a phenyl Grignard reagent, allowing for the incorporation of a functional group into the target structure and affording styrene in about 85% yield within 22 h stirring at room temperature [22]. However, similar yields were obtained when the respective Ni(II) complex was utilized for the very same reaction.

We recently showed that Triphos (2-((diphenylphosphaneyl)methyl)-2-methylpropane-1,3-diyl)bis(diphenylphosphane)- and Triphos^Si (((methylsilanetriyl)tris(methylene))-tris(diphenylphosphane))-derived Ni and Fe complexes are potential noble metal-free platforms to conduct the C–C cross-coupling of aryl iodides and alkynes [23,24]. Furthermore, the Ni(Triphos) complexes were shown to stabilize various oxidation states of Ni (Scheme 2) [25]. Notably, due to the high steric bulk of the Triphos ligand, such Ni complexes neither show any disproportionation of the respective Ni(I) complexes to Ni(0) and Ni(II), nor do they allow for dimerization to give a dinickel complex. The reported reactivity of such Ni(Triphos) complexes pointed our attention towards their application for Kumada cross-coupling reactions. Since Ni(0), Ni(I) and Ni(II) complexes were shown to be reactive complexes and can act as pre-catalysts in the C–C bond coupling reaction under Kumada conditions, we herein explore the role of the oxidation state based on well-defined [(Triphos)Ni^ICl] for the C–C bond formation by UV/Vis and electron paramagnetic resonance (EPR) spectroscopy. The data is furthermore supported by density functional theory (DFT) calculations.

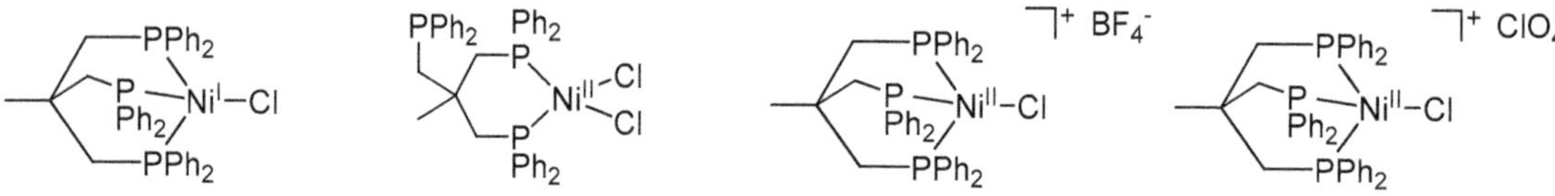

Scheme 2. Nickel complexes [(Triphos)Ni^ICl] (**1**), [(Triphos)Ni^IICl₂] (**2**), [(Triphos)Ni^IICl](BF₄) (**3**) and [(Triphos)Ni^IICl](ClO₄) (**4**).

2. Results and Discussions

Following a recently reported synthetic route, the Ni(I) complex [(Triphos)Ni^ICl] (**1**) was obtained by reduction of the square-planar complex [(Triphos)Ni^IICl₂] (**2**) with cobaltocene in gram scale and good yield (87%). The molecular structure of **1** was previously unequivocally shown by single-crystal X-ray crystallography and the monovalent electronic configuration was confirmed by EPR spectroscopy, as well as its magnetic moment via the Evans method (μ = 1.9 μ_B) [25].

We subsequently investigated the catalytic potential of complex **1** for Kumada-type cross-coupling reactions. As a model reaction, we chose the coupling of para-substituted aryl iodides and phenylmagnesium bromide in THF at room temperature in the presence of 0.5 mol % complex **1**.

Under those conditions, iodobenzene, 4-iodotoluene as well as 4-iodoanisole were successfully coupled to phenylmagnesium bromide and afforded the respective coupling products in 89%, 94% and 86% yield, respectively (Table 1). Likewise, we were able to utilize bromobenzene and 4-bromotoluene with phenylmagnesium bromide, showing that bromo-substituted arenes are suitable for the coupling with complex **1** and can be used instead of the usually expensive iodo-derivatives (Table 1). Here, 1,1'-biphenyl and 4-methyl-1,1'-biphenyl were obtained in 91% and 74% isolated yield, respectively. Following this line, we also tested the application of aryl chlorides and fluorides as substitutes in the cross-coupling reaction procedure. However, the respective coupling products were only obtained in 3% and 5% yield at room temperature, respectively (Table 1). When the reaction temperature was increased to 60 °C, for both aryl chlorides and fluorides, the expected coupling product 1,1'-biphenyl was detected via GC–MS analysis in 100% yield starting from chlorobenzene and in 44% yield for fluorobenzene (Table 1). Thus, aryl chlorides and fluorides are suitable starting materials at higher temperature for Kumada cross-coupling reactions when complex **1** is applied as potential catalyst, but not at room temperature. Notably, in the absence of complex **1**, no formation of any respective bi-aryl product was observed, highlighting the role of the Ni(I) complex **1** as an important pre-catalyst herein. For the latter case, solely starting material was recovered quantitatively.

Table 1. Kumada cross-coupling reaction of aryl halides with phenylmagnesium bromide catalyzed by 0.5 mol % **1** [a].

Aryl Halide	Yield in (%) for X = I	Yield in (%) for X = Br	Yield in (%) for X = Cl	Yield in (%) for X = F
(phenyl)—X	89 [a]	91 [a]	3 [a], 100 [c]	3 [a], 44 [c]
(methylphenyl)—X	94 [a]	74 [a]	0 [b]	–
(methoxyphenyl)—X	86 [a]	–	5	–

[a] Reaction conditions: RArI (2.8 mmol), ArMgBr (5.6 mmol), [Ni(I)] (0.5 mol %), THF (6 mL), RT, 2.5 h. Isolated yields; [b] Coupling product detected by TLC, but not isolated; [c] Reaction conditions: RArI (2.8 mmol), ArMgBr (5.6 mmol), [Ni(I)] (0.5 mol %), THF (6 mL), 60 °C, 2.5 h. Yield determined by GC–MS analysis using 4-methyl-1,1′-biphenyl as an internal standard.

While the coupling of aryl halides and phenyl Grignard reagents proceeds rapidly and affords the coupling products in high yields, we set out to explore the possibility to perform coupling of alkyl halides and/or alkyl Grignard reagents. In a first step, we used phenylethynyl- and methylmagnesium bromide. All such reactions, however, did not yield any of the desired coupling products, and attempts to alter the reaction conditions by changing temperature, reaction time as well as the catalyst concentration failed as well in our hands. Likewise, complex **1** cannot facilitate the conversion of 1,2-diiodoethane and phenylmagnesium bromide. Neither (2-iodoethyl)benzene nor 1,2-diphenylethane were formed under our standard conditions, excluding alkyliodides for application in Kumada cross-coupling reactions with the investigated catalyst. These observations show that the monovalent Ni complex **1** is unsuitable for the coupling of C–C bonds from sp³-centers.

A common limitation of the Kumada cross-coupling reaction is the catalysts' low tolerance towards functional groups within the substrate. To investigate the functional group tolerance of complex **1** as catalyst, we chose different substituted aryl iodides and also heterocyclic iodides as challenging substrates for the Kumada coupling (Table 2). Phenyl iodides with additional alkyl, alkoxy, aryl and fluoride substituents afforded the expected coupling products with yields between 86% and 94% (Table 2, Entries 1–6, 8). Notably, polysubstituted aryl iodides comprising substituents in the *ortho*-position, for example, 2-iodomesitylene and 1-iodonaphthalene, did not show formation of any coupling product (Table 2, Entry 7 and 9). These observations suggest that aryl iodides with substituents directly adjacent to iodine are unsuitable substrates for the cross-coupling reaction utilizing complex **1**, due to steric hindrance of the oxidative addition of the aryl iodide to the Ni(I) center. Likewise, aryl iodides having amino, nitro, nitrile, hydroxyl, carbonyl or carboxylic acid groups did not allow for successful C–C bond formation and resulted in decomposition of complex **1** to a hitherto unknown product mixture (Table 2, Entry 11–17). Similarly, application of heteroaromatic substrates like iodopyridines (Table 2, Entry 18–20) did not result in the formation of the desired products, neither at a higher reaction temperature. Most likely, this lack of reactivity stems from an undesired coordination of the additional functional groups with complex **1**. Thus, Ni(I) complex **1** is only suitable to perform the Kumada coupling of aryl Grignard compounds and alkyl-, alkoxy- and aryl-substituted sterically less-demanding aryl iodides, and has only little tolerance for other functional groups.

Table 2. Kumada cross-coupling reaction of aryl iodides with phenylmagnesium bromide catalyzed by 0.5 mol % **1** [a].

Entry	Product		Yield (%) [b]	Entry	Product	Yield (%) [b]
1		P1	89	11	HO—	0
2		P2	94	12		0
3		P3	86	13		0
4		P4	92	14	H₂N—	0
5	F—	P5	75	15	O₂N—	0
6		P6	96	16	O₂N	0
7				17	NC—	0
8		P7	97	18		0
9			0	19		0
10	I—		0	20		0

[a] Reaction conditions: RArI (2.8 mmol), ArMgBr (5.6 mmol), [Ni(I)] (0.5 mol %), THF (6 mL), RT, 2.5 h;
[b] Isolated yields.

Since complexes featuring a Ni(II) ion were also reported for catalyzing Kumada cross-coupling reactions, we next tested the Ni(II) complexes **2–4** (1 and 0.5 mol % catalyst loading) as potential catalysts in our model reaction, namely the coupling of 4-iodotoluene and phenylmagnesium bromide under otherwise identical conditions as applied above. Ni(II) complex **2** was obtained in 94% yield according to literature-known procedures by complexation of Triphos with NiCl₂, and Ni(II) complexes **3** and **4** were formed via reaction of [Ni(CH₃CN)₆](BF)₂ or Ni(ClO₄)·6H₂O, respectively, to the Ni(II) complex **2** [23,25].

An overview of the observed product yields is provided in Table 3. Yields given for 4-methyl-1,1′-biphenyl were determined by GC–MS analysis using 1,1′-biphenyl as an internal calibration standard. In addition, isolated yields from experiments applying complex **1** are added for comparison. While application of complex **2** generally leads to a complete conversion of 4-iodotoluene to the desired coupling product for both 1 and 0.5 mol % catalyst loading, formation of 4-methyl-1,1′-biphenyl was observed in solely 72% (1 mol %) and 46% (0.5 mol %), as well as 59% (1 mol %) and 41% (0.5 mol %), yields for complexes **3** and **4**, respectively. However, while the amount of 4-methyl-1,1′-biphenyl formed is lowered for **3** and **4**, the results clearly show that Ni(II) complexes can likewise be utilized as potential pre-catalysts for the Kumada reaction, with complexes **1** and **2** showing the highest amounts of isolated product. Here, the choice of the counterions rather than the oxidation state of the nickel center seems to be of more importance. While complexes revealing only halogenide ligands like the tetrahedral Ni(I) complex **1** or the square planar Ni(II) complex **2** showed high activity for the coupling reaction, the formation of 4-methyl-1,1'-biphenyl by complexes **3** and **4** comprising non-coordinating BF_4^- or ClO_4^- counterions is hampered.

Table 3. Ni(I)- and Ni(II)-catalyzed Kumada cross-coupling reaction of 4-iodotoluene with phenylmagnesium bromide.

[Ni] Complex	Catalytic Amount of [Ni]	Yield (%)
1	1 mol %	81 [a], 100 [b]
1	0.5 mol %	94 [a], 100 [b]
2	1 mol %	100 [b]
2	0.5 mol %	100 [b]
3	1 mol %	72 ± 2 [b]
3	0.5 mol %	46 ± 1 [b]
4	1 mol %	59 ± 1 [b]
4	0.5 mol %	41 ± 1 [b]

[a] Isolated yields. [b] Yields determined by GC–MS analysis using 1,1'-biphenyl as an internal standard.

The high reactivity of both of the Ni(I) and Ni(II) complexes pointed our attention towards possible intermediates that are formed upon reaction with either phenylmagnesium bromide or 4-iodotoluene. Thus, we next performed stoichiometric reactions of the different Ni complexes with one equivalent of either reagent in THF. Subsequently, each solution was analyzed by UV/Vis spectroscopy (Figure 1). The UV/Vis spectrum of Ni(I) complex **1** has a broad absorption band at low wavelengths with a maximum at 385 nm (Figure 1A). Spectra of **2**, **3** and **4** differ significantly from the spectrum of **1**. While the UV/Vis spectrum of **2** has an absorption maximum at a wavelength of 465 nm, the spectra of **3** and **4**, which are nearly identical, show an absorption maximum at 425 nm and a broader band with a lower intensity at 670 nm (Figure 1B–D). For complex **1**, showing an absorption band at 385 nm, it is apparent that the addition of 4-iodotoluene did not lead to a change in the complexes' electronic spectrum (Figure 1A). This observation illustrates that the oxidative addition of the aryl iodide is presumably not the first step in the coupling process and no formation of a Ni(III) intermediate takes place. Based on this observation, it can also be excluded that disproportionation of the Ni(I) complex **1** takes place. Addition of phenylmagnesium bromide to **1**, similarly and regardless of whether the experiments were performed in the absence or presence of 4-iodotoluene, led only to minor alterations of the absorption band around 385 nm (Inset Figure 1A). Fundamentally different behavior, however, was observed when the Ni(II) complex **2** was investigated (Figure 1B). While addition of 4-iodotoluene also did not lead to any change of the UV/Vis spectrum, addition of phenylmagnesium bromide led to a significant change in the spectrum; the original band at 465 nm was lost and a new band centered at 385 nm appeared.

Identical trends were observed for the Ni(II) complexes **3** and **4** (Figure 1C,D). Complexes **3** and **4** reveal two characteristic absorption bands at around 430 and 670 nm in THF that disappear after treatment with phenylmagnesium bromide, suggesting an overall similar chemistry as observed for complexes **1** and **2**. It is, however, notable that for complex **3** an additional band centered at 465 nm can be observed that was not observed for the other complexes. The origin of this additional band is hitherto unknown. Altogether it becomes obvious that Ni complexes **1–4** do not react with only the aryl iodide. A brief comparison of the electronic features of complex **1** and complexes **2**, **3** and **4** after treatment with phenylmagnesium bromide clearly suggests similar electronic features. Thus, it becomes clear that phenylmagnesium additionally acts as a reducing agent.

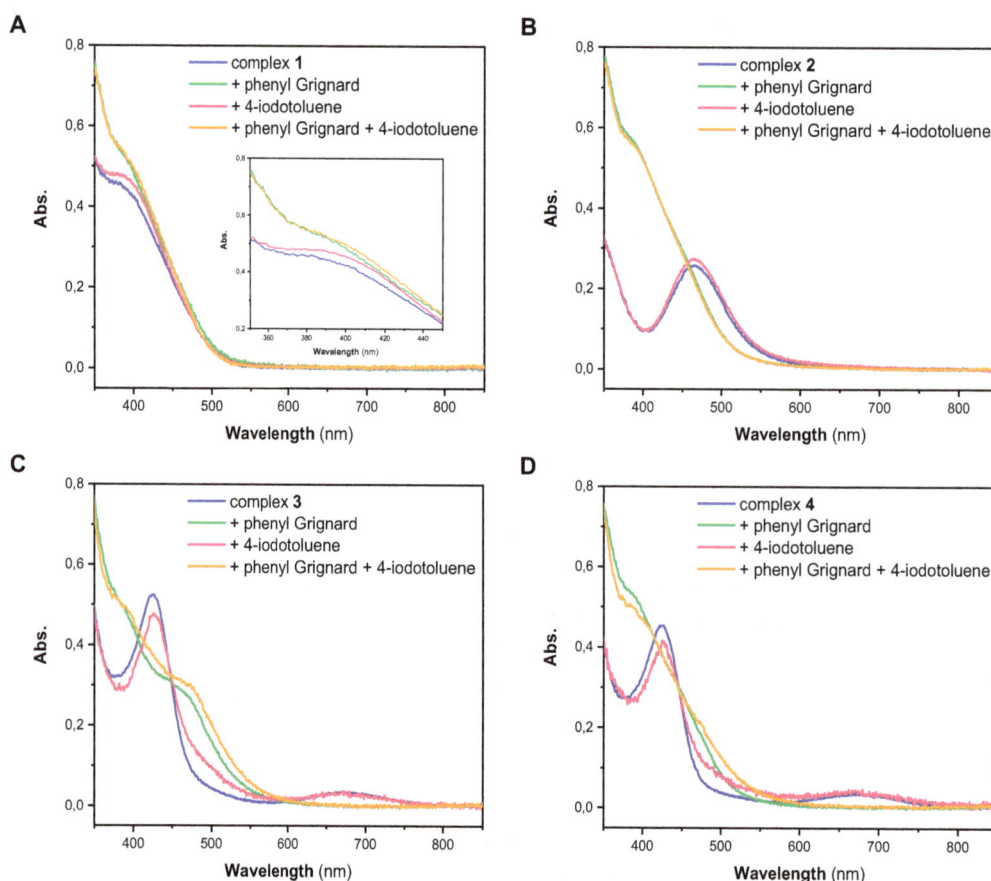

Figure 1. UV/Vis spectra of complexes **1** (**A**), **2** (**B**), 3 (**C**) and **4** (**D**) (blue) and their reaction solutions with 1 equiv. phenylmagnesium bromide (green), 1 equiv. 4-iodotoluene (red) and with both substrates (orange) in THF.

In addition to our UV/Vis studies, we crystallized the products of the different solutions. Notably, crystals were obtained from the reaction of complex **1** with phenylmagnesium bromide as well as of complexes **2**, **3** and **4** with phenylmagnesium bromide and 4-iodotoluene. For all experiments, we obtained a yellow crystalline material suitable for X-ray diffraction. The cell parameters of those crystals closely resembled those of complex **1**, suggesting the formation of a complex with an overall similar structure (Table S1). In addition, the reaction of **4** with phenylmagnesium bromide and 4-iodotoluene afforded a second set of crystalline material with slightly enlarged cell parameters (Table S1). While the quality of this crystal was not sufficiently good, a structural motif was obtained showing a (Triphos)Ni(I) moiety with a coordinated iodine ligand (Figure S1, Table S2).

In order to clarify the formation of a Ni(I) species during C–C bond formation, we further conducted electron paramagnetic resonance (EPR) spectroscopy on the same reaction solutions. This technique is ideal to elucidate the presence of paramagnetic oxidation states of the transition metal. As previously reported, we found a very rich, albeit greatly overlapping, hyperfine structure for complex **1** in THF (Figure S2) [25]. The overall spectrum did not alter upon addition of 4-iodotoluene. Addition of phenylmagnesium bromide to the solution of **1**, however, lead to broadening of the EPR signal and the hyperfine pattern slightly changed, while the g-value remained unaltered. The latter observation again confirms our hypothesis that complex **1** neither undergoes oxidative addition upon addition of aryl iodides nor shows any disproportionation to afford Ni(0) and Ni(II) species.

As was reported before, complexes **2**, **3** and **4** are EPR silent (Figure 2, Figures S3 and S4) and no change was observed upon addition of 4-iodotoluene [23,25]. As expected from the experiments shown above, freeze-quenched reaction solutions of the Ni(II) complexes **2**, **3** or **4** with added phenylmagnesium bromide revealed an EPR spectrum with identical g-values as for complex **1** with phenylmagnesium bromide but with minor changes to the complex hyperfine pattern. Notably, EPR spectra of reaction mixtures of the complexes and the Grignard reagent as well as 4-iodotoluene showed similar features as those without 4-iodotoluene. However, the signal intensity was significantly decreased, suggesting the formation of an EPR-silent species upon addition of 4-iodotoluene. While crystallographic data showed structurally related cell parameters, EPR analysis suggests that as well as [(Triphos)Ni^{I}I] found during crystallization, at least one additional Ni(II) species might have been formed during such experiments.

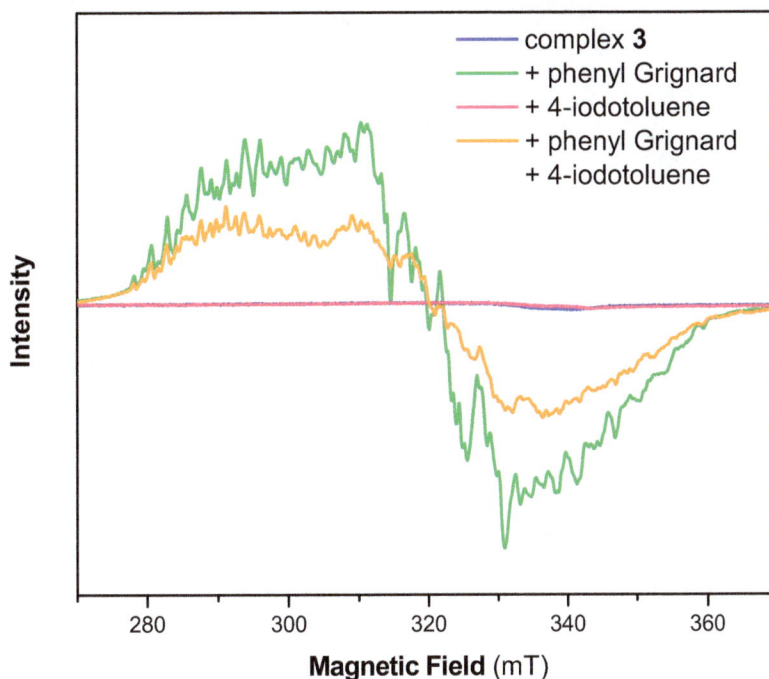

Figure 2. Electron paramagnetic resonance (EPR) spectrum of complex **3** before (blue) and after reaction with 1 equiv. phenyl Grignard (green), 1 equiv. 4-iodotoluene (red) and with both 1 equiv. phenyl Grignard and 1 equiv. 4-iodotoluene (orange) in THF. Experimental conditions: $T = 10$ K, frequency 9.63 GHz, microwave power 2 mW, modulation amplitude 0.3 mT.

As demonstrated by our experiments, the nickel complexes initially react with phenylmagnesium bromide and not with 4-iodotoluene. This finding suggests that in the Ni(I)-driven Kumada cross-coupling of aryl iodides and a Grignard reagent, the Grignard reagent initially acts as a reducing agent for Ni(II) to afford a Ni(I) species. Such a scenario would imply that phenyl radicals originating from the Grignard might form. While we were not able to detect such species in situ, we recognized that after 24 h, solutions of the Ni(II) complexes and phenylmagnesium bromide always contained small amounts of 1,1′-biphenyl (~1%) as analyzed by GC–MS. This assumption is supported by the fact that in the case of an additional treatment of complex **1** with phenylmagnesium bromide, we did not find any 1,1′-biphenyl.

Based on our experimental results we propose the following mechanism for the catalytic cross-coupling reaction catalyzed by the Ni complexes **1–4** (Scheme 3):

(1)   Starting from Ni(II) complexes **2–4**, a reduction to a Ni(I) species by a Grignard reagent as reducing agent occurs as initial reaction step. This reduction likewise leads to formation of 1,1′-biphenyl.

(2)   The Ni(I) species will then undergo transmetalation with a second phenylmagnesium bromide molecule. This proposition stems from the observation that the isolated Ni(I) complex **1** does not react with aryl iodides.

(3) The reactive Ni(I)–phenyl intermediate formed upon transmetalation then reacts under an oxidative addition with the aryl halogenide. It is worth mentioning that the generated Ni(III) species may bear the Triphos ligand either in a κ^2- or κ^3-coordination mode. Occupation of a κ^2-coordination mode seems an attractive option to reduce the steric hindrance at the Ni center and to allow for further reaction steps to proceed.

(4) Subsequently, the coupling product is released by reductive elimination and the Ni(I) species, now bearing an iodide ligand, is regenerated.

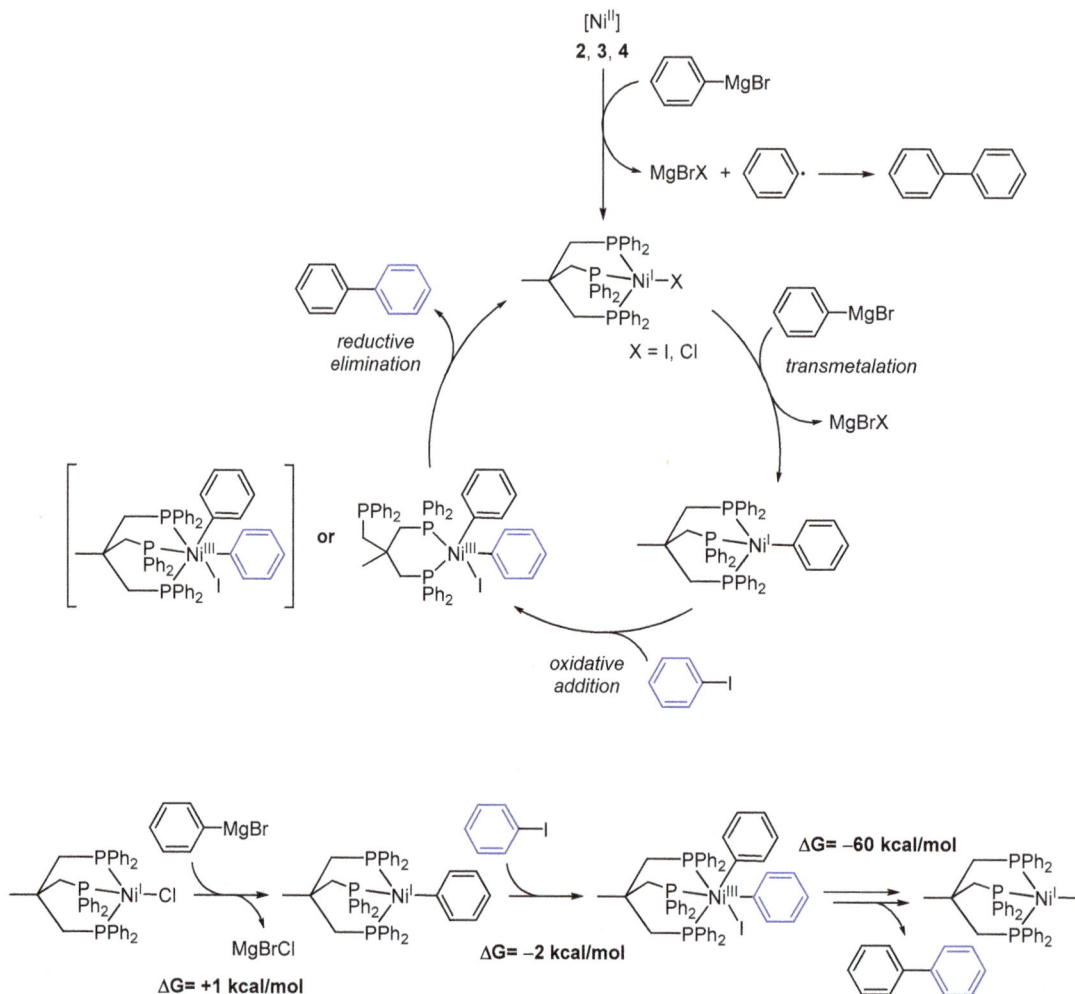

Scheme 3. Proposed mechanism for the Ni-catalyzed Kumada cross-coupling reaction with complexes **1–4**. While possible, DFT calculations suggest the formation of the Ni^{III} intermediate shown in parentheses not to be a significant part of the reaction cycle. Additional reaction scheme for the formation of one 1,1'-biphenyl molecule starting with **1** and ending up with [(Triphos)Ni^{I}I] with calculated ΔG values for each reaction step.

Quantum-chemical calculations of the reaction enthalpies were performed in order to gain support for the proposed cycle. Indeed, the calculations indicated that transmetalation with the Grignard reagent to form [Triphos)Ni^I(phenylate)] is isothermic (+1 kcal/mol) within the accuracy of the calculations, and subsequent reaction with iodobenzene is thermodynamically favored for the κ^2-coordination mode (Scheme 3, lower left) but endothermic for the κ^3-intermediate. As such, κ^2-conformation presents an ideal scaffold for the two phenylates to become co-coordinated to the same metal ion and in spatial vicinity. Moreover, the respective phenyl planes form an angle of 90 degrees in such a way that the two π-systems significantly overlap at the carbenes. The latter likely is of crucial importance as an onset for C–C bond formation. Release of biphenyl and subsequent re-coordination of the phosphine arm is highly exothermic (-60 kcal/mol) and gives rise to the biphenyl final product and [(Triphos)Ni^{I}I], completely in line with experimental observations.

As such, besides giving theoretical support for the proposed mechanism, the calculations additionally underline the importance of the flexibility of the Triphos ligand as a κ^2- or κ^3-coordinated ligand: it can easily harbor a low-spin $3d^8$ Ni(II) halide species as a κ^3-ligand, which is compatible with a preferred 4-coordinate square-planar coordination of low-spin Ni(II). Upon oxidation to Ni(III), $3d^7$, which prefers a five-coordinate square mono-pyramidal coordination, one phosphine easily dissociates in order to stabilize this oxidation state. Upon reductive elimination of biphenyl, re-coordination occurs.

3. Materials and Methods

General procedures. All reactions were performed under a dry N_2 or Ar atmosphere using standard Schlenk techniques or in a glovebox. [(Triphos)Ni^{I}Cl] (**1**), [(Triphos)Ni^{II}Cl$_2$] (**2**), [(Triphos)Ni^{II}Cl](BF$_4$) and [(Triphos)Ni^{II}Cl](ClO$_4$) were synthesized according to previously reported procedures [23,25]. All other compounds were obtained from commercial vendors and used without further purification. All solvents were dried prior to use according to standard methods. ^{1}H-, ^{13}C{^{1}H}-, ^{19}F- and ^{31}P{^{1}H}-NMR spectra were recorded with a Bruker DPX-200 NMR spectrometer or a Bruker DPX-250 NMR spectrometer at room temperature. Peaks were referenced to residual ^{1}H or ^{13}C signals from the deuterated solvent and are reported in parts per million (ppm). Mass spectra were measured with a Shimadzu QP-2010 instrument. UV/Vis spectra were recorded with a Varian Cary 100 spectrometer at room temperature. EPR spectra were recorded at $T = 10$ K on a Bruker Elexsys E500 X-band spectrometer equipped with an Oxford CF935 flow cryostat and a ST9102 resonator. The microwave frequency amounted to 9.636 GHz in all experiments. A microwave power of 2 mW was used. The modulation amplitude of 0.3 mT was chosen to optimize the S/N ratio. A few additional experiments were performed at lower modulation amplitude, but did not reveal additional or more resolved hyperfine structure.

Quantum Chemistry. All calculations were performed with the ORCA program package [26], geometry optimization and frequency analysis was performed by using the BP86 functional [27] and def2-SVP basis set [28] and zeroth order regular approach (zora) [29,30] in order to take into account scalar relativistic effects at nickel and iodine.

General procedure for Kumada coupling reactions. The respective potential catalyst (0.5 mol %) was dissolved in anhydrous THF (6 mL) and the corresponding aryl halide (2.8 mmol) was subsequently added. The solution was stirred for 10 min before the addition of the corresponding Grignard reagent (5.6 mmol). The reaction mixture was stirred for 2.5 h at room temperature. Subsequently, methanol was added to quench the reaction. Silica was added to the solution and the solvent was removed. Purification via column chromatography allowed for isolation of the cross-coupling products P1–P7. ^{1}H, ^{13}C and ^{19}F NMR spectra and GC-MS graphics for characterization of the coupling products are placed in Supplementary Materials.

1,1'-Biphenyl (P1): ^{1}H NMR (200 MHz, CDCl$_3$): $\delta = 7.64–7.59$ (m, 4H, H_{Ar}), 7.50–7.32 (m, 6H, H_{Ar}). ^{13}C NMR (50 MHz, CDCl$_3$): $\delta = 141.4, 128.9, 129.4, 127.3$. GC–MS ($m/z$): Calculated for [C$_{12}H_{10}$]$^+$: 154, found: 154.

4-Methyl-1,1'-biphenyl (P2): ^{1}H NMR (200 MHz, CDCl$_3$): $\delta = 7.62–7.29$ (m, 8H, H_{Ar}), 7.23 (m, 1H, H_{Ar}), 2.39 (s, 3H, CH_3). ^{13}C NMR (50 MHz, CDCl$_3$): $\delta = 141.3, 138.5, 137.2, 129.6, 238.9, 128.8, 127.1, 127.1$. GC–MS ($m/z$): Calculated for [C$_{13}H_{12}$]$^+$: 168, found: 168.

4-Methoxy-1,1'-biphenyl (P3): ^{1}H NMR (200 MHz, CDCl$_3$): $\delta = 7.59–7.50$ (m, 4H, H_{Ar}), 7.47–7.39 (m, 2H, H_{Ar}), 7.35–7.27 (m, 1H, H_{Ar}), 7.03–6.95 (m, 2H, H_{Ar}), 3.86 (s, 3H, CH_3). ^{13}C NMR (50 MHz, CDCl$_3$): $\delta = 159.3, 141.0, 133.9, 128.9, 128.3, 126.9, 126.8, 114.4, 55.5$. GC–MS ($m/z$): Calculated for [C$_{13}H_{12}$O]$^+$: 184, found: 184.

1,4-Diphenylbenzene (P4): ^{1}H NMR (200 MHz, CDCl$_3$): δ = 7.69–7.62 (m, 8H, H_{Ar}), 7.51–7.32 (m, 6H, H_{Ar}). ^{13}C NMR (50 MHz, CDCl$_3$): δ = 140.9, 140.3, 129.0, 127.6, 127.5, 127.2. GC–MS (m/z): Calculated for [C$_{18}$H$_{14}$]$^+$: 230, found: 230.

4-Fluoro-1,1′-biphenyl (P5): ^{1}H NMR (200 MHz, CDCl$_3$): δ = 7.59–7.52 (m, 4H, H_{Ar}), 7.49–7.33 (m, 3H, H_{Ar}), 7.19–7.08 (m, 2H, H_{Ar}). ^{13}C NMR (50 MHz, CDCl$_3$): δ = 162.6 (d, J = 246.2 Hz), 140.4, 137.5 (d, J = 3.3 Hz), 129.0, 128.8 (d, J = 8.0 Hz), 127.4, 127.2, 115.8 (d, J = 21.4 Hz). ^{19}F NMR (235 MHz, CDCl$_3$): –115.9. GC–MS (m/z): Calculated for [C$_{12}$H$_9$F]$^+$: 172, found: 172.

3,5-Dimethyl-1,1′-biphenyl (P6): ^{1}H NMR (200 MHz, CDCl$_3$): δ = 7.62 (d, 2H, H_{Ar}), 7.49–7.35 (m, 5H, H_{Ar}), 7.03 (s, 1H, H_{Ar}), 2.42 (s, 6H, CH_3). ^{13}C NMR (50 MHz, CDCl$_3$): δ = 141.6, 141.4, 138.3, 129.0, 128.7, 127.3, 127.2, 125.2, 21.5. GC–MS (m/z): Calculated for [C$_{14}$H$_{14}$]$^+$: 182, found: 182.

2-Phenylnaphthalene (P7): ^{1}H NMR (200 MHz, CDCl$_3$): δ = 8.05 (s, 1H, H_{Ar}), 7.95–7.85 (m, 3H, H_{Ar}), 7.78–7.71 (m, 3H, H_{Ar}), 7.55–7.35 (m, 5H, H_{Ar}). ^{13}C NMR (50 MHz, CDCl$_3$): δ = 141.3 138.7, 133.8, 132.8, 129.0, 128.6, 128.3, 127.8, 127.6, 127.5, 126.4, 126.1, 125.9, 125.7. GC–MS (m/z): Calculated for [C$_{16}$H$_{12}$]$^+$: 204, found: 204.

4. Conclusions

In conclusion, we herein showed that the Ni(I) complex **1** can indeed serve as a catalyst in C–C bond formation in Kumada cross-coupling reactions. Using attractive reaction conditions with short reaction times of 2.5 h at room temperature and very low catalytic loadings of 0.5 mol %, the coupling products were obtained in good to excellent yields. Owing to the relatively low tolerance of the Ni(I) complex **1** towards functional groups, the substrate scope for Kumada coupling reactions is limited to alkyl-, alkoxy- and aryl-substituted substrates. We also observed that Ni(II) complexes **2–4** are suitable to catalyze the coupling of aryl iodides with phenylmagnesium bromide. Investigating the reaction solutions of stoichiometric reactions of the Ni complexes with the coupling reagents by UV/vis measurements showed that reaction with the Grignard reagents leads to the same catalytic Ni(I) species for all complexes. Furthermore, the Ni complexes did not react under oxidative addition with aryl iodides. EPR measurements confirmed the formation of a Ni(I) species by adding Grignard reagent to the Ni(II) complexes. This is in line with the results from crystallization experiments, where for several reactions employing compounds **2**, **3** or **4**, Ni(I) complex **1** was isolated. Based on our findings, we propose that for Kumada cross-coupling reactions with Ni complexes bearing tripodal phosphine ligand Triphos, a monovalent Ni(I) complex acts as the catalytically active species.

Supplementary Materials:
Cif and cif checked files. Details of the calculations. Figure S1: Structure of the asymmetric unit of compound 1, Figure S2: EPR Spectrum of complex **1** (blue) and its reaction solutions with 1 equiv. phenyl Grignard (green), 1 equiv. 4-iodotoluene (red) and with both 1 equiv. phenyl Grignard and 1 equiv. 4-iodotoluene (orange) in THF, Figure S3: EPR Spectrum of complex **2** (blue) and its reaction solutions with 1 equiv. phenyl Grignard (green), 1 equiv. 4-iodotoluene (red) and with both 1 equiv. phenyl Grignard and 1 equiv. 4-iodotoluene (orange) in THF, Figure S4: EPR Spectrum of complex **4** (blue) and its reaction solutions with 1 equiv. phenyl Grignard (green), 1 equiv. 4-iodotoluene (red) and with both 1 equiv. phenyl Grignard and 1 equiv. 4-iodotoluene (orange) in THF, Figure S5: ^{1}H NMR (200 MHz) and ^{13}C NMR (50 MHz) spectra and GC-MS of 1,1′-biphenyl (**P1**), Figure S6: ^{1}H NMR (200 MHz) and ^{13}C NMR (50 MHz) spectra and GC-MS of 4-methyl-1,1′-biphenyl (**P2**), Figure S7: ^{1}H NMR (200 MHz) and ^{13}C NMR (50 MHz) spectra and GC-MS of 4-methoxy-1,1′-biphenyl (**P3**), Figure S8: ^{1}H NMR (200 MHz), ^{13}C NMR (50 MHz) and ^{19}F NMR (235 MHz) spectra and GC-MS of 4-fluoro-1,1′-biphenyl (**P4**), Figure S9: ^{1}H NMR (200 MHz) and ^{13}C NMR (50 MHz) and GC-MS of 3,5-dimethyl-1,1′-biphenyl (**P5**), Figure S10: ^{1}H NMR (200 MHz) and ^{13}C NMR (50 MHz) and GC-MS of 1,4-biphenylbenzene (**P6**), Figure S11: ^{1}H NMR (200 MHz) and ^{13}C NMR (50 MHz) and GC-MS of 2-phenylnaphthalene (**P7**), Table S1: Cell Parameters of complex **1** and of crystals obtained from the stoichiometric reactions solutions of 1 with 1 equiv. phenylmagnesium bromide, and of **2**, **3** or **4** with both 1 equiv. phenylmagnesium bromide and 1, Table S2: Crystal data and structure refinement of [(Triphos)Ni^{I}I], Data S1: Cartesian coordinates [Angstrom] of geometry-optimized models used in the calculations.

Acknowledgments: This work was supported by the Fonds der Chemischen Industrie (Liebig grant to Ulf-Peter Apfel) and through the Deutsche Forschungsgemeinschaft (Emmy Noether grant to Ulf-Peter Apfel, AP242/2-1).

Author Contributions: Linda Iffland, Anette Petuker and Ulf-Peter Apfel conceived and designed the experiments; Linda Iffland and Anette Petuker performed the experiments. All authors were involved in the analysis of the data; Maurice van Gastel performed DFT calculations and EPR spectroscopy. All authors were involved in writing the paper.

References

1. De Meijere, A.; Diederich, F. *Metal-Catalyzed Cross-Coupling Reactions*; Wiley-VCH: Weinheim, Germany, 2004.

2. Tamao, K.; Sumitani, K.; Kumada, M. Selective carbon–carbon bond formation by cross-coupling of Grignard reagents with organic halides. Catalysis by nickel-phosphine complexes. *J. Am. Chem. Soc.* **1972**, *94*, 4374–4376. [CrossRef]

3. Corriu, R.J.P.; Masse, J.P. Activation of Grignard reagents by transition-metal complexes. A new and simple synthesis of *trans*-stilbenes and polyphenyls. *J. Chem. Soc. Chem. Commun.* **1972**, 144a. [CrossRef]

4. Knappke, C.E.I.; von Wangelin, A.J. 35 years of palladium-catalyzed cross-coupling with Grignard reagents: How far have we come? *Chem. Soc. Rev.* **2011**, *40*, 4948. [CrossRef] [PubMed]

5. Braga, A.A.C.; Ujaque, G.; Maseras, F. Mechanism of Palladium-Catalyzed Cross-Coupling Reactions. In *Computational Modeling for Homogeneous and Enzymatic Catalysis: A Knowledge-Base for Designing Efficient Catalysts*; Morokuma, K., Musaev, D.G., Eds.; Wiley-VCH Verlag GmbH & Co. KGaA: Weinheim, Germany, 2008; pp. 109–130.

6. Rosen, B.M.; Quasdorf, K.W.; Wilson, D.A.; Zhang, N.; Resmerita, A.-M.; Garg, N.K.; Percec, V. Nickel-Catalyzed Cross-Couplings Involving Carbon–Oxygen Bonds. *Chem. Rev.* **2011**, *111*, 1346–1416. [CrossRef] [PubMed]

7. Amatore, C.; Jutand, A. Rates and mechanism of biphenyl synthesis catalyzed by electrogenerated coordinatively unsaturated nickel complexes. *Organometallics* **1988**, *7*, 2203–2214. [CrossRef]

8. Shi, S.; Meng, G.; Szostak, M. Synthesis of Biaryls through Nickel-Catalyzed Suzuki-Miyaura Coupling of Amides by Carbon–Nitrogen Bond Cleavage. *Angew. Chem. Int. Ed.* **2016**, *55*, 6959–6963. [CrossRef] [PubMed]

9. Tsou, T.T.; Kochi, J.K. Mechanism of biaryl synthesis with nickel complexes. *J. Am. Chem. Soc.* **1979**, *101*, 7547–7560. [CrossRef]

10. Lin, C.-Y.; Power, P.P. Complexes of Ni(I): A "rare" oxidation state of growing importance. *Chem. Soc. Rev.* **2017**, *46*, 5347–5399. [CrossRef] [PubMed]

11. Jones, D.G.; McFarland, C.; Anderson, T.J.; Vicic, D.A. Analysis of key steps in the catalytic cross-coupling of alkyl electrophiles under Negishi-like conditions. *Chem. Commun.* **2005**, *33*, 4211–4213. [CrossRef] [PubMed]

12. Jones, D.G.; Martin, J.L.; McFarland, C.; Allen, O.R.; Hall, R.E.; Haley, A.D.; Brandon, R.J.; Konovalova, T.; Desrochers, P.J.; Pulay, P.; et al. Ligand Redox Effects in the Synthesis, Electronic Structure, and Reactivity od an Alkyl-Alkyl Cross-Coupling Catalyst. *J. Am. Chem. Soc.* **2006**, *128*, 13175–13183. [CrossRef] [PubMed]

13. Phapale, V.B.; Buñuel, E.; García-Iglesias, M.; Cárdenas, D.J. Ni-Catalyzed Cascade Formation of $C(sp^3)$–$C(sp^3)$ Bonds by Cyclization and Cross-Coupling Reactions of Iodoalkanes with Alkyl Zinc Halides. *Angew. Chem. Int. Ed.* **2007**, *46*, 8790–8795. [CrossRef] [PubMed]

14. Wu, J.; Nova, A.; Balcells, D.; Brudvig, G.W.; Dai, W.; Guard, L.M.; Hazari, N.; Lin, P.-H.; Pokhrel, R.; Takase, M.K. Nickel(I) Monomers and Dimers with Cyclopentadienyl and Indenyl Ligands. *Chem. Eur. J.* **2014**, *20*, 5327–5337. [CrossRef] [PubMed]

15. Guard, L.M.; Mohadjer Beromi, M.; Brudvig, G.W.; Hazari, N.; Vinyard, D.J. Comparison of dppf-Supported Nickel Precatalysts for the Suzuki-Miyaura Reaction: The Observation and Activity of Nickel(I). *Angew. Chem. Int. Ed.* **2015**, *54*, 13352–13356. [CrossRef] [PubMed]

16. Matsubara, K.; Fukahori, Y.; Inatomi, T.; Tazaki, S.; Yamada, Y.; Koga, Y.; Kanegawa, S.; Nakamura, T. Monomeric Three-Coordinated *N*-Heterocyclic Carbene Nickel(I) Complexes: Synthesis, Structures, and Catalytic Applications in Cross-Coupling Reactions. *Organometallics* **2016**, *35*, 3281–3287. [CrossRef]

17. Page, M.J.; Lu, W.Y.; Poulten, R.C.; Carter, E.; Algarra, A.G.; Kariuki, B.M.; Macgregor, S.A.; Mahon, M.F.; Cavell, K.J.; Murphy, D.M.; et al. Three-Coordinated Nickel(I) Complexes Stabilised by Six-, Seven-, and Eight-Membered Ring *N*-Heterocyclic Carbenes: Synthesis, EPR/DFT Studies and Catalytic Activity. *Chem. Eur. J.* **2013**, *19*, 2158–2167. [CrossRef] [PubMed]

18. Miyazaki, S.; Koga, Y.; Matsumoto, T.; Matsubara, K. A new aspect of nickel-catalyzed Grignard cross-coupling reaction: Selective synthesis, structure, and catalytic behavior of T-shape three-coordinated nickel(I) chloride bearing a bulky NHC ligand. *Chem. Commun.* **2010**, *46*, 1932. [CrossRef] [PubMed]

19. Nagao, S.; Matsumoto, T.; Koga, Y.; Matsubara, K. Monovalent Nickel Complex Bearing a Bulky *N*-Heterocyclic Carbene Catalyzes Buchwald-Hartwig Amination of Aryl Halides under Mild Conditions. *Chem. Lett.* **2011**, *40*, 1036–1038. [CrossRef]

20. Zhang, K.; Conda-Sheridan, M.; Cooke, S.R.; Louie, J. *N*-Heterocyclic Carbene Bound Nickel(I) Complexes and Their Roles in Catalysis. *Organometallics* **2011**, *30*, 2546–2552. [CrossRef] [PubMed]

21. Matsubara, K.; Yamamoto, H.; Miyazaki, S.; Inatomi, T.; Nonaka, K.; Koga, Y.; Yamada, Y.; Veiros, L.F.; Kirchner, K. Dinuclear Systems in Efficient Nickel-Catalyzed Kumada-Tamao-Corriu Cross-Coupling of Aryl Halides. *Organometallics* **2017**, *36*, 255–265. [CrossRef]

22. Marlier, E.E.; Tereniak, S.J.; Ding, K.; Mulliken, J.E.; Lu, C.C. First-Row Transition-Metal Chloride Complexes of the Wide Bite-Angle Diphosphine [iPr]DPDBFphos and Reactivity Studies of Monovalent Nickel. *Inorg. Chem.* **2011**, *50*, 9290–9299. [CrossRef] [PubMed]

23. Petuker, A.; Merten, C.; Apfel, U.-P. Modulating Sonogashira Cross-Coupling Reactivity in Four-Coordinated Nickel Complexes by Using Geometric Control: Modulating Cross-Coupling Reactivity in Ni Complexes. *Eur. J. Inorg. Chem.* **2015**, *2015*, 2139–2144. [CrossRef]

24. Petuker, A.; El-Tokhey, M.; Reback, M.L.; Mallick, B.; Apfel, U.-P. Towards Iron-Catalyzed Sonogashira Cross-Coupling Reactions. *ChemistrySelect* **2016**, *1*, 2717–2721. [CrossRef]

25. Petuker, A.; Mebs, S.; Schuth, N.; Gerschel, P.; Reback, M.L.; Mallick, B.; van Gastel, M.; Haumann, M.; Apfel, U.-P. Spontaneous Si–C bond cleavage in (Triphos[Si])-nickel complexes. *Dalton Trans.* **2017**, *46*, 907–917. [CrossRef] [PubMed]

26. Neese, F. Software update: The ORCA program system, version 4.0: Software update. *Wiley Interdiscip. Rev. Comput. Mol. Sci.* **2017**, e1327. [CrossRef]

27. Becke, A.D. Density-functional exchange-energy approximation with correct asymptotic behavior. *Phys. Rev. A* **1988**, *38*, 3098–3100. [CrossRef]

28. Pantazis, D.A.; Chen, X.-Y.; Landis, C.R.; Neese, F. All Electron Scalar relativistic Basic Sets for Third-Row Transition Metal Atoms. *J. Chem. Theory Comput.* **2008**, *4*, 908–919. [CrossRef] [PubMed]

29. Lenthe, E.; Snijders, J.G.; Baerends, E.J. The zero-order regular approximation for relativistic effects: The effect of spin-orbit coupling in closed shell molecules. *J. Chem. Phys.* **1996**, *105*, 6505–6516. [CrossRef]

30. Van Lenthe, E.; van der Avoird, A.; Wormer, P.E.S. Density functional calculations of molecular hyperfine interactions in zero order regular approximation for relativistic effects. *J. Chem. Phys.* **1998**, *108*, 4783–4796. [CrossRef]

16

Synthesis of a Dichlorodigermasilane: Double Si–Cl Activation by a Ge=Ge Unit

Tomohiro Sugahara [1] (ID)**, Norihiro Tokitoh** [1] **and Takahiro Sasamori** [2,*]

[1] Institute for Chemical Research, Kyoto University, Gokasho, Uji, Kyoto 611-0011, Japan; sugahara@boc.kuicr.kyoto-u.ac.jp (T.S.); tokitoh@boc.kuicr.kyoto-u.ac.jp (N.T.)

[2] Graduate School of Natural Sciences, Nagoya City University, Yamanohata 1, Mizuho-cho, Mizuho-ku, Nagoya, Aichi 467-8501, Japan

* Correspondence: sasamori@nsc.nagoya-cu.ac.jp

Abstract: Halogenated oligosilanes and oligogermanes are interesting compounds in oligosilane chemistry from the viewpoint of silicon-based-materials. Herein, it was demonstrated that a 1,2-digermacyclobutadiene derivative could work as a bis-germylene building block towards double Si–Cl activation to give a halogenated oligometallane, a bis(chlorogermyl)dichlorosilane derivative.

Keywords: Si–Cl activation; germylene; digermene; digermacyclobutadiene

1. Introduction

Halogenated oligosilanes and oligogermanes are attractive compounds as functionalized oligometallanes from the standpoint of oligosilane material chemistry [1–7]. In this regard, Si(II) or Ge(II) species should be an important building block for creating such halogenated oligosilanes/germanes because silylenes (divalent Si(II) species) or germylenes (divalent Ge(II) species) have been known to undergo ready Si–Cl insertion reactions, i.e., Si–Cl activation reactions [8–11]. For example, Iwamoto and Kira reported the facile Si–Cl insertion of the isolable dialkylmetallylenes towards $SiCl_4$ under mild conditions [10]. However, especially in the germanium cases, it is difficult to isolate the insertion products, $>Ge(Cl)SiR_3$, because the insertion reaction of a germylene toward a Si–Cl bond would be reversible in some cases [12]. Thus, the substituents on the Si atom (R of the R_3Si–Cl species) should be bulky and/or electropositive to avoid the α-elimination of R_3Si–Cl from the Ge moiety [8,12–22]. The requirement for the bulkiness on the Si–Cl moiety could make it difficult to create halogenated oligometallanes, such as $>Ge(Cl)–SiCl_2–Ge(Cl)<$ with utilizing the double Si–Cl activation of the Ge(II) species towards $SiCl_4$, because of two unfavorable factors: (i) entropy, two Ge(II) moieties should react with one $SiCl_4$ species; and (ii) the stability of the product, the Cl atoms on the Si atom should promote the α-Si–Cl elimination, i.e., the retro reaction (Scheme 1). In this paper, we chose a 1,2-digermacyclobutadiene derivative [23] as a suitable Ge(II) building block for the double Si–Cl activation of $SiCl_4$ to yield $>Ge(Cl)–SiCl_2–Ge(Cl)<$ species, because the rigid cyclic skeleton should overcome the entropy-disadvantage, and the rigidness of the cyclic skeleton should suppress the α-Si–Cl elimination. Finally, it was found that the stable 1,2-digermacyclobutadiene **1** (1,2-Tbb$_2$-3,4-Ph$_2$-digermacyclobutadiene, Tbb = 2,6-[CH(SiMe$_3$)$_2$]$_2$-4-t-Bu-C$_6$H$_2$, Scheme 2) [24] with $SiCl_4$ afforded the corresponding 1,3-digerma-2-sila-cyclopent-4-ene derivative, the cyclic $>Ge(Cl)–SiCl_2–Ge(Cl)<$ compound. 1,2-Digermacyclobut-1-ene derivative was reacted with $SiCl_4$ to give the double-Si–Cl-insertion product, and the following reduction reaction gave the corresponding $>Ge=Si=Ge<$ species [11]. Although **1** could undergo facile double Si–Cl activation toward $SiCl_4$, neither double Ge–Cl nor C–Cl activation could occur in the reaction of **1** with $GeCl_4/CCl_4$.

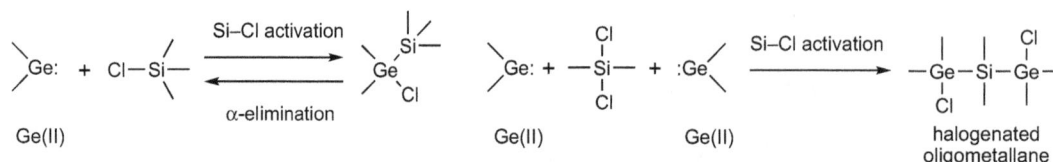

Scheme 1. Depictions for Si–Cl activations of Ge(II) species and Si–Cl α-elimination from a chlorosilylgermane.

2. Results and Discussions

When the stable digermyne **2** bearing bulky aryl substituents, Tbb groups (2,6-[CH(SiMe$_3$)$_2$]$_2$-4-t-Bu-C$_6$H$_2$), was treated with PhC≡CPh (tolan) at room temperature, 1,2-digermacyclobutadiene **1** was isolated as a stable crystalline compound [23–26] via formal [2+2] cycloaddition (Scheme 2). As one can see from the structure of **1**, it is a cyclic 4π-electron conjugated, anti-aromatic compound incorporating Ge(II) moieties. On the basis of theoretical calculations, **1** has considerable –Ge=Ge–C=C– character rather than =Ge–Ge=C–C– [24]. Accodingly, as expected, **1** could work as a building block of the bis-Ge(II) moiety. Reaction of **1** with SiCl$_4$ afforded digermadichlorosilane **3** quantitatively, which could be formed via double Si–Cl insertion reactions of the Ge(II) moieties of the 1,2-digermacyclobutadiene skeleton in **1**. This reaction has been performed under the neat condition at 55 °C because the addition of small amount of SiCl$_4$ or reaction at r.t. afforded very slow conversion of **3**. The obtained dichlorosilane **3** has the >Ge(Cl)–SiCl$_2$–Ge(Cl)< moiety in its 1,3-digerma-2-sila-cyclopent-4-ene skeleton, i.e., **2** should be one of a unique class of compounds of oligohalo-oligometallanes. Thus, **1** was found to work as a bis-germylene building block (>Ge: + :Ge<) towards a double Si–Cl activation.

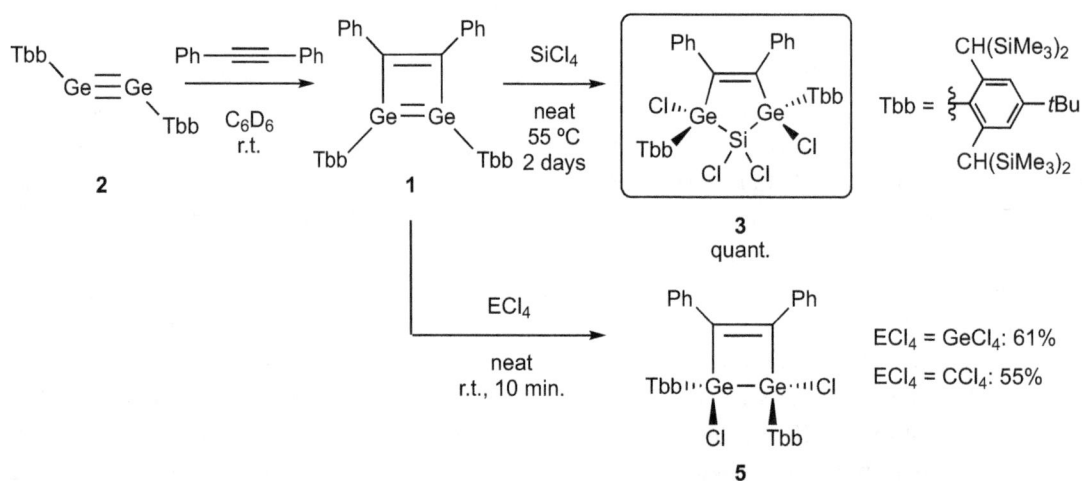

Scheme 2. Preparation of 1,2-digermacyclobutadiene **1**, and its reaction with ECl$_4$ giving digermadichlorosilane **3** (E = Si) and dichlorodigermacyclobutene **5** (E = C, Ge), respectively.

The molecular structure of digermadichlorosilane **3** was definitively determined by X-ray crystallographic analysis (Figure 1). The two Tbb/Cl groups are oriented in (E)-geometry probably due to steric reasons. The five-membered ring skeleton in **3** exhibits the envelope geometry with a deviation of the Si atom from the Ge–C=C–Ge plane by $ca.$ 1.27 Å. While the two Ge–Cl bond lengths are almost the same (Ge1–Cl1: 2.2094(14) Å, Ge2–Cl4: 2.2011(15) Å) within a range of standard deviations, the orientation of the two Cl atoms are slightly different to each other. That is, one of the Cl atom (Cl4) is oriented to outside of the five-membered ring skeleton, but another one (Cl1) is approaching to the central Si atom with the Cl1···Si distance of 3.25 Å, which is far from the other one (Cl4···Si = 3.66 Å) [27]. In addition, the two Cl–Ge–Si angles are considerably different from each other, (Cl1–Ge1–Si = 90.20(8), Cl4–Ge2–Si = 105.40(8)). These asymmetrical structural features indicate weak n(Cl1)$\cdots\sigma^*$(Si–Cl3) interaction. These structural features were reasonably reproduced by the theoretical structural optimization at B3PW91/6-311G(2d) [28]. The theoretically-optimized structure of the less hindered model **3'**, which has Me groups instead of Tbb groups, exhibits a completely planar five-membered skeleton with C_2 symmetry. Thus, these structural features observed in **3** could be due to the steric congestion.

In the expectation of obtaining the Ge analogue of **3**, digermadichlorogermane **4**, the reaction of **1** with GeCl$_4$ was attempted. As a result, the expected product, **4**, was not obtained, but the 1,2-dichloro-1,2-digermacyclobut-3-ene **5** was obtained as a predominant product even under the conditions of using only a small amount of GeCl$_4$ in the dark [29]. In addition, the reaction of **1** with CCl$_4$ also furnished the formation of **5** without any formation of the CCl$_2$-insertion product **6**. 1,2-Dichloro-1,2-digermacyclobutene **5** showed considerable stability in the air, and it can object to further purification by silica gel column chromatography. Although the reaction mechanism for the formation of **5** by the reaction of **1** with GeCl$_4$ or CCl$_4$ was not clear at present, the formation of **5** is most likely interpreted in terms of the double-chlorination of **1** with the elimination of ECl$_2$ (E = Ge or C) moiety.

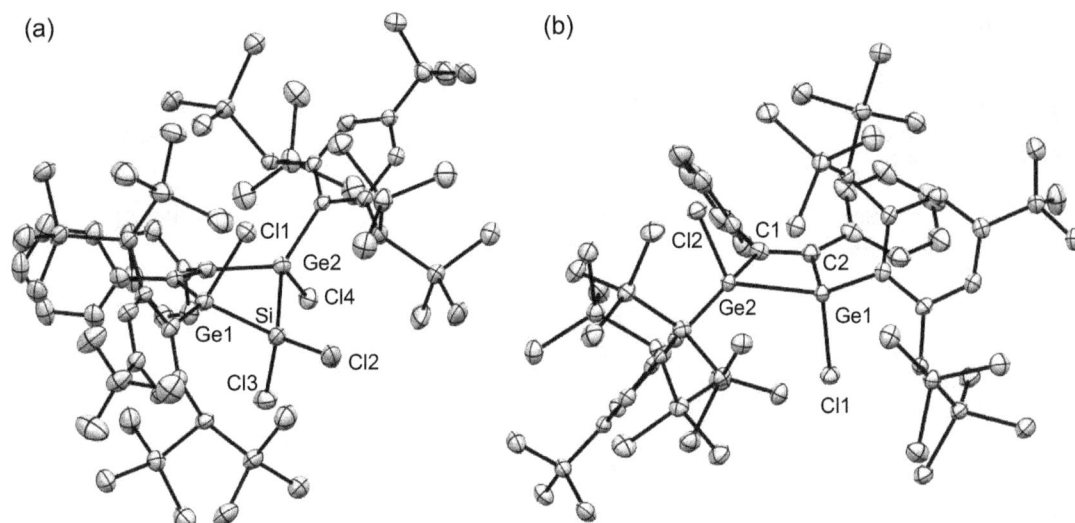

Figure 1. Molecular structures of (**a**) digermadichlorosilane **3** and (**b**) dichlorodigermacyclobutene **5** with atomic displacement parameters set at 50% probability. All hydrogen atoms and solvent molecules (THF and benzene) were omitted for clarity and only selected atoms are labeled. Selected bond lengths (Å) and angles (deg.): (**a**) **3**: Ge1–Si, 2.3734(16); Ge2–Si, 2.3938(15); Ge1–Cl1, 2.2094(14); Ge2–Cl4: 2.2011(15); Si–Cl2, 2.052(2); Si–Cl3, 2.053(2); Ge1–Si–Ge2, 91.15(5); Cl1–Ge1–Si, 90.29(5); Cl4–Ge2–Si, 105.40(6); Cl2–Si–Cl3, 106.04(9); (**b**) **5**: Ge1–Ge2, 2.4694(6); Ge1–Cl1, 2.2098(11); Ge2–Cl2, 2.2049(11); Ge1–C2, 1.984(4); Ge2–C1, 1.996(4); C2–Ge1–Ge2, 74.24(13); Ge1–Ge2–C1, 73.02(12); Ge2–C1–C2, 106.8(3); C1–C2–Ge1, 105.6(3).

The difference of the products in the reaction of **1** with ECl$_4$ (E = Si, Ge, and C) between E = Si and E = Ge, C cases should be of great interest. Although we could not draw a definitive conclusion, we performed the thermodynamic energy calculations (free energies) on the reaction of **1** with ECl$_4$ (E = Si, Ge, C) to give the insertion products, **3**, **4**, and **6**, or the chlorination products, **5** and Cl$_2$E: (calculated as 1/2 Cl$_2$E=ECl$_2$) at the B3PW91/6-311G(2d) level of theory (Scheme 3) [28]. In the case of E = Si, the formation of **3** should be exothermic by 2.3 kcal/mol, and that of **5** with Cl$_2$Si=SiCl$_2$ was estimated as an endothermic reaction by 4.5 kcal/mol. However, in the case of E = Ge or C, the formation of **5** with Cl$_2$E=ECl$_2$ was thermodynamically favorable (E = Ge: ΔG = −27 kcal/mol, E = C: ΔG = −81 kcal/mol) relative to the formation of **4** or **6** (E = Ge (**4**): ΔG = −24 kcal/mol, E = C

(**6**): $\Delta G = -63$ kcal/mol). Thus, thermodynamic energy difference between cases of E = Si, Ge, and C could give us some hints on the difference of the reaction products, though the reasonable reaction mechanisms are not clear at present.

The structure of 1,2-dichloro-1,2-digermacyclobut-3-ene **5** was revealed by the X-ray crystallographic analysis. The two Tbb/Cl moieties are oriented in (*E*)-geometries, in the digermacyclobutene skeleton in **5**. The Ge–Ge bond length is 2.4694(6) Å, which is within a range of singly-bonded Ge–Ge distances. The lengths of the two Ge–Cl bonds are almost identical as Ge1–Cl1 = 2.2098(11) Å and Ge2–Cl2 = 2.2049(11) Å, which are similar to those of **3**. The Ge1–C2 and Ge2–C1 (1.984(4), 1.996(4) Å) bond lengths in the digermacyclobutene skeleton of **5** are slightly longer, and shorter relative to those of the only example of the previously reported chlorinated 1,2-digerma-3-cyclobutadiene derivative **7** (Ge–Cl: 2.145(2)–2.150(2), Ge–C: 1.998(6), 2.002(6) Å) (Scheme 4) [30]. Interestingly, reduction of the isolated **5** with lithium naphthalenide was found to reproduce 1,2-digermacyclobutadiene **1** quantitatively, as evidenced by the ^{1}H NMR spectra.

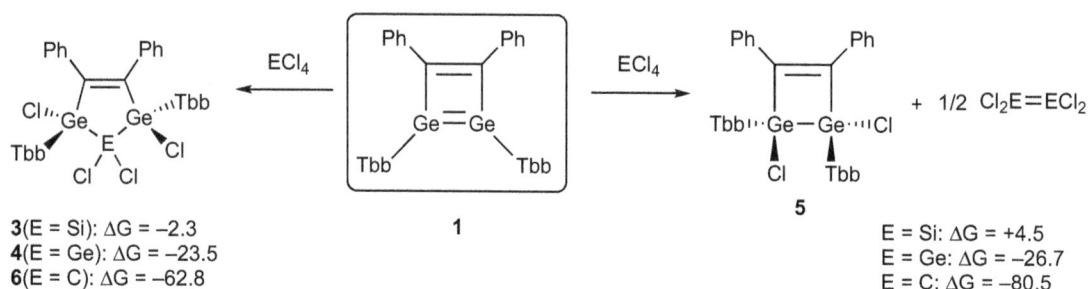

3(E = Si): $\Delta G = -2.3$
4(E = Ge): $\Delta G = -23.5$
6(E = C): $\Delta G = -62.8$

E = Si: $\Delta G = +4.5$
E = Ge: $\Delta G = -26.7$
E = C: $\Delta G = -80.5$

Scheme 3. Theoretical calculations on ΔG values (in kcal/mol) in the reactions of **1** with ECl$_4$ (E = Si, Ge, C) to give insertion products (**3**, **4**, **6**) or chlorinated product **5**.

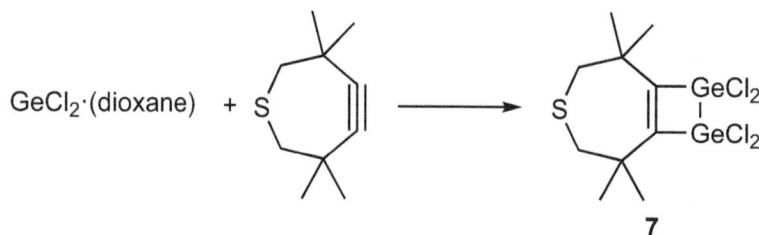

Scheme 4. Reported reaction of GeCl$_2$·(dioxane) with the highly strained alkyne to give the first example of chlorinated 1,2-digerma-3-cyclobutadiene derivative **7** [30].

3. Materials and Methods

3.1. General Information

All manipulations were carried out under an argon atmosphere using either a Schlenk line techniques or glove boxes. Solvents were purified by the Ultimate Solvent System, Glass Contour Company (Laguna Beach, CA, USA) [31]. ^{1}H, ^{13}C, and ^{29}Si NMR spectra were measured on a JEOL AL-300 spectrometer ([1]H: 300 MHz, [13]C: 75 MHz, [29]Si: 59 MHz). Signals arising from residual C$_6$D$_5$H (7.15 ppm) in the C$_6$D$_6$ were used as an internal standard for the ^{1}H NMR spectra, and that of C$_6$D$_6$ (128.0 ppm) for the ^{13}C NMR spectra, and external SiMe$_4$ 0.0 ppm for the ^{29}Si NMR spectra. High-resolution mass spectra (HRMS) were measured on a Bruker micrOTOF focus-Kci mass spectrometer (on ESI-positive mode). All melting points were determined on a Büchi Melting Point Apparatus M-565 and are uncorrected. 1,2-digermacyclobutadiene **1** was prepared according to literature procedure [24].

3.2. Experimental Details

3.2.1. Reaction of 1,2-Tbb$_2$-1,2-Digermacyclobutadiene 1 with an Excess of SiCl$_4$

A solution of 1,2-Tbb$_2$-1,2-digermacyclobutadiene 1 (56.0 mg, 0.046 mmol) in SiCl$_4$ (1.0 mL, 8.8 mmol, excess) was treated at 55 °C for 48 h, and the color of the dark red solution disappeared. After removal of residual SiCl$_4$ under the reduced pressure, the residue was recrystallized from THF at room temperature to give compound 3 as colorless crystals in quantitative yield (64.2 mg, 0.046 mmol).

Data for 3: colorless crystals, m.p. = 68.7–69.7 °C (dec.); ^{1}H NMR (300 MHz, C$_6$D$_6$, r.t.): δ 0.11 (s, 36H, SiMe$_3$), 0.32 (s, 36 H, SiMe$_3$), 1.24 (s, 18H, *t*-Bu), 2.51 (bs, 4H, CH), 6.70–6.76 (m, 2H, ArH), 6.84–6.90 (m, 8H, ArH), 7.22 (d, 4H, *J* = 7.2 Hz, ArH); ^{13}C NMR (75 MHz, C$_6$D$_6$, 298 K): δ 1.80 (Si$\underline{Me}$$_3$), 2.17 (Si$\underline{Me}$$_3$), 30.96 ($\underline{C}Me_3$), 30.99 ($\underline{C}$H), 34.32 ($\underline{C}Me_3$), 124.20 ($\underline{Ar}$H), 126.87 ($\underline{Ar}$H), 127.81 ($\underline{Ar}$H), 130.97 ($\underline{Ar}$H), 132.28 ($\underline{Ar}$), 140.70 ($\underline{Ar}$), 150.13 ($\underline{Ar}$), 151.14 ($\underline{Ar}$), 162.66 ($\underline{C}$Ar); ^{29}Si NMR (59 MHz, C$_6$D$_6$, 298 K): δ 3.61 ($\underline{Si}$Me$_3$), 3.84 ($\underline{Si}$Me$_3$), 28.83 (Ge$\underline{Si}$Ge); MS (DART-TOF, positive mode): m/z calcd. for C$_{62}$H$_{109}$^{35}Cl$_4$^{74}Ge$_2$Si$_9$ 1393.3630 ([M + H]$^+$), found 1393.3681 ([M + H]$^+$).

3.2.2. Reaction of 1,2-Tbb$_2$-1,2-Digermacyclobutadiene 1 with an Excess of GeCl$_4$

A C$_6$D$_6$ solution of 1,2-Tbb$_2$-1,2-digermacyclobutadiene 1 (38.3 mg, 0.0313 mmol) was treated with an excess amount of GeCl$_4$ (0.3 mL, 2.6 mmol) at room temperature. After stirring of the reaction mixture for 10 min, the solvent and GeCl$_4$ were removed under reduced pressure. The residue was recrystallized from benzene at room temperature to give compound 5 as main product in 61% yield (24.6 mg, 0.0190 mmol).

Data for 5: colorless crystals, m.p. 90.4–91.4 °C; ^{1}H NMR (300 MHz, C$_6$D$_6$, r.t.): δ 0.13 (s, 36H, SiMe$_3$), 0.27 (s, 36 H, SiMe$_3$), 1.26 (s, 18H, *t*-Bu), 2.58 (s, 4H, CH), 6.88–6.93 (m, 6H, ArH), 7.00 (t, 4H, *J* = 7.2 Hz, ArH), 7.39 (d, 4H, *J* = 7.2 Hz, ArH); ^{13}C NMR (75 MHz, C$_6$D$_6$, 298 K): δ 1.83 (Si$\underline{Me}$$_3$), 1.86 (Si$\underline{Me}$$_3$), 30.26 ($\underline{C}Me_3$), 31.03 ($\underline{C}$H), 34.39 ($\underline{C}Me_3$), 124.16 ($\underline{Ar}$H), 128.00 ($\underline{Ar}$H), 128.67 ($\underline{Ar}$H), 129.79 ($\underline{Ar}$H), 133.71 ($\underline{Ar}$), 139.21 ($\underline{Ar}$), 150.45 ($\underline{Ar}$), 151.29 ($\underline{Ar}$), 167.29 ($\underline{C}$Ar); MS (DART-TOF, positive mode): m/z calcd. for C$_{62}$H$_{109}$^{35}Cl$_2$^{74}Ge$_2$Si$_8$ 1295.4484 ([M + H]$^+$), found 1295.4492 ([M + H]$^+$).

3.2.3. Reaction of 1,2-Tbb$_2$-1,2-Digermacyclobutadiene 1 with an Excess of CCl$_4$

A C$_6$D$_6$ solution of 1,2-Tbb$_2$-1,2-digermacyclobutadiene 1 (32.9 mg, 0.0269 mmol) was treated with an excess amount of CCl$_4$ (0.2 mL, 2.1 mmol) at room temperature. After stirring of the reaction mixture for 10 min, the solvent and CCl$_4$ were removed under reduced pressure. The residue was recrystallized from benzene at room temperature to give compound 3 as main product in 55% yield (22.8 mg, 0.0175 mmol).

3.3. Computational Methods

The level of theory and the basis sets used for the structural optimization are contained within the main text. Frequency calculations confirmed minimum energies for all optimized structures. All calculations were carried out using the *Gaussian 09* program package [28]. Computational time was generously provided by the Supercomputer Laboratory in the Institute for Chemical Research of Kyoto University.

3.4. X-ray Crystallographic Analysis

Single crystals of [3·(thf)] and [5·2(benzene)] were obtained from recrystallization from THF and benzene, respectively. Intensity data were collected on a RIGAKU Saturn70 CCD system with VariMax Mo Optics using Mo Kα radiation (λ = 0.71075 Å). The structures were solved by a direct method (SIR2004 [32]) and refined by a full-matrix least square method on F^2 for all reflections (*SHELXL*-97 [33]). All hydrogen atoms were placed using AFIX instructions, while all other atoms were refined anisotropically. Supplementary crystallographic data were deposited at the Cambridge Crystallographic Data Centre (CCDC; under reference numbers: CCDC-1578241 and 1578242 for [**3**·(thf)] and [**5**·2(benzene)], respectively) and can be obtained free of charge via https://www.ccdc.cam.ac.uk/structures/. X-ray crystallographic data for [**3**·(thf)] and [**5**·2(benzene)]. Data for [**3**·(thf)] ($C_{66}H_{116}Cl_4Ge_2OSi_9$): M = 1465.37, triclinic, $P–1$ (no.2), a = 12.6367(7) Å, b = 16.9170(6) Å, c = 20.4887(10) Å, α= 91.3815(14)°, β = 105.252(2)°, γ = 109.642(3)°, V = 3949.2(3) Å^3, Z = 2, $D_{calc.}$ = 1.232 g·cm^{-3}, μ = 1.070 mm^{-1}, $2\theta_{max}$ = 51.0°, measd./unique refls. = 83580/14641 ($R_{int.}$ = 0.1095), param = 767, GOF = 1.117, R_1 = 0.0683/0.1122 [$I{>}2\sigma(I)$/all data], wR_2 = 0.1188/0.1359 [$I{>}2\sigma(I)$/all data], largest diff. peak and hole 1.681 and -0.592 e.Å^{-3} (CCDC-1578241). Data for [**5**·2(benzene)] ($C_{74}H_{120}Cl_2Ge_2Si_8$): M = 1450.49, triclinic, $P–1$ (no.2), a = 11.6792(2) Å, b = 15.7581(3) Å, c = 24.7906(5) Å, α = 76.2640(10)°, β = 88.0800(10)°, γ = 70.2510(10)°, V = 4165.89(14) Å^3, Z = 2, $D_{calc.}$ = 1.156 g·cm^{-3}, μ = 0.937 mm^{-1}, $2\theta_{max}$ = 50.0°, measd./unique refls. = 64887/14555 ($R_{int.}$ = 0.0810), param = 805, GOF = 1.289, R_1 = 0.0637/0.0804 [$I{>}2\sigma(I)$/all data], wR_2 = 0.1247/0.1311 [$I{>}2\sigma(I)$/all data], largest diff. peak and hole 0.983 and -0.689 e.Å^{22123} (CCDC-1578242).

4. Conclusions

It was demonstrated that 1,2-digermacyclobutadiene **1** could work as a bis-germylene building block (>Ge: + :Ge<) towards double Si–Cl activation in the reaction of **1** with SiCl$_4$ to give the halogenated oligometallane, bis(chlorogermyl)dichlorosilane **3**. Conversely, GeCl$_4$ and CCl$_4$ were found to work as double-chlorinating reagents towards **1** giving dichlorodigermacyclobutene **5**. Thus, **1** would be an interesting building block for oligohalo-oligometallanes.

Acknowledgments: This work was partially supported by the following grants: Grant-in-Aid for Scientific Research (B) (no. 15H03777), Grant-in-Aid for Challenging Exploratory Research (no. 15K13640), Scientific Research on Innovative Areas, "New Polymeric Materials Based on Element-Blocks" (#2401) (no. 25102519), "Stimuli-Responsive Chemical Species for the Creation of Functional Molecules" (#2408) (no. 24109013), and the project of Integrated Research on Chemical Synthesis from the Japanese Ministry of Education, Culture, Sports, Science, and Technology (MEXT), as well as by the "Molecular Systems Research" project of the RIKEN Advanced Science Institute and the Collaborative Research Program of the Institute for Chemical Research, Kyoto University. We would like to thank Toshiaki Noda and Hideko Natsume at Nagoya University for the expert manufacturing of custom-tailored glasswares. T. Sugahara would like to thank the Japan Society for the Promotion of Science (JSPS) for a fellowship (no. JP16J05501).

Author Contributions: Takahiro Sasamori and Norihiro Tokitoh conceived and designed the experiments; Tomohiro Sugahara performed the experiments; Takahiro Sasamori and Tomohiro Sugahara performed the XRD analysis and wrote the paper; and Takahiro Sasamori performed theoretical calculations.

References

1.   Miller, R.D.; Michl, J. Polysilane High Polymers. *Chem. Rev.* **1989**, *89*, 1359–1410. [CrossRef]

2.   Amadoruge, M.L.; Weinert, C.S. Singly Bonded Catenated Germanes: Eighty Years of Progress. *Chem. Rev.* **2008**, *108*, 4253–4294. [CrossRef] [PubMed]

3.   Mochida, K.; Chiba, H. Synthesis, absorption characteristics and some reactions of polygermanes. *J. Organomet. Chem.* **1994**, *473*, 45–54. [CrossRef]

4.   Lickiss, P.D.; Smith, C.M. Silicon derivatives of the metals of groups 1 and 2. *Coord. Chem. Rev.* **1995**, *145*, 75–124. [CrossRef]

5.   Sekiguchi, A.; Lee, V.Y.; Nanjo, M. Lithiosilanes and their application to the synthesis of polysilane dendrimers. *Coord. Chem. Rev.* **2000**, *210*, 11–45. [CrossRef]

6.   Kyushin, S.; Matsumoto, H. Ladder Polysilanes. *Adv. Organomet. Chem.* **2003**, *49*, 133–166. [CrossRef]

7.   Weinert, C.S. Syntheses, structures and properties of linear and branched oligogermanes. *Dalton Trans.* **2009**, 1691–1699. [CrossRef] [PubMed]

8. Drost, C.; Hitchcock, P.B.; Lappert, M.F. Unprecedented Oxidative Chlorosilylation Addition Reactions to a Diarylgermylene and -stannylene. *Organometallics* **2002**, *21*, 2095–2100. [CrossRef]

9. Iwamoto, T.; Masuda, H.; Kabuto, C.; Kira, M. Trigermaallene and 1,3-Digermasilaallene. *Organometallics* **2005**, *24*, 197–199. [CrossRef]

10. Kira, M.; Iwamoto, T.; Ishida, S.; Masuda, H.; Abe, T.; Kabuto, C. Unusual Bonding in Trisilaallene and Related Heavy Allenes. *J. Am. Chem. Soc.* **2009**, *131*, 17135–17144. [CrossRef] [PubMed]

11. Sugahara, T.; Sasamori, T.; Tokitoh, N. Highly Bent 1,3-Digerma-2-silaallene. *Angew. Chem. Int. Ed.* **2017**, *56*, 9920–9923. [CrossRef] [PubMed]

12. Mallela, S.P.; Geanangel, R.A. New Cyclic and Acyclic Silicon–Germanium and Silicon–Germanium–Tin Derivatives. *Inorg. Chem.* **1994**, *33*, 1115–1120. [CrossRef]

13. Mallela, S.P.; Geanangel, R.A. Preparation and structural characterization of new derivatives of digermane bearing tris(trimethylsilyl)silyl substituents. *Inorg. Chem.* **1991**, *30*, 1480–1482. [CrossRef]

14. Ichinohe, M.; Sekiyama, H.; Fukaya, N.; Sekiguchi, A. On the Role of *cis,trans*-(t-Bu$_3$SiGeCl)$_3$ in the Reaction of GeCl$_2$·Dioxane with Tri-*tert*-butylsilylsodium: Evidence for Existence of Digermanylsodium *t*-Bu$_3$SiGe(Cl)$_2$Ge(Cl)(Na)Si*t*-Bu$_3$ and Digermene *t*-Bu$_3$Si(Cl)Ge=Ge(Cl)Si*t*-Bu$_3$. *J. Am. Chem. Soc.* **2000**, *122*, 6781–6782. [CrossRef]

15. Fukaya, N.; Sekiyama, H.; Ichinohe, M.; Sekiguchi, A. Photochemical Generation of Chlorine-substituted Digermenes and Their Rearrangement to Germylgermylenes. *Chem. Lett.* **2002**, 802–803. [CrossRef]

16. Sekiguchi, A.; Ishida, Y.; Fukaya, N.; Ichinohe, M. The First Halogen-Substituted Cyclotrigermenes: A Unique Halogen Walk over the Three-Membered Ring Skeleton and Facial Stereoselectivity in the Diels-Alder Reaction. *J. Am. Chem. Soc.* **2002**, *124*, 1158–1159. [CrossRef] [PubMed]

17. Lee, V.Y.; Yasuda, H.; Ichinohe, M.; Sekiguchi, A. SiGe$_2$ and Ge$_3$: Cyclic Digermenes that Undergo Unexpected Ring-Expansion Reactions. *Angew. Chem. Int. Ed.* **2005**, *44*, 6378–6381. [CrossRef] [PubMed]

18. Lee, V.Y.; Yasuda, H.; Ichinohe, M.; Sekiguchi, A. Heavy cyclopropene analogues R$_4$SiGe$_2$ and R$_4$Ge$_3$ (R = SiMe*t*Bu$_2$)—New members of the cyclic digermenes family. *J. Organomet. Chem.* **2007**, *692*, 10–19. [CrossRef]

19. Wagler, J.; Brendler, E.; Langer, T.; Pöttgen, R.; Heine, T.; Zhechkov, L. Ylenes in the M^{II}→Si^{IV} (M=Si, Ge, Sn) Coordination Mode. *Chem. Eur. J.* **2010**, *16*, 13429–13434. [CrossRef] [PubMed]

20. Al-Rafia, S.M.; Malcolm, A.C.; McDonald, R.; Ferguson, M.J.; Rivard, E. Trapping the Parent Inorganic Ethylenes H$_2$SiGeH$_2$ and H$_2$SiSnH$_2$ in the Form of Stable Adducts at Ambient Temperature. *Angew. Chem. Int. Ed.* **2011**, *50*, 8354–8357. [CrossRef] [PubMed]

21. Katir, N.; Matioszek, D.; Ladeira, S.; Escudié, J.; Castel, A. Stable *N*-Heterocyclic Carbene Complexes of Hypermetallyl Germanium(II) and Tin(II) Compounds. *Angew. Chem. Int. Ed.* **2011**, *50*, 5352–5355. [CrossRef] [PubMed]

22. Lee, V.Y.; Ito, Y.; Yasuda, H.; Takanashi, K.; Sekiguchi, A. From Tetragermacyclobutene to Tetragermacyclobutadiene Dianion to Tetragermacyclobutadiene Transition Metal Complexes. *J. Am. Chem. Soc.* **2011**, *133*, 5103–5108. [CrossRef] [PubMed]

23. Cui, C.; Olmstead, M.M.; Power, P.P. Reactivity of Ar'GeGeAr' (Ar' = C$_6$H$_3$-2,6-Dipp$_2$, Dipp = C$_6$H$_3$-2,6-*i*Pr$_2$) toward Alkynes: Isolation of a Stable Digermacyclobutadiene. *J. Am. Chem. Soc.* **2004**, *126*, 5062–5063. [CrossRef] [PubMed]

24. Sugahara, T.; Guo, J.-D.; Sasamori, T.; Karatsu, Y.; Furukawa, Y.; Ferao, A.E.; Nagase, S.; Tokitoh, N. Reaction of a Stable Digermyne with Acetylenes: Synthesis of a 1,2-Digermabenzene and a 1,4-Digermabarrelene. *Bull. Chem. Soc. Jpn.* **2016**, *89*, 1375–1384. [CrossRef]

25. Tashkandi, N.Y.; Pavelka, L.C.; Caputo, C.A.; Boyle, P.D.; Power, P.P.; Baines, K.M. Addition of alkynes to digermynes: Experimental insight into the reaction pathway. *Dalton Trans.* **2016**, *45*, 7226–7230. [CrossRef] [PubMed]

26. Zhao, L.; Jones, C.; Frenking, G. Reaction Mechanism of the Symmetry-Forbidden [2+2] Addition of Ethylene and Acetylene to Amido-Substituted Digermynes and Distannynes Ph$_2$N-EE-NPh$_2$, (E = Ge, Sn): A Theoretical Study. *Chem. Eur. J.* **2015**, *21*, 12405–12413. [CrossRef] [PubMed]

27. Tillmann, J.; Meyer, L.; Schweizer, J.I.; Bolte, M.; Lerner, H.-W.; Wagner, M.; Holthausen, M.C. Chloride-Induced Aufbau of Perchlorinated Cyclohexasilanes from Si$_2$Cl$_6$: A Mechanistic Scenario. *Chem. A-Eur. J.* **2014**, *20*, 9234–9239. [CrossRef] [PubMed]

28. *Gaussian 09 Program*; Gaussian, Inc.: Wallingford, CT, USA, 2009.

29. Ohtaki, T.; Ando, W. Dichlorodigermacyclobutanes and Digermabicyclo[2.2.0]hexanes from the Reactions of [Tris(trimethylsilyl)methyl]chlorogermylene with Olefins. *Organometallics* **1996**, *15*, 3103–3105. [CrossRef]

30. Espenbetov, A.A.; Struchkov Yu, T.; Kolesnikov, S.P.; Nefedov, O.M. Crystal and Molecular Structure of $\Delta^{1,7}$2,2,6,6,-Tetramethyl-4-thia-8,8,9,9-tetrachloro-8,9-digermabicyclo[5.2.0]nonene; The First Representative of 1,2-Digermacyclobutenes. *J. Organomet. Chem.* **1984**, *275*, 33–37. [CrossRef]

31. Pangborn, A.B.; Giardello, M.A.; Grubbs, R.H.; Rosen, R.K.; Timmers, F.J. Safe and Convenient Procedure for Solvent Purification. *Organometallics* **2004**, *15*, 1518–1520. [CrossRef]

32. Burla, M.C.; Caliandro, R.; Camalli, M.; Carrozzini, B.; Cascarano, G.L.; De Caro, L.; Giacovazzo, C.; Polidori, G.; Spagna, R. *SIR2004*: An improved tool for crystal structure determination and refinement. *J. Appl. Cryst.* **2005**, *38*, 381–388. [CrossRef]

33. Sheldrick, G.M. A short history of *SHELX*. *Acta Crystallogr. Sect. A* **2008**, *64*, 112–122. [CrossRef] [PubMed]

17

Dehydrogenation of Surface-Oxidized Mixtures of 2LiBH$_4$ + Al/Additives (TiF$_3$ or CeO$_2$)

Juan Luis Carrillo-Bucio, Juan Rogelio Tena-García and Karina Suárez-Alcántara *

Morelia Unit of the Materials Research Institute of the National Autonomous University of Mexico, Antigua Carretera a Pátzcuaro 8701, Col. Ex Hacienda de San José de la Huerta, Morelia 58190, Michoacán, Mexico; jlcarrillob@iim.unam.mx (J.L.C.-B.); juanrogelio_tenagarcia@iim.unam.mx (J.R.T.-G.)

* Correspondence: karina_suarez@iim.unam.mx

Abstract: Research for suitable hydrogen storage materials is an important ongoing subject. LiBH$_4$–Al mixtures could be attractive; however, several issues must be solved. Here, the dehydrogenation reactions of surface-oxidized 2LiBH$_4$ + Al mixtures plus an additive (TiF$_3$ or CeO$_2$) at two different pressures are presented. The mixtures were produced by mechanical milling and handled under welding-grade argon. The dehydrogenation reactions were studied by means of temperature programmed desorption (TPD) at 400 °C and at 3 or 5 bar initial hydrogen pressure. The milled and dehydrogenated materials were characterized by scanning electron microscopy (SEM), X-ray diffraction (XRD), and Fourier transformed infrared spectroscopy (FT-IR) The additives and the surface oxidation, promoted by the impurities in the welding-grade argon, induced a reduction in the dehydrogenation temperature and an increase in the reaction kinetics, as compared to pure (reported) LiBH$_4$. The dehydrogenation reactions were observed to take place in two main steps, with onsets at 100 °C and 200–300 °C. The maximum released hydrogen was 9.3 wt % in the 2LiBH$_4$ + Al/TiF$_3$ material, and 7.9 wt % in the 2LiBH$_4$ + Al/CeO$_2$ material. Formation of CeB$_6$ after dehydrogenation of 2LiBH$_4$ + Al/CeO$_2$ was confirmed.

Keywords: hydrogen storage; borohydrides; reactive mixtures

1. Introduction

LiBH$_4$ is an outstanding material regarding its hydrogen content (18.4 wt %) [1]. However, its dehydrogenation temperature is too high for any practical application in hydrogen storage. Pure LiBH$_4$ presents two dehydrogenation steps; one minor at approximately 320 °C and the main one at 500 °C [1]. To reduce the dehydrogenation temperature, improve reaction kinetics or reversibility, LiBH$_4$ has been mixed with several compounds in different proportions. The list includes but is not limited to other borohydrides such as: Ca(BH$_4$)$_2$ [2], NaBH$_4$ [3], or Mg(BH$_4$)$_2$ [4], other complex hydrides such as LiNH$_2$ [5], alanates of Li or Na [6,7], binary hydrides such as MgH$_2$ [8–10], CaH$_2$ [11–13], TiH$_2$ [12], halide salts such as LiCl [14], oxides [15], scaffolds [16], and metals such as Mg, Ti, V, Cr, Sc, or Al [12,17]. The main characteristics of these mixtures are collated in Table 1.

The last system, the 2LiBH$_4$ + Al, is of potential interest. Siegel et al. [18] anticipated, based on first-principles calculations, that the reaction:

$$2LiBH_4 + Al \rightarrow AlB_2 + 2LiH + 3H_2 \tag{1}$$

would release 8.6 wt % hydrogen at 277 °C and $p(H_2) = 1$ bar. Experimentally, Zhang et al. performed the dehydrogenation of 2LiBH$_4$ + Al at a pressure of 0.001 bar H$_2$ and from roughly 325–525 °C [19]. Zhang et al. demonstrated a hydrogen release of approximately 4 wt % by means of isothermal dehydrogenation at 375 °C, meanwhile complete dehydrogenation was obtained up to 577 °C by means of thermogravimetric measurement [19]. Kang et al. achieved the release of 7.2 wt % hydrogen when the mixture was catalyzed with TiF_3, at a pressure of 0.001 bar, 450 °C, and 3 h [20]. Hansen et al. [17] demonstrated that the dehydrogenation reaction of $2LiBH_4 + 3Al$ occurs at $p(H_2) = 10^{-2}$ bar when the material is heated up to 500 °C; and that the re-hydrogenation demonstrated only partial reversibility. Other examples include the experiments of Ravnsbaek et al., where the dehydrogenation reaction of $LiBH_4 + Al$ (1:0.5) was characterized by in-situ synchrotron radiation powder X-ray diffraction from room temperature to 500 °C and dynamic vacuum [21]. All these experiments have in common a very low or vacuum dehydrogenation pressure (summarized in Table 1). The vacuum dehydrogenation pressure is inadequate for any practical application. Additionally, it is well-known that the hydrogen pressure can affect the dehydrogenation products of $LiBH_4$ mixtures. For example with the system $2LiBH_4 + MgH_2 \rightarrow 2LiH + MgB_2 + 4H_2$, a correlation was demonstrated between the dehydrogenation pressure and re-hydrogenation [22]. With the $2LiBH_4$–Al system, Yang et al. proposed that a desorption backpressure of 3 bar could contribute to the formation of AlB_2 [12]; which would improve, in principle, the reversibility of hydrogenation/dehydrogenation reactions. A study of the dehydrogenation backpressure with $2LiBH_4$–Al systems has been poorly explored.

Among the additives for improving reaction kinetics or reversibility; TiF_3 is the material most commonly used, and it is almost mandatory to test TiF_3 in all new mixtures. In their part, oxides such as TiO_2, ZrO_2, Nb_2O_5 or MoO_3 have resulted in successful accelerators for hydrogen desorption reactions [23,24]. CeO_2 is not a commonly used additive for hydrogenation/dehydrogenation reactions. However, interactions of hydrogen with CeO_2 have been reported [25]. The possible effects of CeO_2 as an additive for hydrogen storage materials deserve research.

Conventionally, hydrogen storage materials are produced and handled in a high or ultra-high purity inert atmosphere, i.e., high-purity argon. This is done to avoid the deactivation of materials caused by the formation of thick oxide films that impede hydrogen diffusion from/to the bulk of the storage material. However, some reports have expressed that the use of high-purity inert atmosphere can be relaxed [26] and that allowing some surface oxidation can be helpful to the dehydrogenation kinetics [27].

Here, the dehydrogenation behavior is presented of $2LiBH_4 + Al$ added with 5 wt % of TiF_3 or CeO_2, and surface-modified (oxidized) by the effects of impurities in welding-grade argon. The dehydrogenation reactions were performed at 3 bar and 5 bar of hydrogen. The effects of the borohydride–Al mixture, additives, and surface-oxide effects are discussed.

Table 1. Reported hydrogen desorption conditions for several $LiBH_4$ mixtures.

Material and/or Proposed Reaction, and Reported ΔH^0 (If Available) (kJ/mol H_2)	Desorption Conditions p (bar) and T (°C)	Comments
$LiBH_4 \rightarrow LiH + B + 3/2H_2$ [1]	p: not specified T: 320 °C and 500 °C	Multi-step dehydrogenation reaction
$LiBH_4 \rightarrow Li + B + 2H_2$ 95.1 kJ/mol H_2 [28]	p: 1 bar T: 25 °C	From standard formation enthalpy of $LiBH_4$
$xLiBH_4 + (1-x)\,Ca(BH_4)_2$ [2]	p: not specified T: 370 °C for $x = 0.4$	$x = 0 - 1$, eutectic melting at 200 °C
$0.62LiBH_4$-$0.38NaBH_4$ [3]	p: not specified T: onset at 287 °C, peaks at 488 °C and 540 °C	Multi-step dehydrogenation reaction
$xLiBH_4 + (1-x)\,Mg(BH_4)_2$ [4]	p: 5 bar T: 170 °C and 215 °C	$x = 0 - 1$, eutectic melting at 180 °C Multi-step dehydrogenation reaction
$LiBH_4 + 2LiNH_2 \rightarrow Li_3BN_2 + 4H_2$ 23 kJ/mol H_2 [5]	p: 100–0.01 bar T: 430 °C	From pressure composition isotherm.
$LiBH_4 + LiAlH_4$ [6]	p: 0.2 bar T: 118 °C and 210 °C	2:1 mixture, two-step dehydrogenation. Dehydrogenation temperature reduced if TiF_3 addition.
$LiBH_4 + NaAlH_4$ [7]	p: 1 bar He T: from room temperature up to 210 °C for the doped systems and 110–250 °C for the undoped systems.	Molar ratios 1:1, 2:3 and 1:3; with and without $TiCl_3$ additive. Multi-step dehydrogenation reaction.

Table 1. *Cont.*

Material and/or Proposed Reaction, and Reported ΔH^0 (If Available) (kJ/mol H_2)	Desorption Conditions p (bar) and T (°C)	Comments
$2LiBH_4 + MgH_2 \rightarrow 2LiH + MgB_2$ + $4H_2$ 50.4 kJ/mol H_2 [10]	p: 3 bar H_2 T: 350–400 °C	Multi-step dehydrogenation reaction
$6LiBH_4 + CaH_2 \leftrightarrow 6LiH + CaB_6$ + $10H_2$ 59 kJ/mol H_2 [11]	p: 1.3 bar flowing He T: onset at 150 °C, maximum at 350 °C, finished at 450 °C	-
$LiBH_4 + TiH_2$ [12]	p: not specified (argon) T: ~410 °C	-
$LiBH_4 + LiCl$ (1:1) to give $Li(BH_4)_{1-x}Cl_x$ ($x \approx 0.23$) [14]	p: not specified (argon) T: 300–550 °C	Cl^- to BH_4^- substitution at $LiBH_4$
$2LiBH_4 + Al \rightarrow 2LiH + AlB_2$ + $3H_2$ 57.9 kJ/mol H_2 [18]	277 °C [18]	Theoretical desorption temperature
	Dehydrogenation: p: 0.001 bar H_2 T: 325 °C to 525 °C [19]	H_2 release of about 4 wt % Multi-step dehydrogenation reaction
	p: 0.001 bar T: 450 °C [20]	Catalyzed with TiF_3
	p: dynamic vacuum T: up to 500 °C [21]	Formation of $Li_xAl_{1-x}B_2$

2. Results

2.1. Characterization of As-Milled Materials

Scanning Electron Microscopy. Some of the descriptions below contain remarks about SEM images that are shown in the Supplementary Materials. Also for comparison purposes, the SEM images of $LiBH_4$ and Al without milling are presented in the Supplementary Materials. Non-milled $LiBH_4$ is composed of large crystals embedded in an amorphous phase. Non-milled Al is composed of particles of approximately 50 μm, heavily agglomerated and forming flakes of 1 mm (Supplementary Materials). Figure 1 presents the most representative SEM images of the as-milled materials. The as-milled $2LiBH_4 + Al$ material, Figure 1a, formed a three-dimensional, porous structure. Interestingly the surface of $2LiBH_4 + Al$ is covered with crystals of approximately 2–3 μm. The material $2LiBH_4 + Al/TiF_3$, Figure 1b, also presented a three-dimensional structure. The material consisted of elongated crystals of 10 μm length and 2 μm width. Elemental analysis by energy-dispersive X-ray spectroscopy (EDS) of those crystals revealed a B-rich phase, i.e., the $LiBH_4$. The addition of CeO_2, Figure 1c, to the base material, produced an agglomerated, spherical, and compacted material of about 50 μm in diameter, with some surface formations approximating flake-shape. Here, spots of CeO_2 are clearly distinguishable from the base material. In principle, the morphology characteristics of hydrogen storage materials must be reduced particle size, low agglomeration, homogeneity of component materials, and the formation of porous structures allowing the inflow/outflow of hydrogen while maintaining good thermal conductivity. SEM images showed interesting morphologies of the $LiBH_4 + Al/additives$, depending on the additive material.

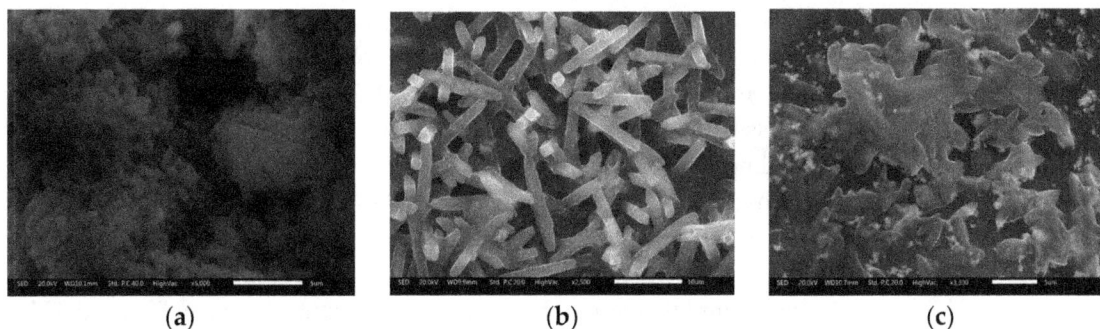

(a)   (b)   (c)

Figure 1. Scanning electron microscopy (SEM) images of the as-milled materials: (**a**) $2LiBH_4 + Al$; (**b**) $2LiBH_4 + Al/TiF_3$; (**c**) $2LiBH_4 + Al/CeO_2$.

Powder X-ray diffraction. Figure 2 shows the X-ray diffraction patterns of as-milled materials. In the three studied materials, the presence of Al is evidenced by the strong diffraction peaks. By contrast, the $LiBH_4$ presented attenuated peaks as as-milled materials. $LiBH_4$ diffraction peaks in Figure 2 are a mixture of orthorhombic (main) and hexagonal (minor) phases to a different degree in the different mixtures. The material $2LiBH_4$ + Al/CeO_2 presented strong peaks of CeO_2, in agreement with the CeO_2 particles found in the SEM images. The material added with TiF_3 did not present the characteristic peaks of TiF_3. Some researchers have proposed the in-situ formation of $Ti(BH_4)_3$ by the reaction of $LiBH_4$ and TiF_3, followed by a rapid decomposition of $Ti(BH_4)_3$ [29]. Reactions to form other stable fluorine salts such as TiF_2 or AlF_3 cannot be discarded. These side reactions during ball-milling can be the reason for the absence of TiF_3 diffraction peaks. The materials also presented a small amount of aluminum oxide. Refinement of the Al and CeO_2 phases (not including $LiBH_4$ phases, $R_{wp} \approx 7$) produced the following Al-crystal sizes: Al in $2LiBH_4$ + Al: 309.6 ± 16.7 nm; Al in $2LiBH_4$ + Al/TiF_3: 129.0 ± 5.0 nm; Al in $2LiBH_4$ + Al/CeO_2: 142.8 ± 9.2 nm. The CeO_2 crystal size was determined as 120.3 ± 4.2 nm. All these facts point to the formation of fine mixtures of $2LiBH_4$ + Al/additive and an influence of the additives on the crystalline characteristics of the samples.

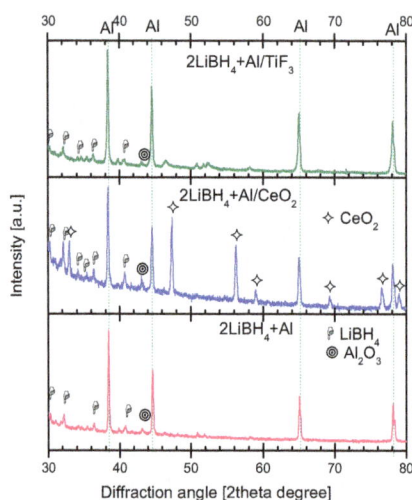

Figure 2. X-ray diffraction patterns of the as-milled materials: (**a**) $2LiBH_4$ + Al; (**b**) $2LiBH_4$ + Al/TiF_3; (**c**) $2LiBH_4$ + Al/CeO_2.

Fourier Transformed Infrared Spectroscopy. An effect of the ball milling is the increase of the local pressure and temperature during ball-ball or ball-vial collisions that induce the decomposition of sensitive materials. In the case of the $2LiBH_4$ + Al mixtures, it is intended to decrease the decomposition temperature during dehydrogenation, but not to decompose during ball-milling. Thus FT-IR was performed to check the "survival" of $LiBH_4$ after ball-milling. Borohydrides present two regions of interest B–H bending ($1000–1600$ cm^{-1}) and H–B–H stretching ($2000–2500$ cm^{-1}) [29,30]. Both IR active modes are presented in all the as-milled samples (Supplementary Materials), indicating sufficient thermal stability during ball milling.

X-ray photoelectron spectroscopy. XPS results are presented in Figure 3. Frame 3a presents the Li 1s XPS spectra. The Li 1s XPS spectrum of $LiBH_4$ with a clean surface was reported to present a peak at 57.1 eV; meanwhile, the oxygen-exposed samples presented a contribution (shoulder) of Li_2O at 55.5 eV [27]. The studied materials present the main peak in a range between 57.2 and 57.5 eV. Because of that, the main contribution of Li binding is not the Li–O interactions. Frame 3(b) presents the B 1s XPS spectra. The reported peak for pure $LiBH_4$ is located at 188.3–188.4 eV [31]. After exposing $LiBH_4$ to oxygen or moisture, a new peak located at 191.5–191.4 eV emerged and it was attributed to $LiBO_2$ [31]. The studied materials presented both peaks, the $LiBH_4$ and the $LiBO_2$ at 188.3–188.0 eV and 191.9–191.3 eV, respectively. The peak intensity ratio of the $LiBH_4$ and $LiBO_2$ is not homogeneous throughout the set of studied materials. The mixture of $2LiBH_4 + Al$ presented the lowest intensity of the $LiBO_2$ peak. Frame c of Figure 3 presents the O 1s XPS spectra; as a reference, the O 1s spectra at the Al raw-material was included. In that spectrum, the peak at 533.6 eV can be related to an Al_2O_3 layer [32]. The XPS spectra of $LiBH_4$, $2LiBH_4 + Al/TiF_3$, and $2LiBH_4 + Al/CeO_2$ presented two main peaks; one at 533.9 eV and other at 532.6 eV. This last peak can be attributed to the $LiBO_2$ [32]. The $LiBO_2$ peak is very much attenuated in the $2LiBH_4 + Al$, in agreement with the result at the B 1s spectra. The Al 2p spectrum of the as-received Al and as-milled $2LiBH_4 + Al$/additive materials is shown in the Supplementary Materials. The (as-received) Al curve presents the characteristic metallic and oxide peaks of Al [33]. Meanwhile, no signal above noise was detected for the mixtures, indicating Al segregation to sub-surface layers.

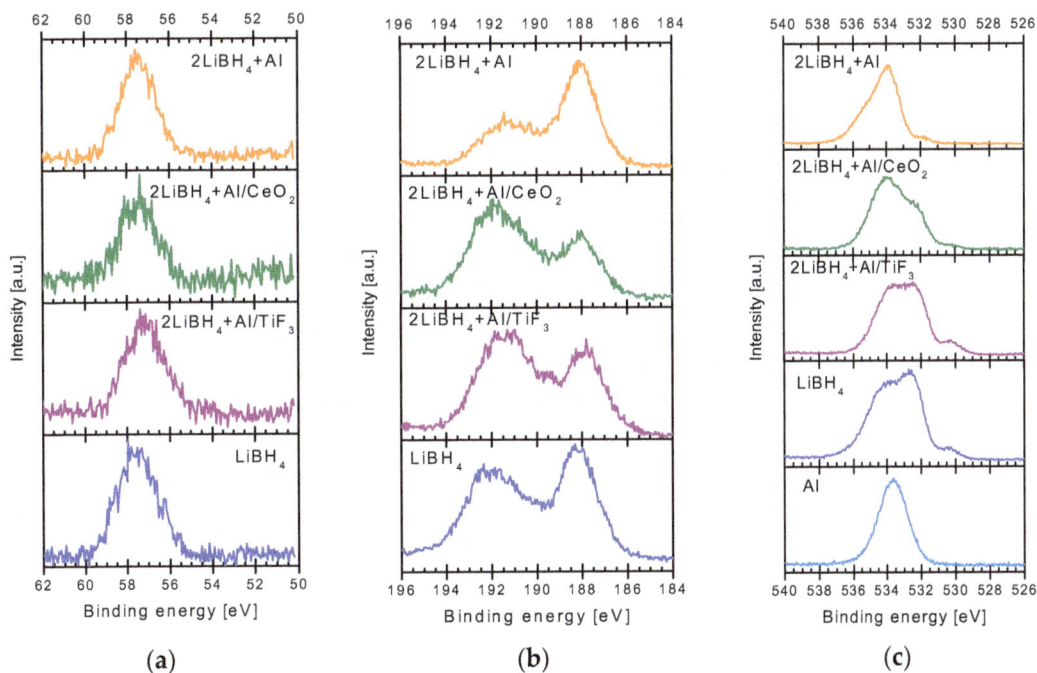

Figure 3. X-ray photoelectron spectroscopy (XPS) spectra of as-milled $2LiBH_4 + Al$, $2LiBH4 + Al/TiF_3$, and $2LiBH_4/CeO_2$. (**a**) Li 1s edge; (**b**) B 1s edge; (**c**) O 1s edge.

2.2. Dehydrogenation Reactions

Figure 4 presents the dehydriding reactions traced by thermal-programed control. For two frames, the sample temperature also was plotted. The dehydrogenations of $2LiBH_4 + Al$ and $2LiBH_4 + Al/TiF_3$ at 3 bar and 5 bar initial pressure are presented in Figure 4a. The dehydrogenations were performed up to 400 °C. The dehydrogenations occurred as multiple-step reactions. The first step with both materials was observed to start at 110 °C and produced a small hydrogen release. In the $2LiBH_4 + Al$ material, the second step was observed at 210 °C, and a slight change in reaction rate at roughly 325 °C. The dehydrogenation reactions in the $2LiBH_4 + Al$ material were essentially finished at 400 °C and reached a hydrogen release of −5.8 wt % for the 3 bar initial pressure reaction and −6.0 wt % for the 5 bar initial pressure reaction. The dehydrogenations of $2LiBH_4 + Al/TiF_3$ at 3 bar and 5 bar initial pressure also are multiple-step reactions. At both initial pressures, the dehydrogenation first step onset is situated at about 110 °C. At the 3 bar dehydrogenation, the second reaction step is located at 240 °C. The last step initiated at 340 °C, after a clear-defined plateau. At 3 bar initial pressure the release of hydrogen accounted for −9.3 wt %. The situation is very different for the same material at 5 bar initial pressure, here the onset of the second dehydrogenation step was observed at 200 °C. A slowdown of the dehydrogenation rate was observed at 300 °C. Finally, the dehydrogenation finished after reaching 400 °C, releasing −6.0 wt %. This last quantity is greater than the theoretical value of −8.6 wt %, thus a release of B-compounds is probable. It means that higher initial pressure in combination with lower final temperature needs to be explored for this material.

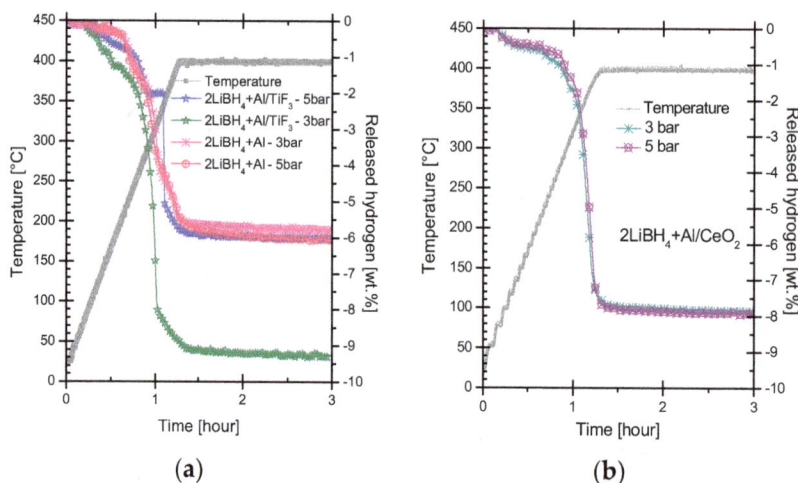

Figure 4. Temperature programed hydrogen desorption of (**a**) $2LiBH_4 + Al$, and $2LiBH_4 + Al/TiF_3$; (**b**) $2LiBH_4 + Al/CeO_2$ at 3 and 5 bar initial pressure of hydrogen.

Figure 4b presents the dehydrogenation reactions of $2LiBH_4 + Al/CeO_2$ at 3 bar and 5 bar initial hydrogen pressure up to 400 °C. The dehydrogenations reactions are similar at both pressures and they are also multistep reactions. The onset temperature of the first dehydrogenation step is below 100 °C. The dehydrogenation reactions presented the second step starting at 220 °C. The material $2LiBH_4 + Al/CeO_2$ released −7.9 wt % of hydrogen in both cases. In the material $2LiBH_4 + Al/CeO_2$, dehydrogenation at 350 °C was achieved and can be observed in the Supplementary Materials. The dehydrogenation is a multi-step reaction with a hydrogen release of −5.6 wt % at 5 bar and −6.5 wt % at 3 bar after 3 h of reaction.

2.3. Characterization of Dehydrogenated Materials

Figure 5 presents the most representative SEM images of the $2LiBH_4 + Al$, $2LiBH_4 + Al/TiF_3$, and $2LiBH_4 + Al/CeO_2$ materials dehydrogenated at 5 bar, other SEM images are available in the Supplementary Materials. In general, dehydrogenated materials presented important agglomerations of about 200 μm size. Thus an important increase in particle size after dehydrogenation was observed. The agglomerated materials were composed of an amorphous matrix with crystalline zones of cubic or hexagonal morphologies. Elemental analysis of those crystals revealed the presence of B and Al as major components; and Ti, F, Ce, and O as minor components, accordingly with the additive. An important observation is that the atomic composition of Al and B in the crystals of all the tested materials was not consistent. The elemental analysis was performed in small areas on the crystals, and the results range from roughly the expected 1:2 atomic ratio of AlB_2 to 1:12 atomic ratio of AlB_{12}. The loss of available B for the re-hydrogenation reaction is a common drawback for all borohydrides and their mixtures. B can be lost as borane compounds (BH_3, B_2H_6, etc.), form B-clusters or other boranes that are unreactive at moderate pressures and temperature for further re-hydrogenation reaction. Elemental analysis during SEM data collection indicates good B retention with the dehydrogenated samples.

Due to the detection limit of elemental analysis with the SEM microscope, Li-compounds were not located. The SEM images of the materials with CeO_2 deserve some attention; the as-milled materials presented clear CeO_2 bright spots (Figure 1c); meanwhile, the dehydrogenated materials (Supplementary Materials) presented a more homogeneous distribution of Ce along the sample image. Another interesting point was a tendency to formation of crystals with higher dehydrogenation pressure. A greater quantity of crystals was observed at samples dehydrogenated at 5 bar initial pressure than at 3 bar initial pressure.

Figure 5. SEM images of the dehydrogenated (DHH) materials: (**a**) $2LiBH_4$ + Al-DHH 5 bar, 400 °C; (**b**) $2LiBH_4$ + Al/TiF$_3$-DHH 5 bar, 400 °C; (**c**) $2LiBH_4$ + Al/CeO$_2$-DHH 5 bar, 400 °C.

Powder X-ray diffraction patterns of the dehydrogenated (labeled as DHH) materials are presented in Figure 6. The diffractograms of the dehydrogenated materials at 3, and 5 bar are presented in Frame (a,b) respectively. In general, the exact determination of the reaction extent based purely on the X-ray diffraction (XRD) results is difficult: this is due to Al and LiH sharing the same crystal symmetry and close crystal cell dimensions. Thus the peaks of the reactant and the reaction product overlap, particularly at low diffraction angles. Additionally, the (101) peak of AlB$_2$ (P6MMM) is pretty close to the Al (200) and LiH (200) peak, these peaks also overlap. The rest of the AlB$_2$ peaks are appreciable when zooming the plot with data-processing software; otherwise, these peaks appear rather small. All the dehydrogenated materials presented minor peaks of Li$_2$O and LiOH. The dehydrogenated $2LiBH_4$ + Al and $2LiBH_4$ + Al/TiF$_3$ presented Al$_2$O$_3$. The oxides and hydroxide could evolve/crystalize from the initial surface oxides present in the as-milled materials and the hydrogen atmosphere during heating. The dehydrogenated 3 bar- and 5 bar-$2LiBH_4$ + Al/CeO$_2$ materials are very interesting regarding the oxides content; they no longer present peaks of CeO$_2$ after the dehydrogenation reaction. CeO$_2$ reacted to form CeB$_6$ (COD-5910033).

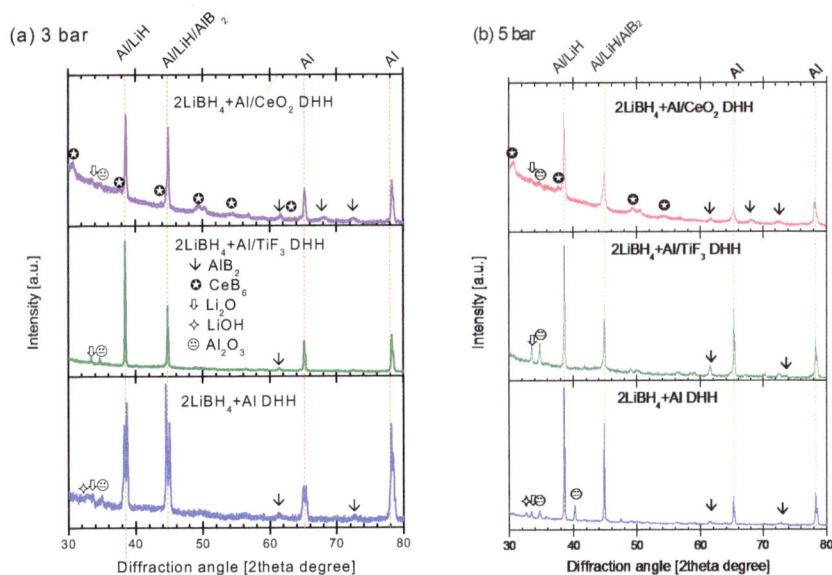

Figure 6. X-ray diffraction patterns of dehydrogenated materials. Frame (**a**) after dehydrogenation at 3 bar initial hydrogen pressure; Frame (**b**) after dehydrogenation at 5 bar initial hydrogen pressure.

FT-IR spectra of all dehydrogenated products are presented in Figure 7. This figure confirms the incomplete dehydrogenation of the materials after reactions. The dehydrogenated materials presented differences in the H–B–H bending region compared to the as-milled materials. First, a shift of the peaks to higher wavenumbers was observed, from 1100–1200 cm^{-1} in as-milled materials to 1400–1560 cm^{-1} in dehydrogenated materials. Secondly, the intensity of these peaks was different with the additive; the 2LiBH$_4$ + Al material presented the highest peak intensity meanwhile the materials added with TiF$_3$ and CeO$_2$ presented reduced signal intensity. The B–H stretching region (2000–2500 cm^{-1}) also presented shifts in wavenumber, however, this shift was not as marked as the H–B–H region. It amounts to approximately 75 cm^{-1}. The materials dehydrogenated at 5 bar presented more intense H–B–H peaks than the materials dehydrogenated at 3 bar. This result points to a different reaction mechanism.

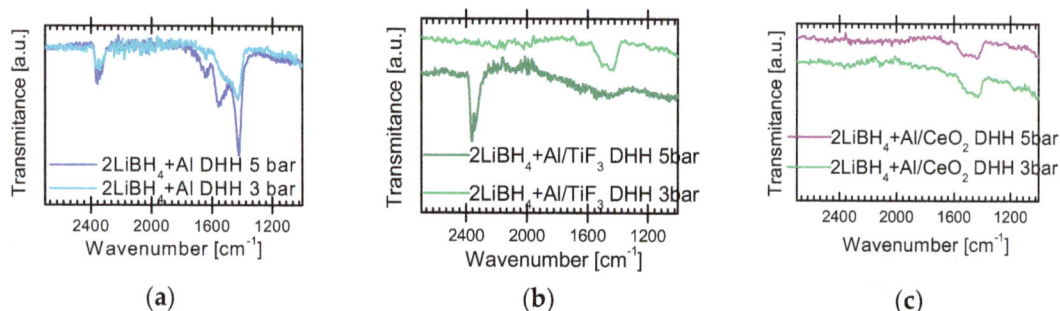

Figure 7. Fourier Transformed Infrared Spectroscopy (FT-IR) of dehydrogenated (DHH) samples: (a) 2LiBH$_4$ + Al-DHH; (b) 2LiBH$_4$ + Al/TiF$_3$; (c) 2LiBH$_4$ + Al/CeO$_2$-DHH.

3. Discussion

Milling effects. A proper integration of LiBH$_4$ and Al in the mixtures was observed after the ball-milling. The milling of Al is not an easy process. Al has the tendency of sintering instead of dispersing unless using proper milling conditions [34]. The milling conditions used here represent a good balance between the optimum milling of Al and reduced degradation of LiBH$_4$.

Protective atmosphere effects. Customarily; the storage, production, processing, and handling of hydrogen storage materials are performed with a high purity argon atmosphere, with oxygen and moisture levels below 0.1 ppm. This would raise significantly the cost of the hydrogen storage materials if commercialization is intended. In our experiments the argon purity was not so high, the supplier-guaranteed oxygen and moisture levels were 10 ppm. This allowed surface oxidation as demonstrated by means of XPS experiments. Kato et al. [27] demonstrated that the surface oxidation of LiBH$_4$ produces Li surface segregation, the formation of Li$_2$O over LiBH$_4$, reduction of diborane desorption, and enhancement of the rate of hydrogen desorption. XPS suggests the formation of LiBO$_2$ rather than Li$_2$O on the surface of our samples and the segregation of Al to sub-surface layers. As a result of the surface oxidation, a reduction of the dehydrogenation temperature was observed here and compared with similar materials carefully protected against surface oxidation [17–22]. A reduction of the dehydrogenation activation barrier was proposed by Kang et al. [35] if LiBH$_4$ or the intermediate LiBH donate one electron (each), to a catalyst on their surfaces. Oxygen is well-known as an electron acceptor. Thus surface oxidation could reduce the activation barrier and help to reach thermodynamic-predicted temperatures for the dehydrogenation reaction.

Dehydrogenation kinetics. The dehydrogenations of the 2LiBH$_4$ + Al, 2LiBH$_4$ + Al/TiF$_3$, and 2LiBH$_4$ + Al/CeO$_2$ materials are multi-step reactions. The multistep nature of the dehydrogenation reaction of the 2LiBH$_4$ + Al is shared with the dehydrogenation of LiBH$_4$ [36] and the RHC LiBH$_4$ + MgH$_2$ [22,37]. Reports of LiBH$_4$-Al dehydrogenation in several molar proportions also described a multistep mechanism for the dehydrogenation reaction [17,19–21,38]. In the materials presented here, the first dehydrogenation step occurred at low temperature, i.e., 100–110 °C. And the main dehydrogenation step occurred between 200–300 °C, finishing at 400 °C. The main dehydrogenation temperature interval is close to the temperature predicted by Siegel et al. for the dehydrogenation reaction of $2LiBH_4 + Al$, 277 °C [18]. A reduction of activation barrier can be responsible for the reduced dehydrogenation temperatures, as pointed out above. Another good point concerning all the studied materials is that the dehydrogenation is rapid, completed within 1 h, mainly during the heating period.

The effect of dehydrogenation pressure. It must be pointed out that this is the first report of dehydrogenation of $2LiBH_4 + Al$/additive mixtures at non-vacuum pressures, i.e., at fuel cell compatible working pressures. Thus the possibility of using these mixtures has been demonstrated for disposable or non-rechargeable hydrogen storage applications as long as further re-hydrogenation can be proved. In general, our materials dehydrogenated at 3 bar presented a better hydrogen release quantity. Further research will be conducted to prove if re-hydrogenation is possible.

The effect of the additives. TiF_3 is a well-known additive in the hydrogen storage area, it is almost mandatory to try TiF_3 as an additive in hydrogen storage systems. Meanwhile, CeO_2 is not commonly used as an additive. CeO_2 has been anticipated to interact with hydrogen [25], and a revision of its effects as additive deserve attention. However, few examples have been published. Ceria could uptake small amounts of hydrogen below 391 °C [39]. Even more, Lin et al. [39] demonstrated an interfacial effect of $CeH_{2.73}/CeO_2$ functioning as "hydrogen pump" and reducing the hydrogen desorption temperature of MgH_2–Mg_2NiH_4–$CeH_{2.73}/CeO_2$. In that work [39], $CeH_{2.73}$ and CeO_2 were formed during successive hydrogenation and oxidation reactions. $LiBH_4$ or $LiBH_4 + 1/2MgH_2$ were mixed with $CeCl_3$, CeF_3, or CeH_2 to improve the hydrogenation/dehydrogenation kinetics, reversibility, or to reduce dehydrogenation temperature [40–44]. In the present work, CeO_2 demonstrated its effectiveness reducing the dehydrogenation temperature. The characterization of dehydrogenated materials by powder X-ray diffraction demonstrated the formation of CeB_6. This boride has been observed after dehydrogenation of $LiBH_4$ or $LiBH_4 + 1/2MgH_2$ destabilized with $CeCl_3$, CeF_3, or CeH_2 [40–44]. On the other hand, X-ray diffraction did not show additional formation of Li or Al-oxides by liberation of the oxygen of CeO_2. The expected formation of oxides is low: for example, a mass balance indicates that if all the oxygen of CeO_2 forms Li_2O, the lithium oxide would be 1.7 wt %. To conclude this part, CeO_2 additive was effective in reducing the dehydrogenation temperature and producing good hydrogen release.

4. Materials and Methods

4.1. Sample Preparation

All reactives were purchased from Sigma-Aldrich and used without further purification. The Al was granular, with a particle size roughly of 1 mm and 99.7% purity. The $LiBH_4$ purity was ≥95%, meanwhile, the purity of TiF_3 and CeO_2 were 99% and 99.995% respectively. The molar ratio of $LiBH_4$ and Al was 2:1. The amount of the additive in each sample was 5 wt %. The mixtures of $2LiBH_4 + Al$ + additives were produced by mechanical milling. The milling was performed in batches of 1 gram of mixture as needed for performing reactions or characterization. The milling was performed in a planetary mill (Across-International) with a rotation of the main plate of 2400 rpm. The milling vials were machinated in stainless steel 316 L with an internal volume of 100 mL, with bolted lids. The milling balls were of yttrium-stabilized zirconium oxide (1 cm diameter). The powder to ball ratio was 1:15. The total milling time was 5 h divided into periods of 1 h milling and 10 min resting. In each cycle of milling-pause, the rotation direction of the planetary mill was inverted. The handling and storage of materials were performed inside a glove box filled with welding grade argon.

4.2. Characterization of the Ball-Milled and Dehydrogenated Materials

Scanning electron microscopy (SEM) images were obtained in a JSM-IT300 microscope (JEOL, Tokio, Japan). Samples were dispersed on carbon tape over a Cu sample holder. The SEM samples were prepared inside the argon glove box and transferred to the microscope by means of a glove bag to avoid oxidation; however, slight oxidation could have been possible. SEM images were obtained by backscattered or secondary electrons and 10 kV or 20 kV of acceleration voltage, accordingly to each sample characteristics.

Powder X-ray diffraction (PXRD) characterization was performed in a BrukerB8 diffractometer (Cu Kα = 1.540598 Å, Bruker AXS, Karlsruhe, Germany). The powders of as-milled and dehydrogenated materials were compacted in a dedicated sample-holder, then they were covered with a Kapton foil for protection against ambient oxygen and moisture. Data processing and phase identification were performed with Diffract Suite Eva (Version 4.2.1.10, Bruker AXS, Karlsruhe, Germany, 2016) or MAUD software (Version 2.55, Trento University, Trento, Italy, 2015). ICSD (Inorganic Crystal Structure Database, Karlsruhe) or COD (Crystallography Open Database) databases were used for phase identification.

Fourier transformed infrared spectroscopy characterization was performed in a Varian 640-IR, FT-IR Spectrometer (Agilent Technologies, Santa Clara, CA, USA). The studied materials were compacted in KBr pellets. The KBr was purchased from Sigma-Aldrich and dried just before pellet preparation. About 2.5 mg of each material was dispersed in 50 mg of dry KBr. FT-IR data was collected in attenuated total reflection (ATR) mode.

X-ray photoelectron spectroscopy analyses were performed only on the as-milled materials. XPS experiments were accomplished in an ultra-high vacuum (UHV) system Scanning XPS microprobe PHI 5000 Versa Probe II (Physical Electronics, Minneapolis, United States of America), with an Al Kα X-ray source (photon energy of 1486.6 eV) monochromatic at 25.4 W, and an multichannel detector. The surface of the samples was etched for 5 min with 1 kV Ar$^+$ at 55.56 nA·mm^{-2}. The XPS spectra were obtained at 45° to the normal surface in the constant pass energy mode, E_0 = 117.40 and 11.75 eV for survey surface and high-resolution narrow scan, respectively. The peak positions were referenced to the background silver 3d$_{5/2}$ photo-peak at 368.2 eV, having an FWHM of 0.56 eV, and C 1s hydrocarbon groups at 285.0 eV, Au 4f$_{7/2}$ in 84.0 eV central peak core level position. XPS characterization was performed on the as-milled materials and raw materials NaBH$_4$ and Al as necessary. Powder samples were compacted in an adequate sample holder and transferred to the equipment by means of a glove bag.

4.3. Dehydrogenation Reaction

Dehydrogenations of 2LiBH$_4$ + Al + additive materials were performed in a Sievert's-type reactor. This reactor was designated and constructed by the research group. It consists basically of twins of a sample holder and a reference holder, a well-known-volume reservoir of H$_2$ for sample and a well-known-volume reservoir of H$_2$ for reference, high precision pressure transducers for sample and reference, and delicate control of the reservoirs, reference-holder and sample-holder temperatures. The registered temperatures and pressures versus time were converted to hydrogen release in wt % with the following formula [45]:

$$\text{wt}\,\%(H_2) = 100 \times \frac{M(H_2) \times \Delta p \times V_{sample}}{m \times R \times T_{sample} \times Z_{fact}} + 100 \times \frac{M(H_2) \times \Delta p \times V_{reservoir}}{m \times R \times T_{reservoir} \times Z_{fact}}, \tag{2}$$

where $M(H_2)$ is the hydrogen molar mass (2.01588 g·mol^{-1}). $\Delta p = \Delta p_{sample} - \Delta p_{reference}$ is in bar, where Δp_{sample} and $\Delta p_{reference}$ mean the actual pressure minus the initial pressure of sample and reference. At zero-time both sample and reference initial pressures are equal. This performs as a differential pressure transducer and helps reducing small (1 × 10^{-3} bar) variations of the pressures caused by thermal effects. The V and T are the volume and temperature of the reservoir and sample holder, in cm^3 and Kelvin degrees respectively; m is the sample mass in g; R is the gas constant (83.14459 cm^3·bar·K^{-1}·mol^{-1}) and, Z_{fact} is the hydrogen compressibility factor [46,47]. It is necessary to mention that the sample holder and reference holder volume account for less than 1% of the reservoirs volume, meeting the appropriate conditions for hydrogen sorption/desorption and Sieverts law.

Samples were transferred to/from the Sieverts-type reactor without oxygen contact by means of a closing valve at the sample holder. Dehydrogenations were performed by a temperature-controlled process. The dehydrogenation initial pressure was fixed manually at 3 or 5 bar, the reservoir temperature was fixed at 40 °C. The sample temperature was raised from room temperature to 350 °C or 400 °C with a heating rate of 5 or 6 °C/min. The total dehydrogenation time was 3 h. Then, the system was cooled down and the remaining hydrogen was released. The dehydrogenation reaction was marked by a significant and sudden increase of the registered pressure beyond the temperature effects. The gases, hydrogen and argon, used during the dehydrogenation experiments were of chromatographic and high purity grade.

5. Conclusions

The 2LiBH4 + Al/additives mixtures were prepared by ball milling. The milling conditions were optimized for the integration of Al and preservation of 2LiBH$_4$. The use of additives TiF$_3$ and CeO$_2$ produced different morphologies with as-milled materials. The studied materials presented a significant reduction of the dehydrogenation temperature that can be related to the surface-oxidation. The surface oxidation was the result of the use of welding-grade argon. The dehydrogenation reactions were observed to take place in two main steps, with onsets at 100 °C and 200–300 °C. The maximum released hydrogen was 9.3 wt % in the 2LiBH$_4$ + Al/TiF$_3$ material, and 7.9 wt % in the 2LiBH$_4$ + Al/CeO$_2$ material. Formation of CeB$_6$ after dehydrogenation of 2LiBH$_4$ + Al/CeO$_2$ was confirmed.

Supplementary Materials:
Figure S1: SEM image of LiBH$_4$ (not ball-milled), Figure S2: SEM image of Al (not ball-milled), Figure S3: SEM of as-milled 2LiBH$_4$ + Al material, Figure S4: SEM of as-milled 2LiBH$_4$ + Al/TiF$_3$ material, Figure S5: SEM of as-milled 2LiBH$_4$ + Al/CeO$_2$ material, Figure S6: FT-IR of the as-milled 2LiBH$_4$ + Al, 2LiBH$_4$ + Al/TiF$_3$ and 2LiBH$_4$/CeO$_2$, Figure S7: Al 1s XPS of the as-milled 2LiBH$_4$ + Al, 2LiBH$_4$ + Al/TiF$_3$ and 2LiBH$_4$/CeO$_2$, Figure S8: SEM of dehydrogenated 2LiBH$_4$ + Al at 5 bar 400 °C, Figure S9: SEM of dehydrogenated 2LiBH$_4$ + Al at 3 bar 400 °C, Figure S10: SEM and EDS of dehydrogenated 2LiBH$_4$ + Al/TiF$_3$ at 5 bar 400 °C, Figure S11: SEM and EDS of dehydrogenated 2LiBH$_4$ + Al/TiF$_3$ at 3 bar 400 °C, Figure S12: SEM of dehydrogenated 2LiBH$_4$ + Al/CeO$_2$ DHH 5 bar 400 °C, Figure S13: SEM of dehydrogenated 2LiBH$_4$ + Al/CeO$_2$ DHH 3 bar 400 °C, Figure S14: Dehydrogenation of 2LiBH$_4$ + Al/CeO$_2$ (3 bar and 5 bar, 350 °C), Figure S15: SEM of dehydrogenated 2LiBH$_4$ + Al/CeO$_2$ DHH 5 bar 350 °C, Figure S16: EDS of dehydrogenated 2LiBH$_4$ + Al/CeO$_2$ DHH 5 bar 350 °C.

Acknowledgments: The present work was supported by SENER-CONACyT Project 215362 "Investigación en mezclas reactivas de hidruro: nanomateriales para almacenamiento de hidrógeno como vector energético". Authors want to thank Omar Solorza-Feria and CINVESTAV-DF for facilitating FT-IR characterization. Authors would also like to thank Orlando Hernández for SEM characterization. XRD was performed at Universidad Autónoma Metropolitana-Iztapalapa, we thank Federico González García. Authors would like to thank Sandra Rodil and Lázaro Huerta Arcos for XPS characterization at IIM-UNAM.

Author Contributions: The findings in this manuscript are part of Juan Luis Carrillo-Bucio master's degree work. Juan Rogelio Tena-García performed the XRD experiments and interpretation. The manuscript was written with contributions of all authors. All authors have given approval to the final version of the manuscript.

References

1. Züttel, A.; Wenger, P.; Rentsch, S.; Sudan, P.; Mauron, P.; Emmenegger, C. LiBH$_4$ a new hydrogen storage material. *J. Power Sources* **2003**, *118*, 1–7. [CrossRef]

2. Lee, J.Y.; Ravnsbæk, D.; Lee, Y.S.; Kim, Y.; Cerenius, Y.; Shim, H.J.; Jensen, T.R.; Hur, N.H.; Cho, Y.W. Decomposition reactions and reversibility of the LiBH$_4$–Ca(BH$_4$)$_2$ composite. *J. Phys. Chem. C* **2009**, *113*, 15080–15086. [CrossRef]

3. Liu, Y.; Reed, D.; Paterakis, C.; Contreras Vasquez, L.; Baricco, M.; Book, D. Study of the decomposition of a 0.62LiBH$_4$–0.38NaBH$_4$ mixture. *Int. J. Hydrog. Energy* **2017**, *42*, 22480–22488. [CrossRef]

4. Gil-Bardají, E.; Zhao-Karger, Z.; Boucharat, N.; Nale, A.; van Setten, M.J.; Lohstroh, W.; Reohm, E.; Catti, M.; Fichtner, M. LiBH$_4$–Mg(BH$_4$)$_2$: A physical mixture of metal borohydrides as hydrogen storage material. *J. Phys. Chem. C* **2011**, *115*, 6095–6101. [CrossRef]

5. Aoki, M.; Miwa, K.; Noritake, T.; Kitahara, G.; Nakamori, Y.; Orimo, S.; Towata, S. Destabilization of LiBH$_4$ by mixing with LiNH$_2$. *Appl. Phys. A* **2005**, *80*, 1409–1412. [CrossRef]

6. Mao, J.F.; Guo, Z.P.; Liua, H.K.; Yu, X.B. Reversible hydrogen storage in titanium-catalyzed LiAlH$_4$–LiBH$_4$ system. *J. Alloys Compd.* **2009**, *487*, 434–438. [CrossRef]

7. Shi, Q.; Yu, X.; Feidenhans, R.; Vegge, T. Destabilized LiBH$_4$–NaAlH$_4$ mixtures doped with titanium based catalysts materials research. *J. Phys. Chem. C* **2008**, *112*, 18244–18248. [CrossRef]

8. Vajo, J.J.; Skeith, S.L.; Mertens, F. Reversible storage of hydrogen in destabilized LiBH$_4$. *J. Phys. Chem. B* **2005**, *109*, 3719–3722. [CrossRef] [PubMed]

9. Barkhordarian, G.; Klassen, T.; Dornheim, M.; Bormann, R. Unexpected kinetic effect of MgB$_2$ in reactive hydride composites containing complex borohydrides. *J. Alloys Compd.* **2007**, *440*, L18–L21. [CrossRef]

10. Bösenberg, U.; Doppiu, S.; Mosegaard, L.; Barkhordarian, G.; Eigen, N.; Borgschulte, A.; Jensen, T.R.; Cerenius, Y.; Gutfleisch, O.; Klassen, T.; et al. Hydrogen sorption properties of MgH$_2$–LiBH$_4$ composites. *Acta Mater.* **2007**, *55*, 3951–3958. [CrossRef]

11. Pinkerton, F.E.; Meyer, M.S. Reversible hydrogen storage in the lithium borohydride—Calcium hydride coupled system. *J. Alloys Compd.* **2008**, *464*, L1–L4. [CrossRef]

12. Yang, J.; Sudik, A.; Wolverton, C. Destabilizing LiBH$_4$ with a Metal (M = Mg, Al, Ti, V, Cr, or Sc) or Metal Hydride (MH$_2$ = MgH$_2$, TiH$_2$, or CaH$_2$). *J. Phys. Chem. C* **2007**, *111*, 19134–19140. [CrossRef]

13. Li, Y.; Li, P.; Qu, X. Investigation on LiBH$_4$–CaH$_2$ composite and its potential for thermal energy storage. *Sci. Rep.* **2017**, *7*, 41754–41758. [CrossRef] [PubMed]

14. Arnbjerg, L.M.; Ravnsbæk, D.B.; Filinchuk, Y.; Vang, R.T.; Cerenius, Y.; Besenbacher, F.; Jørgensen, J.E.; Jakobsen, H.J.; Jensen, T.R. Structure and dynamics for LiBH$_4$–LiCl solid solutions. *Chem. Mater.* **2009**, *21*, 5772–5782. [CrossRef]

15. Yu, X.B.; Grant, D.M.; Walker, G.S. Dehydrogenation of LiBH$_4$ destabilized with various oxides. *J. Phys. Chem. C* **2009**, *113*, 17945–17949. [CrossRef]

16. Lee, H.S.; Hwang, S.J.; To, M.; Lee, Y.S.; Whan Cho, Y. Discovery of fluidic LiBH$_4$ on scaffold surfaces and its application for fast co-confinement of LiBH$_4$–Ca(BH$_4$)$_2$ into mesopores. *J. Phys. Chem. C* **2015**, *119*, 9025–9035. [CrossRef]

17. Hansen, B.R.S.; Ravnsbæk, D.B.; Reed, D.; Book, D.; Gundlach, C.; Skibsted, J.; Jensen, T.R. Hydrogen storage capacity loss in a LiBH$_4$–Al composite. *J. Phys. Chem. C* **2013**, *117*, 7423–7432. [CrossRef]

18. Siegel, D.J.; Wolverton, C.; Ozoliņš, V. Thermodynamic guidelines for the prediction of hydrogen storage reactions and their application to destabilized hydride mixtures. *Phys. Rev. B* **2007**, *76*, 134102–134106. [CrossRef]

19. Zhang, Y.; Tian, Q.; Zhang, J.; Liu, S.S.; Sun, L.X. The dehydrogenation reactions and kinetics of 2LiBH$_4$–Al composite. *J. Phys. Chem. C* **2009**, *113*, 18424–18430. [CrossRef]

20. Kang, X.D.; Wang, P.; Ma, L.P.; Cheng, H.M. Reversible hydrogen storage in LiBH$_4$ destabilized by milling with Al. *Appl. Phys. A* **2007**, *89*, 963–966. [CrossRef]

21. Ravnsbæk, D.B.; Jensen, T.R. Mechanism for reversible hydrogen storage in LiBH$_4$–Al. *J. Appl. Phys.* **2012**, *111*, 112621–112628. [CrossRef]

22. Bosenberg, U.; Ravnsbæk, D.B.; Hagemann, H.; D'Anna, V.; Bonatto-Minella, C.; Pistidda, C.; van Beek, W.; Jensen, T.R.; Bormann, R.; Dornheim, M. Pressure and temperature influence on the desorption pathway of the LiBH$_4$–MgH$_2$ composite system. *J. Phys. Chem. C* **2010**, *114*, 15212–15217. [CrossRef]

23. Saldan, I.; Llamas-Jansa, I.; Hino, S.; Frommen, C.; Hauback, B.C. Synthesis and thermal decomposition of Mg(BH$_4$)$_2$–TMO (TMO=TiO$_2$; ZrO$_2$; Nb$_2$O$_5$; MoO$_3$) composites. *IOP Conf. Ser. Mater. Sci. Eng.* **2015**, *77*, 012041–012047. [CrossRef]

24. Saldan, I.; Frommen, C.; Llamas-Jansa, I.; Kalantzopoulos, G.N.; Hino, S.; Arstad, B.; Heyn, R.; Zavorotynska, O.; Deledda, S.; Sørby, M.H.; et al. Hydrogen storage properties of γ-Mg(BH$_4$)$_2$ modified by MoO$_3$ and TiO$_2$. *Int. J. Hydrog. Energy* **2015**, *40*, 12286–12293. [CrossRef]

25. Sohlberg, K.; Pantelides, S.T.; Pennycook, S.J. Interactions of hydrogen with CeO$_2$. *J. Am. Chem. Soc.* **2001**, *123*, 6609–6611. [CrossRef] [PubMed]

26. Suárez-Alcántara, K.; Palacios-Lazcano, A.F.; Funatsu, T.; Cabañas-Moreno, J.G. Hydriding and dehydriding in air-exposed Mg–Fe powder mixtures. *Int. J. Hydrog. Energy* **2016**, *41*, 23380–23387. [CrossRef]

27. Kato, S.; Bielmann, M.; Borgschulte, A.; Zakaznova-Herzog, V.; Remhof, A.; Orimo, S.I.; Zuttel, A. Effect of the surface oxidation of $LiBH_4$ on the hydrogen desorption mechanism. *Phys. Chem. Chem. Phys.* **2010**, *12*, 10950–10955. [CrossRef] [PubMed]

28. JANAF Thermochemical Tables. Available online: http://kinetics.nist.gov/janaf/ (accessed on 15 July 2017).

29. Fang, Z.Z.; Ma, L.P.; Kang, D.; Wang, P.J.; Wang, P.; Cheng, H.M. In situ formation and rapid decomposition of $Ti(BH_4)_3$ by mechanical milling $LiBH_4$ with TiF_3. *Appl. Phys. Lett.* **2009**, *94*. [CrossRef]

30. D'Anna, V.; Spyratou, A.; Sharma, M.; Hagemann, H. FT-IR spectra of inorganic borohydrides. *Spectrochim. Acta Part A Mol. Biomol. Spectrosc.* **2014**, *128*, 902–906. [CrossRef] [PubMed]

31. Xiong, Z.; Cao, L.; Wang, J.; Mao, J. Hydrolysis behavior of $LiBH_4$ films. *J. Alloys Compd.* **2017**, *698*, 495–500. [CrossRef]

32. Haeberle, J.; Henkel, K.; Gargouri, H.; Naumann, F.; Gruska, B.; Arens, M.; Tallarida, M.; Schmeißer, D. Ellipsometry and XPS comparative studies of thermal and plasma enhanced atomic layer deposited Al_2O_3-films. *J. Nanotechnol.* **2013**, *4*, 732–742. [CrossRef] [PubMed]

33. Van den Brand, J.; Sloof, W.G.; Terryn, H.; de Wit, J.H.W. Correlation between hydroxyl fraction and O/Al atomic ratio as determined from XPS spectra of aluminum oxide layers. *Surf. Interface Anal.* **2004**, *36*, 81–88. [CrossRef]

34. Ramezani, M.; Neitzert, T. Mechanical milling of aluminum powder using planetary ball milling process. *J. Achiev. Mater. Manuf. Eng.* **2012**, *55*, 790–798.

35. Kang, J.K.; Kim, S.Y.; Han, Y.S.; Muller, R.P.; Goddard, W.A., III. A candidate $LiBH_4$ for hydrogen storage: Crystals structures and reaction mechanisms of intermediate phases. *Appl. Phys. Lett.* **2005**, *87*. [CrossRef]

36. Züttel, A.; Rentsch, S.; Fischer, P.; Wenger, P.; Sudan, P.; Mauron, P.; Emmenegger, C. Hydrogen storage properties of LiBH4. *J. Alloys Compd.* **2003**, *356–357*, 515–520. [CrossRef]

37. Kim, K.B.; Shim, J.H.; Park, S.H.; Choi, I.S.; Oh, K.H.; Cho, Y.W. Dehydrogenation reaction pathway of the $LiBH_4$–MgH_2 composite under various pressure conditions. *J. Phys. Chem. C* **2015**, *119*, 9714–9720. [CrossRef]

38. Friedrichs, O.; Kim, J.W.; Remhof, A.; Buchter, F.; Borgschulte, A.; Wallacher, D.; Cho, Y.W.; Fichtner, M.; Oh, K.H.; Zuttel, A. The effect of Al on the hydrogen sorption mechanism of $LiBH_4$. *Phys. Chem. Chem. Phys.* **2009**, *11*, 1515–1520. [CrossRef] [PubMed]

39. Lin, H.J.; Tang, J.J.; Yu, Q.; Wang, H.; Ouyang, L.Z.; Zhao, Y.J.; Liu, J.W.; Wang, W.H.; Zu, M. Symbiotic $CeH_{2.73}$/CeO_2 catalyst: A novel hydrogen pump. *Nano Energy* **2014**, *9*, 80–87. [CrossRef]

40. Gennari, F.C.; Fernández Albanesi, L.; Puszkiel, J.A.; Arneodo Larochette, P. Reversible hydrogen storage from $6LiBH_4$–MCl_3 (M = Ce, Gd) composites by in-situ formation of MH_2. *Int. J. Hydrog. Energy* **2011**, *36*, 563–570. [CrossRef]

41. Liu, B.H.; Zhang, B.J.; Jiang, Y. Hydrogen storage performance of $LiBH_4$ + $1/2MgH_2$ composites improved by Ce-based additives. *Int. J. Hydrog. Energy* **2011**, *36*, 5418–5424. [CrossRef]

42. Jin, S.A.; Lee, Y.S.; Shim, J.H.; Cho, Y.W. Reversible hydrogen storage in $LiBH_4$–MH_2 (M = Ce, Ca) composites. *J. Phys. Chem. C* **2008**, *112*, 9520–9524. [CrossRef]

43. Gennari, F.C. Destabilization of $LiBH_4$ by MH_2 (M = Ce, La) for hydrogen storage: Nanostructural effects on the hydrogen sorption kinetics. *Int. J. Hydrog. Energy* **2011**, *36*, 15231–15238. [CrossRef]

44. Zhang, B.J.; Liu, B.H.; Li, Z.P. Destabilization of $LiBH_4$ by (Ce, La)(Cl, F)$_3$ for hydrogen storage. *J. Alloys Compd.* **2011**, *509*, 751–757. [CrossRef]

45. Peschke, M. Wasserstoffspeicherung in Reaktiven Hydrid-Kompositen Einfluss von fluorbasierten Additiven auf das Sorptionsverhalten des MgB_2–LiH-Systems. Diplomarbeit (Undergraduate Thesis), GKSS-Forschungszentrum Geesthacht GmbH Helmut-Schmidt-Universität/Universität der Bundeswehr, Hamburg, Germany, April 2009.

46. Züttel, A.; Borgschulte, A.; Schlapbach, L.; Chorkendorf, I.; Suda, S. Properties if hydrogen. In *Hydrogen as A Future Energy Carrier*; Züttel, A., Borgschulte, A., Schlapbach, L., Eds.; John Wiley & Sons: Wienheim, Germany, 2008; pp. 71–94. ISBN 978-3-527-30817-0.

47. Hemmes, H.; Driessen, A.; Driessen, R. Thermodynamic properties of hydrogen at pressures up to 1 Mbar and Temperatures between 100 and 1000 K. *J. Phys. C Solid State Phys.* **1986**, *19*, 3571–3585. [CrossRef]

18

Lewis Base Complexes of Magnesium Borohydride: Enhanced Kinetics and Product Selectivity upon Hydrogen Release

Marina Chong [1], Tom Autrey [1,2],* ⓘ and Craig M. Jensen [2],*

[1]   Pacific Northwest National Laboratory, 902 Battelle Blvd, Richland, WA 99352, USA; marinac628@gmail.com
[2]   Department of Chemistry, University of Hawaii at Manoa, 2545 McCarthy Mall, Honolulu, HI 96822, USA
*   Correspondence: tom.autrey@pnnl.gov (T.A.); Jensen@hawaii.edu (C.M.J.)

Abstract: Tetrahydofuran (THF) complexed to magnesium borohydride has been found to have a positive effect on both the reactivity and selectivity, enabling release of H_2 at <200 °C and forms $Mg(B_{10}H_{10})$ with high selectivity.

Keywords: hydrogen storage; Lewis base adducts; borohydride

1. Introduction

Over the past decade, there has been a significant international effort involving chemists, materials scientists and physicists to discover and demonstrate a solid-state hydrogen storage material that would enable a fuel cell electric vehicle 5 min refueling time and a 500 km driving range. Only a few of the thousands of materials investigated have garnered as much interest as $Mg(BH_4)_2$ [1–12]. The high gravimetric density of H_2, ca. 14.7 wt % H_2 and thermodynamics for H_2 release lie in the narrow window required for reversibility under moderate pressure and temperature. The dehydrogenation of the borohydride to MgB_2 has a calculated ΔH_0 of 38.6 kJ/(mol H_2) and ΔS of 111.5 J/(K·mol H_2), predicting a plateau pressure of 1 bar H_2 of 73 °C [13]. These thermodynamic properties together with the borohydride's high gravimetric hydrogen density, and demonstrated hydrogen cycling compatibility [1,9] suggest its application as a reversible hydrogen carrier for PEM fuel cell applications. Two critical challenges remaining are (i) the slow rates of hydrogen release and (ii) the thermodynamic stability of the major dehydrogenation product, magnesium dodecaborane, $Mg(B_{12}H_{12})$, occasionally referred to as the dead-end for reversibility.

At temperatures greater than 460 °C the borohydride releases ~14 wt % hydrogen, giving mixture of products, i.e., MgB_2, MgH_2, Mg and amorphous boron, depending on reaction conditions [6,10,11,14,15]. Hydrogenation of this product mixture at 400 bar H_2 and 270 °C results in the uptake of 6.1 wt % hydrogen [5]. NMR studies concluded that $MgB_{12}H_{12}$, forming at temperatures greater than 250 °C, is a thermodynamic endpoint, preventing re-hydrogenation to $Mg(BH_4)_2$ [4]. On the other hand, the reversal of MgB_2 to $Mg(BH_4)_2$ occurs, albeit, under extreme conditions of 950 bar H_2 and 400 °C [9]. This demonstrated that reversibility can be achieved, however, under conditions that are impractical for commercial hydrogen storage applications. Similarly, the lithium, sodium, and potassium salts of $B_{12}H_{12}{}^{2-}$ have been hydrogenated, in the presence of metal hydrides, to the corresponding borohydride under 1000 bar of H_2 at 500 °C [16]. Whether the pathway for the hydrogenation of MgB_2 to $Mg(BH_4)_2$ involves $MgB_{12}H_{12}$ remains an open question.

The use of additives to enhance kinetics of hydrogen release from $Mg(BH_4)_2$ has been the subject of several investigations [17–20]. An early study found that $TiCl_3$ lowered the onset temperature of hydrogen release from 262 to 88 °C [17]. More recently, significant reductions in the onset temperature of hydrogen release have been observed upon the addition of $NbCl_5$ and a Ti–Nb nanocomposite [18]; metal fluorides such as CaF_2, ZnF_2, TiF_3, and NbF_5 [18–20] and $ScCl_3$ [19]. Hydrogen release is also induced by mechanically milling $Mg(BH_4)_2$ with TiO_2 resulting in release of 2.4 wt % H_2 at 271 °C while undergoing reversible dehydrogenation to $Mg(B_3H_8)_2$ [20]. Alternatively, the thermal dehydrogenation of $Mg(BH_4)_2$ has been shown to be accelerated in eutectic mixtures with $LiBH_4$ [21–23]. Another study claimed that the addition of LiH to $Mg(BH_4)_2$ induced hydrogen evolution at temperatures as low as 150 °C and enabled the cycling of 3.6 wt % H_2 through 20 cycles at 180 °C [24].

2. Results

The high temperature and pressure required for reversibility led us to explore the decomposition pathways at lower temperatures. The decomposition of $Mg(BH_4)_2$ over a prolonged period (5 weeks) under 1 bar nitrogen at 200 °C yields $Mg(B_3H_8)_2$ as the major product [1]. While formation of the $B_3H_8{}^-$ anion has been recognized from thermal condensation studies of $BH_4{}^-$ in solution [25], this finding provided evidence that an analogous process may take place during solid state decomposition contrary to theoretical predictions. Furthermore, under 120 bar hydrogen pressure and 250 °C, the $Mg(B_3H_8)_2$ intermediate was completely converts back to $Mg(BH_4)_2$ after 48 h. The subsequent hydrogenation of independently synthesized $Mg(B_3H_8)_2 \cdot THF$ (THF = tetrahydofuran), where attempts to remove the solvent were unsuccessful, then demonstrated that quantitative re-hydrogenation to $Mg(BH_4)_2$ could be achieved under 50 bar H_2 and 5 h at 200 °C [26]. We concluded that the faster rate exhibited by the solvated sample resulted from a phase change induced by the coordination of the THF. Studies of borohydrides and boranes in the context of hydrogen storage, have typically focused on complete solvent removal. The presence of residual solvent is generally considered problematic and the various synthetic routes to $Mg(BH_4)_2$ often call for rigorous efforts to obtain a pure, solvent-free product. However, our findings suggested that the solvent coordination might have the beneficial effect of enhancing dehydrogenation kinetics. Only a handful of studies have explored the dehydrogenation of $Mg(BH_4)_2$ coordinated to a solvent, the majority of which have highlighted nitrogen donors [27–30].

Our observation of the kinetic enhancement of the hydrogenation of $Mg(B_3H_8)_2$ to $Mg(BH_4)_2$ prompted us to further explore how solvent coordination affects hydrogen release temperatures. We have examined the effect of dimethyl sulfide (DMS), diethyl ether (Et_2O), triethylamine (TEA), diglyme (Digly), dimethoxy ethane (DME) and THF, encompassing a range of Lewis basicity, on the decomposition of $Mg(BH_4)_2$. Alternative syntheses, complex polymorphism, predicted thermodynamic properties, and attempts to improve the hydrogen cycling capacity of $Mg(BH_4)_2$ have been widely explored and reviews of these activities were recently published [20,31]. However, the solid-state chemistry of the interconversion of the borane intermediates involved in these systems remains largely unexplored. Therefore, a unique aspect of this work has been the direct observation and characterization of the borane products and metastable reaction intermediates by MAS and solution phase ^{11}B NMR studies.

Table 1. Ligand ratios in synthesized solvates, determined by ^{1}H NMR.

Solvate	Mg:Ligand Ratio
$Mg(BH_4)_2 \cdot DMS$ [§]	1:0.34
$Mg(BH_4)_2 \cdot TEA$	1:1.8
$Mg(BH_4)_2 \cdot Et_2O$	1:0.36
$Mg(BH_4)_2 \cdot Digly$	1:1.18
$Mg(BH_4)_2 \cdot DME$	1:2.2
$Mg(BH_4)_2 \cdot THF$	1:2.8

[§] The dimethyl sulfide (DMS) solvate was obtained through the synthetic protocol as described by Zanella et al. [32]. The DMS is weakly bond to the magnesium cation and readily removed by heating.

The TEA, Et_2O, Digly, DME and THF solvates of $Mg(BH_4)_2$, were prepared by adding an excess of solvent to $Mg(BH_4)_2$ at room temperature. Subsequently the solvent was removed *en vacuo* to obtain a crystalline solid. The stoichiometry of the solvates was determined from the relative integrated intensities of the signals observed in the ^1H NMR spectra as summarized in Table 1. Where we could find crystal structure information for solvates of $Mg(BH_4)_2$, the stoichiometry of solvate to Mg cation determine by NMR in our work is slightly greater than reported for $Mg(BH_4)_2 \cdot DME$ 1:1.5 and slightly lower for $Mg(BH_4)_2 \cdot THF$ 1:3 [33].

Unsolvated $Mg(BH_4)_2$ and solvate powders were dehydrogenated via combinatorial screening equipment made by Unchained Labs® (Pleasanton, CA, USA), consisting of a 24 well plate design. Heating of the samples was conducted in a screening pressure reactor at 180 °C for 24 h under N_2 flow. Product ratios determined by ^{11}B NMR are shown in Table 2. Entry 1 shows the low reactivity of unsolvated $Mg(BH_4)_2$ at 180 °C with 93% BH_4^- remaining. This result is typical of the slow kinetics of dehydrogenation for borohydride complexes at temperatures below 300 °C. Dehydrogenation of the TEA complex favored formation of $B_3H_8^-$, along with a trace amount of $B_{10}H_{10}^{2-}$. The ether additives showed higher levels of dehydrogenation at 180 °C. Another difference found with the ether complexes is the observation of $B_{10}H_{10}^{2-}$ as the major product, suggesting either a competing dehydrogenation path or that the presence of these ether ligands encourages further reactivity of the $B_3H_8^-$ to form more deeply dehydrogenated products. Of the ether solvates, DME and THF provided the highest conversion of BH_4^- with $B_{10}H_{10}^{2-}$ as the major product. Only small amounts of $B_{12}H_{12}^{2-}$, demonstrating that the decomposition was ~10× more selective for $B_{10}H_{10}^{2-}$ than $B_{12}H_{12}^{2-}$, much higher than the ~1.5× selectivity exhibited by the Digly solvate. These findings motivated further exploration of the dehydrogenation reaction.

Table 2. Distribution of products of $Mg(BH_4)_2$ solvates determined from integration of ^{11}B NMR peaks in mol % after dehydrogenation at 180 °C, 24 h, 1 atm N_2. The balance of products consist of trace quantities of boric acid due to hydrolysis of unstable polyboranes.

Sample	$B_{10}H_{10}^{2-}$	$B_3H_8^-$	$B_{12}H_{12}^{2-}$	BH_4^-
$Mg(BH_4)_2$		3		93
$Mg(BH_4)_2 \cdot TEA$	2	6		89
$Mg(BH_4)_2 \cdot Et_2O$	4	4		88
$Mg(BH_4)_2 \cdot Digl$	5	2	3	82
$Mg(BH_4)_2 \cdot DME$	46	14	4	30
$Mg(BH_4)_2 \cdot THF$	31	12	3	39

3. Discussion

A recent study asserted that closo-boranes are secondary products formed upon aqueous workup of the low temperature dehydrogenation reactions [34]. To determine if formation of $B_{10}H_{10}^{2-}$ occurs directly in the solid state reaction, the ^{11}B VT MAS NMR spectrum of dehydrogenated (1 atm N_2 at 180 °C for 24 h) $Mg(BH_4)_2 \cdot THF$ was obtained (Figure 1). At room temperature, the observed resonances were broad, typical of solid state spectra for quadrupolar nuclei. Heating the sample to 160 °C sharpened the BH_4^- peak and the resonances for $B_{10}H_{10}^{2-}$ at −2 and −27 ppm could be resolved. The peaks at −2 and −27 ppm assigned to the basal and apical boron in $B_{10}H_{10}$ based on the 1:4 ratio integration ratio in the solid state spectrum at 160 °C. At 20 °C the peaks are barely perceptible from the baseline. The sample is subsequently dissolved in a mixture of THF/D_2O for solution NMR analysis. A solution phase spectrum of the dehydrogenated complex was obtained for comparison of product distribution and line width after dissolution in THF/D_2O (Figure 1c). The same high selectivity for $B_{10}H_{10}^{2-}$ is observed in both solution (−2, −30 ppm) and solid-state NMR with respective yields of 19% and 18%.

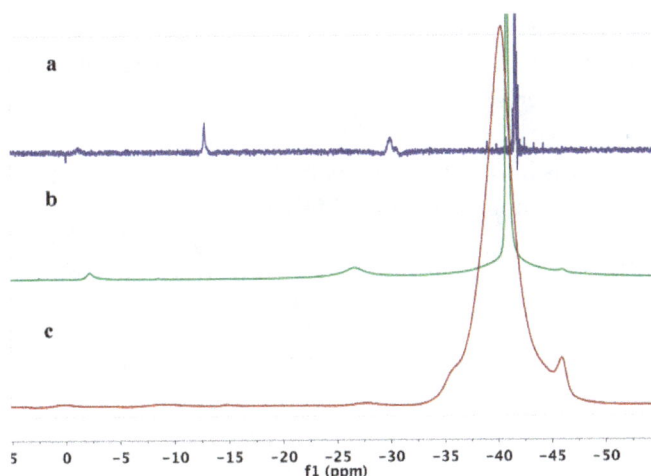

Figure 1. ^{11}B NMR spectra of Mg(BH$_4$)$_2$·THF (tetrahydofuran) dehydrogenated (**a**) solution phase dissolved in 1:2 THF:D$_2$O, (**b**) solid state collected at 160 °C and (**c**) solid state collected at 20 °C. Experimental set-up described in references [35,36].

In situ VT MAS ^{11}B NMR studies of the Mg(BH$_4$)$_2$·THF complex provides additional insight. As seen in Figure S1, the room temperature spectrum contains resonances for both unsolvated and solvated Mg(BH$_4$)$_2$ at −41 and −44 ppm respectively. See Figure S2 for a reference spectrum of unsolvated Mg(BH$_4$)$_2$. The downfield shift of the THF solvated BH$_4$$^-$ complex is comparable to the downfield shift reported for Mg(BH$_4$)$_2$·4NH$_3$ [37]. Upon heating the two peaks collapse into a single narrow peak at 90 °C. We interpret the narrow line width (FWHM = 32 Hz) as being indicative of a fluid phase. This is consistent with the observation of the melting of the THF solvate between 80–100 °C in a melting point apparatus and similar to the m.p. of 90 °C reported for Mg(BH$_4$)$_2$·2NH$_3$ [36].

A comparison of the IR spectra of the solvated and unsolvated Mg(BH$_4$)$_2$ complex is shown in Figure 2. The single prominent stretch observed in the B–H stretching region between 2300–2500 cm^{-1} [38] for Mg(BH$_4$)$_2$ is indicative of lack of directional bonding between the Mg cation and the tetrahedral environment of BH$_4$$^-$. The additional coordination of THF molecules results in the BH$_4$$^-$ also bonding in a mono or bidentate mode to the Mg cation. This lowering of symmetry leads to a spectrum with a number of overlapping bands occur between 2000–2500 cm^{-1}. The modified coordination may play a role in the dehydrogenation mechanism and energetics.

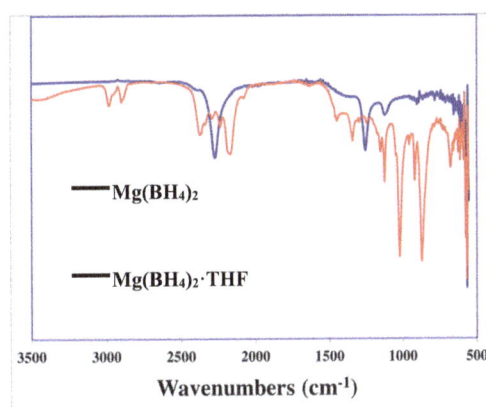

Figure 2. Attenuated Total Reflectance-Infrared spectra of unsolvated Mg(BH$_4$)$_2$ blue spectra with simple B–H stretching region and Mg(BH$_4$)$_2$·THF red spectrum with complex B–H stretching frequency.

The melting of the THF adduct is also likely to be a contributing factor to the enhanced kinetics. However, the onset of dehydrogenation occurs at temperatures above the melting point of the

THF complex. The THF may also reduce the activation energy of clustering to form more deeply dehydrogenated products by altering the coordination mode between Mg^{2+} and BH_4^- through donation of electron density or steric interactions. The high selectivity for $MgB_{10}H_{10}$ over $MgB_{12}H_{12}$ is surprising. Either THF influences the reaction pathway, i.e., lower the barrier for a branching point that pushes the reaction towards $MgB_{10}H_{10}$ formation or THF flips the thermodynamic stability of the closoboranes making $MgB_{10}H_{10}$ more stable than $MgB_{12}H_{12}$.

4. Materials and Methods

All sample preparation and storage was conducted either in a nitrogen glovebox or on a Schlenk line. Solvents were dried over molecular sieves and verified by NMR for purity before use.

4.1. Synthesis of Mg(BH₄)₂

Magnesium borohydride was synthesized following a method described by Zanella et al. Di-*n*-butylmagnesium (Sigma Aldrich, Milwaukee, WI, USA) was added dropwise to borane-dimethylsulfide complex (Sigma Aldrich) in toluene according to the reaction scheme:

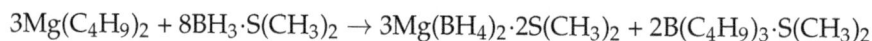

$$3Mg(C_4H_9)_2 + 8BH_3 \cdot S(CH_3)_2 \rightarrow 3Mg(BH_4)_2 \cdot 2S(CH_3)_2 + 2B(C_4H_9)_3 \cdot S(CH_3)_2$$

The mixture was allowed to stir at room temperature for a minimum of 3 h and subsequently filtered, washed with toluene, and dried *en vacuo* at room temperature for 6 h and then at 75 °C overnight. The product, a fine white powder, was found to consist of >95% α-$Mg(BH_4)_2$ by XRD analysis.

4.2. Synthesis of Solvent Adducts of Mg(BH₄)₂

The TEA, Et_2O, Digly, and THF solvates of $Mg(BH_4)_2$, were typically prepared by adding an excess of solvent to $Mg(BH_4)_2$ at room temperature and stirring for 30 min. Excess solvent was then removed *en vacuo* either at room temperature or up to 45 °C for higher boiling point solvents, for as long as needed to obtain a crystalline solid. The DMS adduct was obtained during the synthesis of $Mg(BH_4)_2$ as described above prior to removal of the DMS by heating.

4.3. Characterization of Mg(BH₄)₂ Adducts and Decomposition Products by Solution NMR

Powders were typically dissolved in a 1:2 mixture of THF:deuterium oxide (D_2O) and analyzed within 10 min on a Varian 300 MHz spectrometer with ^{11}B chemical shifts referenced to $BF_3 \cdot Et_2O$ (δ = 0 ppm) and ^1H referenced to TMS (δ = 0 ppm). ^{11}B was measured at 96.23 MHz and ^1H was measured at 299.95 MHz. A relaxation delay of 15 s was used for all ^{11}B analyses with a 90° pulse width of 6 µs. An external standard was added to the quartz NMR tubes to determine the solubility of the powder in the THF/D_2O mixture. The standard consisted of an aqueous solution of sodium tetraphenylborate ($NaBPh_4$) sealed in a glass capillary. Calculation of percent composition of decomposed products was based on peak areas.

4.4. Solid State NMR

Sample powders were packed into 4 mm zirconium oxide rotors and spun at 12 kHz on a Varian 500 MHz spectrometer (Varian, Palo Alto, CA, USA) equipped with a HX 4 mm probe.

4.5. In Situ NMR

The characterization of $Mg(BH_4)_2 \cdot THF$ during heating to 200 °C was conducted by variable temperature (VT) solid state magic angle spin (MAS) NMR in a Varian 500 MHz spectrometer 5 mm HXY probe. ^1H and ^{11}B shifts were referenced to tetramethylsilane at 0 ppm and lithium borohydride at −41.6 ppm and measured at 499.87 and 160.37 MHz respectively. ^1H and ^{11}B spectra were obtained with a 2 s and 5 s relaxation delay and 90° pulse width of 6 µs. The sample powder was packed in a

5 mm zirconia rotor under 1 atm N_2 with a Teflon spacer and then capped with a customized plastic bushing capable of withstanding pressures up to 200 bar. The details of the rotor design are given in detail elsewhere [32,33] and have been modified to accommodate 5 mm rotors. The rotors were spun at 5 kHz at room temperature and subsequently heated at a rate of about 6 °C/min and held at specific temperatures during the ramp at which ^{11}B and ^{1}H spectra were obtained. The duration of the analyses at the set temperatures was approximately 45 min.

5. Conclusions

In summary, characterization of the dehydrogenation products arising from $Mg(BH_4)_2 \cdot THF$ complex by solution and solid-state NMR shows that the dehydrogenation mechanism is highly selective for $B_{10}H_{10}^{2-}$ over $B_3H_8^-$ (theoretical H_2 release 8.1 wt % vs. 2.5 wt % in the absence of solvates) and $B_{12}H_{12}^{2-}$, a kinetic dead end. The dehydrogenation of $Mg(BH_4)_2$ at temperatures below 200 °C and potential for cycling between $Mg(BH_4)_2$ and $MgB_{10}H_{10}$ have significant implications for hydrogen storage applications. Further studies into optimizing the reaction through modification of ligand to Mg ratios are currently underway.

Acknowledgments: The authors gratefully acknowledge research support from the Hydrogen Materials—Advanced Research Consortium (HyMARC), established as part of the Energy Materials Network under the U.S. Department of Energy, Office of Energy Efficiency and Renewable Energy, Fuel Cell Technologies Office. We authors thank Junzhi Yang for preliminary experimental results on solvated magnesium borane complexes, Heather Job for assistance with the combinatorial decomposition experiments, Gary Edvenson for insightful discussion on borane cluster chemistry and David Hoyt and Sarah Burton from the Environmental Molecular Science Laboratory (EMSL) for assistance with the solid state NMR. EMSL is a DOE Office of Science User Facility sponsored by the Office of Biological and Environmental Research and located at Pacific Northwest National Laboratory (PNNL). PNNL a multi-program national laboratory operated by Battelle for the U.S. Department of Energy under Contract DE-AC05-76RL01830.

Author Contributions: Marina Chong, Tom Autrey and Craig Jensen conceived and designed the experiments; Marina Chong performed the experiments; Marina Chong, Tom Autrey and Craig Jensen contributed to analyzing the data and writing the paper.

References

1.　Chong, M.; Karkamkar, A.; Autrey, T.; Orimo, S.; Jalisatgi, S.; Jensen, C.M. Reversible dehydrogenation of magnesium borohydride to magnesium triborane in the solid state under moderate conditions. *Chem. Commun.* **2011**, *47*, 1330–1332. [CrossRef]

2.　Filinchuk, Y.; Richter, B.; Jensen, T.R.; Dmitriev, V.; Chernyshov, D.; Hagemann, H. Porous and dense magnesium borohydride frameworks: Synthesis, stability, and reversible absorption of guest species. *Angew. Chem. Int. Ed.* **2011**, *50*, 11162–11166. [CrossRef]

3.　Hanada, N.; Chłopek, K.; Frommen, C.; Lohstroh, W.; Fichtner, M. Thermal decomposition of $Mg(BH_4)_2$ under He flow and H_2 pressure. *J. Mater. Chem.* **2008**, *18*, 2611–2614. [CrossRef]

4.　Hwang, S.-J.; Bowman, R.C., Jr.; Reiter, J.W.; Rijssenbeek, J.; Soloveichik, G.; Zhao, J.-C.; Kabbour, H.; Ahn, C.C. NMR confirmation for formation of $[B_{12}H_{12}]^{2-}$ complexes during hydrogen desorption from metal borohydrides. *J. Phys. Chem. C* **2008**, *112*, 3164–3169. [CrossRef]

5.　Li, H.-W.; Kikuchi, K.; Sato, T.; Nakamori, Y.; Ohba, N.; Aoki, M.; Miwa, K.; Towata, S.; Orimo, S. Synthesis and hydrogen storage properties of a single-phase magnesium borohydride $Mg(BH_4)_2$. *Mater. Trans.* **2008**, *49*, 2224–2228. [CrossRef]

6.　Matsunaga, T.; Buchter, F.; Mauron, P.; Bielman, M.; Nakamori, Y.; Orimo, S.; Ohba, N.; Miwa, K.; Towata, S.; Züttel, A. Hydrogen storage properties of $Mg[BH_4]_2$. *J. Alloys Compd.* **2008**, *459*, 583–588. [CrossRef]

7.   Nakamori, Y.; Miwa, K.; Ninomiya, A.; Li, H.-W.; Ohba, N.; Towata, S.; Züttel, A.; Orimo, S. Correlation between thermodynamical stabilities of metal borohydrides and cation electronegativites: First-principles calculations and experiments. *Phys. Rev. B* **2006**, *74*, 045126. [CrossRef]

8.   Ozolins, V.; Majzoub, E.H.; Wolverton, C. First-principles prediction of a ground state crystal structure of magnesium borohydride. *Phys. Rev. Lett.* **2008**, *100*, 135501. [CrossRef]

9.   Severa, G.; Ronnebro, E.; Jensen, C.M. Direct hydrogenation of magnesium boride to magnesium borohydride: Demonstration of >11 weight percent reversible hydrogen storage. *Chem. Commun.* **2010**, *46*, 421–423. [CrossRef]

10. Van Setten, M.J.; de Wijs, G.A.; Fichtner, M.; Brocks, G.A. A Density Functional Study of alpha-Mg(BH$_4$)$_2$. *Chem. Mater.* **2008**, *20*, 4952–4956. [CrossRef]

11. Yan, Y.; Li, H.-W.; Nakamori, Y.; Ohba, H.; Miwa, K.; Towata, S.; Orimo, S. Differential scanning calorimetry measurements of magnesium borohydride Mg(BH$_4$)$_2$. *Mater. Trans.* **2008**, *49*, 2751–2752. [CrossRef]

12. Zavorotynska, O.; Deledda, S.; Hauback, B.C. Kinetics studies of the reversible partial decomposition reaction in Mg(BH$_4$)$_2$. *Int. J. Hydrog. Energy* **2016**, *41*, 9885–9892. [CrossRef]

13. Zhang, Y.; Majzoub, E.; Ozoliņš, V.; Wolverton, C. Theoretical prediction of metastable intermediates in the decomposition of Mg(BH$_4$)$_2$. *J. Phys. Chem. C* **2012**, *116*, 10522–10528. [CrossRef]

14. Chłopek, K.; Frommen, C.; Léon, A.; Zabara, O.; Fichtner, M. Synthesis and properties of magnesium tetrahydroborate, Mg(BH$_4$)$_2$. *J. Mater. Chem.* **2007**, *17*, 3496–3503. [CrossRef]

15. Li, H.-W. Dehydriding and rehydriding processes of well-crystallized Mg(BH$_4$)$_2$ accompanying with formation of intermediate compounds. *Acta Mater.* **2008**, *56*, 1342–1347. [CrossRef]

16. White, J.L.; Newhouse, R.J.; Zhang, J.Z.; Udovic, T.J.; Stavila, V. Understanding and mitigating the effects of stable dodecahydro-closo-dodecaborate intermediates on hydrogen-storage reactions. *J. Phys. Chem. C* **2016**, *120*, 25725–25731. [CrossRef]

17. Li, H.-W.; Kikuchi, K.; Nakamori, Y.; Miwa, K.; Towata, S.; Orimo, S. Effects of ball milling and additives on dehydriding behaviors of well-crystallized Mg(BH$_4$)$_2$. *Scr. Mater.* **2007**, *57*, 679–682. [CrossRef]

18. Bardají, E.G.; Hanada, N.; Zabara, O.; Fichtner, M. Effect of several metal chlorides on the thermal decomposition behaviour of α-Mg(BH$_4$)$_2$. *Int. J. Hydrog. Energy* **2011**, *36*, 12313–12318. [CrossRef]

19. Newhouse, R.J.; Stavila, V.; Hwang, S.-J.; Klebanoff, L.; Zhang, J.Z. Reversibility and improved hydrogen release of magnesium borohydride. *J. Phys. Chem. C* **2010**, *114*, 5224–5234. [CrossRef]

20. Saldan, I.; Frommen, C.; Llamas-Jansa, I.; Kalantzopoulos, G.N.; Hino, S.; Arstad, B.; Heyn, R.H.; Zavorotynska, O.; Deledda, S.; Sørby, M.H.; et al. Hydrogen storage properties of γ–Mg(BH$_4$)$_2$ modified by MoO$_3$ and TiO$_2$. *Int J. Hydrog. Energy* **2015**, *40*, 12286–12293. [CrossRef]

21. Bardají, E.G.; Zhao-Karger, Z.; Boucharat, N.; Nale, A.; van Setten, M.J.; Lohstroh, W.; Röhm, E.; Catti, M.; Fichtner, M. LiBH$_4$−Mg(BH$_4$)$_2$: A physical mixture of metal borohydrides as hydrogen storage material. *J. Phys. Chem. C* **2011**, *115*, 6095–6101. [CrossRef]

22. Hagemann, H.; D'Anna, V.; Rapin, J.-P.; Černý, R.; Filinchuk, Y.; Kim, K.C.; Sholl, D.S.; Parker, S.T. New fundamental experimental studies on α-Mg(BH$_4$)$_2$ and other borohydrides. *J. Alloys Compd.* **2011**, *509*, S688–S690. [CrossRef]

23. Nale, A.; Catti, M.; Bardají, E.G.; Fichtner, M. On the decomposition of the 0.6LiBH$_4$−0.4Mg(BH$_4$)$_2$ eutectic mixture for hydrogen storage. *Int. J. Hydrog. Energy* **2011**, *36*, 13676–13682. [CrossRef]

24. Yang, J.; Fu, H.; Song, P.; Zheng, J.; Li, X. Reversible dehydrogenation of Mg(BH$_4$)$_2$–LiH composite under moderate conditions. *Int. J. Hydrog. Energy* **2012**, *37*, 6776–6783. [CrossRef]

25. Muetterties, E.L.; Knoth, W.H. *Polyhedral Boranes*; Marcel Dekker: New York, NY, USA, 1968.

26. Chong, M.; Matsuo, M.; Orimo, S.; Autrey, T.; Jensen, C.M. Selective reversible hydrogenation of Mg(B$_3$H$_8$)$_2$/MgH$_2$ to Mg(BH$_4$)$_2$: Pathway to reversible borane-based hydrogen storage? *Inorg. Chem.* **2015**, *54*, 4120–4125. [CrossRef]

27. Chen, J.; Chua, Y.S.; Wu, H.; Xiong, Z.; He, T.; Zhou, W.; Ju, X.; Yang, M.; Wu, G.; Chen, P. Synthesis, structures and dehydrogenation of magnesium borohydride–ethylenediamine composites. *Int. J. Hydrog. Energy* **2015**, *40*, 412–419. [CrossRef]

28. Yang, Y.; Liu, Y.; Zhang, Y.; Li, Y.; Gao, M.; Pan, H. Hydrogen storage properties and mechanisms of Mg(BH$_4$)$_2$·2NH$_3$–xMgH$_2$ combination systems. *J. Alloys Compd.* **2014**, *585*, 674–680. [CrossRef]

29. Zhao, S.; Xu, B.; Sun, N.; Sun, Z.; Zeng, Y.; Meng, L. Improvement in dehydrogenation performance of $Mg(BH_4)_2 \cdot 2NH_3$ doped with transition metal: First-principles investigation. *Int. J. Hydrog. Energy* **2015**, *40*, 8721–8731. [CrossRef]

30. Soloveichik, G.; Her, J.H.; Stephens, P.W.; Gao, Y.; Rijssenbeek, J.; Andrus, M.; Zhao, J.-C. Ammine magnesium borohydride complex as a New material for hydrogen storage: Structure and properties of $Mg(BH_4)_2 \cdot 2NH_3$. *Inorg. Chem.* **2008**, *47*, 4290–4298. [CrossRef]

31. Zavorotynska, O.; El-Kharbachi, A.; Deledda, S.; Hauback, B.C. Recent progress in magnesium borohydride $Mg(BH_4)_2$: Fundamentals and applications for energy storage. *Int. J. Hydrog. Energy* **2016**, *41*, 14387–14404. [CrossRef]

32. Zanella, P.; Crociani, L.; Masciocchi, N.; Giunchi, G. Facile high-yield synthesis of pure, crystalline $Mg(BH_4)_2$. *Inorg. Chem.* **2007**, *46*, 9039–9041. [CrossRef]

33. Wegner, W.; Jaroń, T.; Dobrowolski, M.A.; Dobrzycki, Ł.; Cyrański, M.K.; Grochala, W. Organic derivatives of $Mg(BH_4)_2$ as precursors towards MgB_2 and novel inorganic mixed-cation borohydrides. *Dalton Trans.* **2016**, *45*, 14370–14377. [CrossRef]

34. Yan, Y.; Remhof, A.; Rentsch, D.; Züttel, A. The role of $MgB_{12}H_{12}$ in the hydrogen desorption process of $Mg(BH_4)_2$. *Chem. Commun.* **2015**, *51*, 700–702. [CrossRef]

35. Hoyt, D.W.; Turcu, R.V.F.; Sears, J.A.; Rosso, K.; Burton, S.D.; Felmy, A.R.; Hu, J.-Z. High-pressure magic angle spinning nuclear magnetic resonance. *J. Magn. Res.* **2011**, *212*, 378–385. [CrossRef]

36. Turcu, R.V.F.; Hoyt, D.W.; Rosso, K.M.; Sears, J.A.; Loring, J.S.; Felmy, A.R.; Hu, J.-Z. Rotor design for high pressure magic angle spinning nuclear magnetic resonance. *J. Magn. Res.* **2013**, *226*, 64–69. [CrossRef]

37. Gao, L.; Guo, Y.H.; Li, Q.; Yu, X.B. The comparison in dehydrogenation properties and mechanism between $MgCl_2(NH_3)/LiBH_4$ and $MgCl_2(NH_3)/NaBH_4$ systems. *J. Phys. Chem. C* **2010**, *114*, 9534–9540. [CrossRef]

38. Marks, T.J.; Kolb, J.R. Covalent transition metal, lanthanide, and actinide tetrahydroborate complexes. *Chem. Rev.* **1977**, *77*, 263–293.

19

Field-Induced Single-Ion Magnet Behaviour in Two New Cobalt(II) Coordination Polymers with 2,4,6-Tris(4-pyridyl)-1,3,5-triazine

Dong Shao [1] (iD)**, Le Shi** [1]**, Hai-Yan Wei** [2,*] **and Xin-Yi Wang** [1,*] (iD)

[1] State Key Laboratory of Coordination Chemistry, Collaborative Innovation Center of Advanced Microstructures, School of Chemistry and Chemical Engineering, Nanjing University, Nanjing 210023, China; shaodong1130@163.com (D.S.); leshiyun1994@163.com (L.S.)

[2] Jiangsu Key Laboratory of Biofunctional Materials, School of Chemistry and Materials Science, Nanjing Normal University, Nanjing 210023, China

* Correspondence: weihaiyan@njnu.edu.cn (H.-Y.W.), wangxy66@nju.edu.cn (X.-Y.W.)

Abstract: We herein reported the syntheses, crystal structures, and magnetic properties of a two-dimensional coordination polymer $\{[Co^{II}(TPT)_{2/3}(H_2O)_4][CH_3COO]_2 \cdot (H_2O)_4\}_n$ (**1**) and a chain compound $\{[Co^{II}(TPT)_2(CHOO)_2(H_2O)_2]\}_n$ (**2**) based on the 2,4,6-Tris(4-pyridyl)-1,3,5-triazine (TPT) ligand. Structure analyses showed that complex **1** had a cationic hexagonal framework structure, while **2** was a neutral zig-zag chain structure with different distorted octahedral coordination environments. Magnetic measurements revealed that both complexes exhibit large easy-plane magnetic anisotropy with the zero-field splitting parameter $D = 47.7$ and 62.1 cm^{-1} for **1** and **2**, respectively. This magnetic anisotropy leads to the field-induced slow magnetic relaxation behaviour. However, their magnetic dynamics are quite different; while complex **1** experienced a dominating thermally activated Orbach relaxation at the whole measured temperature region, **2** exhibited multiple relaxation pathways involving direct, Raman, and quantum tunneling (QTM) processes at low temperatures and Orbach relaxation at high temperatures. The present complexes enlarge the family of framework-based single-ion magnets (SIMs) and highlight the significance of the structural dimensionality to the final magnetic properties.

Keywords: single-ion magnet; cobalt(II); magnetic anisotropy; coordination polymers

1. Introduction

Single-molecule magnets (SMMs) have attracted continuous interest in the field of molecular magnetism since the discovery of magnetic bistability in Mn_{12} acetate clusters [1]. These macroscopic compounds have opened a new window to the microscopic world of the quantum and have great potential in high-density information storage and quantum computing [2]. For a long time, metal clusters with high spin states have been highly pursued. However, the magnetisation reversal barriers (U_{eff}) and blocking temperatures (T_B) have not been effectively enhanced for years. Recent research on mononuclear lanthanide and transition metal based SMMs, also termed as single-ion magnets (SIMs), revealed that high magnetic anisotropy is vitally important to the performance of SMMs [3]. As a result, SIMs have aroused a growing interest in the field of molecular magnetism and numerous related studies have been reported [4].

Since the first report of field-induced slow magnetic relaxation behaviour in a mononuclear Fe^{II} complex in 2010 by Long and co-workers [5], transition-metal-based SIMs have experienced a fast expansion in the literature, especially for those based on Co(II) ions. Intense efforts have been devoted to the design and synthesis of isolated 3d SIMs with coordination numbers ranging from two to eight in various geometries [6]. Furthermore, the field-induced SIM behaviour has also been explored in some one-dimensional (1D) [7,8], two-dimensional (2D) [9–16], and even three-dimensional (3D) Co^{II} coordination frameworks [17,18], as long as the magnetic coupling between the spin centres can be ignored. The magnetic anisotropic spin centres in the frameworks can be well isolated by organic linkers or diamagnetic cyanometallates, and thus can be regarded as isolated single-ion magnetic centres. In this sense, the coordination frameworks offer potentially excellent platforms for their structural stability, diversity, and flexibility, and can be used to control the magnetic anisotropy and even the magnetic relaxation processes [19].

Recently, our group has been continuously focused on the single-ion magnetism of Co^{II} ions with pentagonal bipyramidal (PBP) geometry, and we have reported a series of mononuclear complexes [20,21], as well as 1D and 2D coordination polymers based on the PBP Co^{II} building blocks [8–10]. The influence of axial ligands and magnetic exchange interactions has been carefully studied in these systems. We noticed that PBP mononuclear Co^{II} complexes are roubst SIM systems with easy-plane magnetic anisotropy, and negligible magnetic interactions can maintain the field-induced SIM behaviour in these frameworks, while weak magnetic interactions suppress the slow magnetic relaxation.

To continue our study, a classic ligand, 2,4,6-Tris(4-pyridyl)-1,3,5-triazine (TPT), which can act as a three-connected node (Figure 1), was assembled with Co^{II} ions and a rigid macrocyclic pentadentate ligand dibenzo-15-crown-5-ether with the aim of preparing more isolated Co^{II} SIMs in multi-dimensional frameworks. However, due to unknown reasons, the Co^{II} centres were not chelated by the macrocyclic ligand in the self-assembly process. Serendipitously, the Co^{II} ions were bridged by the TPT ligands, forming two new coordination polymers with different structures. Herein, we reported the syntheses, crystal structures, and magnetic properties of a 2D cationic framework compound $\{[Co^{II}(TPT)_{2/3}(H_2O)_4][CH_3COO]_2 \cdot (H_2O)_4\}_n$ (1) and a neutral 1D chain compound $\{[Co^{II}(TPT)_2(CHOO)_2(H_2O)_2]\}_n$ (2). Magnetic measurements revealed that both complexes exhibit large easy-plane magnetic anisotropy and field-induced slow magnetic relaxation behaviour.

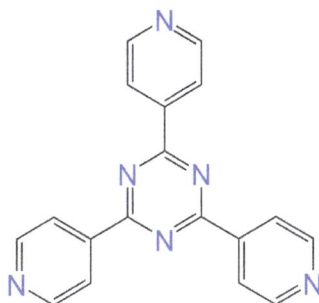

Figure 1. The TPT (2,4,6-Tris(4-pyridyl)-1,3,5-triazine) ligand.

2. Results and Discussion

2.1. Crystal Structure Description

Single-crystal X-ray diffraction analyses revealed that complex 1 crystallises in the trigonal space group $R\bar{3}m$ and has a 2D hexagonal planar structure (Table S1, Figure 2b). It should be noted that a similar compound $[Co(NCS)_2(H_2O)_{0.65}(MeOH)_{0.35}(TPT)_2] \cdot 2.4(H_2O)$ (CoNCS) has been structurally studied in detail by Murugesu and co-workers, and it has a Shubnikov hexagonal plane net [17]. The asymmetric unit of 1 contains one lattice water molecule, one acetate anion, and a 25% occupied Co^{II} ion coordinated by one water molecule and one 50% occupied $(TPT)_{1/6}$ ligand (Figure S1). The Co^{II} centre in 1 resides in a slightly elongated octahedral geometry consisting of four equatorial O atoms from water molecules and two N atoms from bridging TPT ligands in the axial positions (Figure 1a).

The average Co–O and Co–N bond lengths are 2.071(2) and 2.144(4) Å, respectively. The continuous shape measure value (CShM) related to the ideal octahedron (O_h symmetry) for the Co^{II} centre was calculated to be 0.101 using the Shape 2.1 program [22]. As for the ligand TPT, it is on the 3-fold axis and acts as a three-connected node using all three pyridine groups to coordinate three Co^{II} centres. Bridged by the three-connected bridging ligand, a 2D honeycomb (6, 3) framework was formed. The shortest intralayer Co⋯Co distance is 13.347 Å. These layers are further packed alternatingly along the c axis in a staggered hcb (Shubnikov hexagonal plane net) mode (Figure S2). Thus, although there is very large opening window in one single layer, there are no significant open channels in the structure of **1**. The shortest interlayer Co⋯Co distance is 8.293 Å. As we can see, the Co^{II} ions in **1** are well separated by the organic ligands, which will lead to a very weak (if not completely absent) magnetic interaction between these spin centres. This neglible magnetic interaction will not quench the SIM behaviour of **1**, as will be discribed later.

As for compound **2**, it has a 1D chain structure bridged by two of the three pyridines of the TPT ligand (Figure 2d). Though similar Co^{II} complexes based on a TPT ligand with a zigzag chain structure have been reported in the literature [23,24], the dynamic magnetic properties of these chains have never been studied. High magnetic anisotropy was revealed in the compound $[Co(Piv)_2(TPT)(C_2H_5OH)_2]_n$ (**CoPiv**) by Eremenko and co-workers, which showed a spin state switching after desolvation [23]. Complex **2** crystallises in the monoclinic space group $C2/c$ and the Co^{II} centre is also in a distorted octahedral geometry with a N_2O_4 coordination environment, where the N and O atoms are provided by the TPT ligands, coordinated water molecules, and formate groups, respectively (Figure S3 and Figure 2c). We have to point out that the formate anions should be formed by the hydrolysis of the N,N′-dimethylformamide (DMF) solvent under the hydrothermal condition, which has been reported widely in the literature [25]. The average Co–O and Co–N bond lengths are 2.091(2) and 2.176(1) Å, which are slightly larger than those of compound **1** and close to those of compound **CoPiv**. The CShM value is calculated to be 0.276, suggesting a more distorted octahedron. Different from that in **1**, the TPT ligand in **2** serves as a ditopic bridge, linking the $Co(H_2O)_2(CHOO)_2$ core to form a neutral 1D chain. The interchain Co⋯Co distance is 12.913 Å, while the shortest interchain Co⋯Co distance is of 8.293 Å. As in **1**, these long Co⋯Co distances separate the Co^{II} ions efficiently and make compound **2** another candidate to be a SIM system with a framework structure (Figure S4).

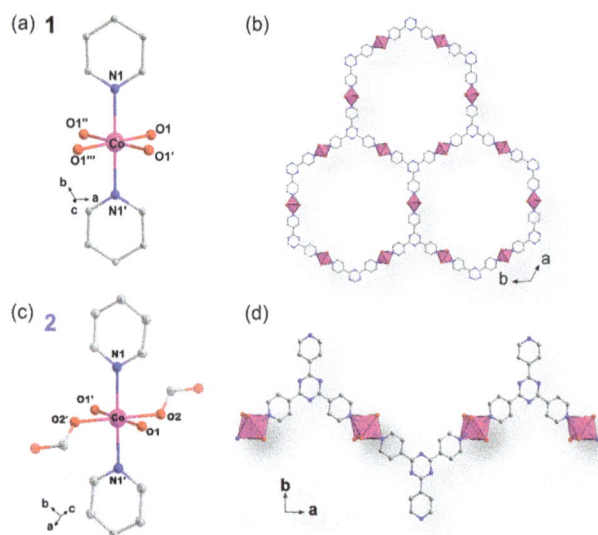

Figure 2. The local coordination environment of Co^{II} ions in **1** (**a**) and **2** (**c**); (**b**) the two-dimensional (2D) honeycomb structure of **1** along the c direction; (**d**) the one-dimensional (1D) chain structure along the a axis. All H atoms and solvent water molecules were omitted for clarity. Symmetry transformations used to generate equivalent atoms: #1 $-x + y + 1$, y, z; #2 $x - y$, $-y$, $-z + 2$; #3 $-x + 1$, $-y$, $-z + 2$ for **1**; #1 $-x + 1/2$, $-y + 3/2$, $-z$; #2 $-x$, y, $-z + 1/2$ for **2**.

2.2. Magnetic Properties

Variable-temperature magnetic susceptibilities for compounds **1** and **2** in the temperature range of 2–300 K were performed on ground single crystal samples under an external direct current (dc) field of 1000 Oe (Figure 3). At 300 K, the $\chi_M T$ values for **1** and **2** are 2.826 and 3.103 cm$^3\cdot$mol$^{-1}\cdot$K, respectively, which are significantly higher than the spin-only value for a high-spin Co^{II} ion with $g = 2$ (1.875 cm$^3\cdot$mol$^{-1}\cdot$K). These high values can be attributed to the orbit contribution and significant spin-orbit coupling (SOC) and are comparable to the reported values (2.1–3.8 cm$^3\cdot$mol$^{-1}\cdot$K) for high-spin Co^{II} centres in the distorted octahedral geometry [11–18]. Upon cooling, the $\chi_M T$ curves decrease very slowly down to the vicinity of 50 K and then abruptly decrease to 2 K, which is typical of anisotropic mononuclear Co^{II} complexes [4]. The $\chi_M T$ values of **2** in the high temperature region are obviously larger than those of **1**, indicating that **2** might possess a higher magnetic anisotropy. The field-dependent magnetisations for **1** and **2** were also measured at 2, 3, and 5 K with fields up to 70 kOe. The largest magnetisation values at 70 kOe reach 2.22, 2.21, and 2.14 μ_B for **1**, and 2.43, 2.42, and 2.33 μ_B for **2** at 2, 3, and 5 K, respectively (Figure 3). The lack of saturation of the magnetisation also suggests the presence of appreciable magnetic anisotropy in these 1D and 2D coordination polymers.

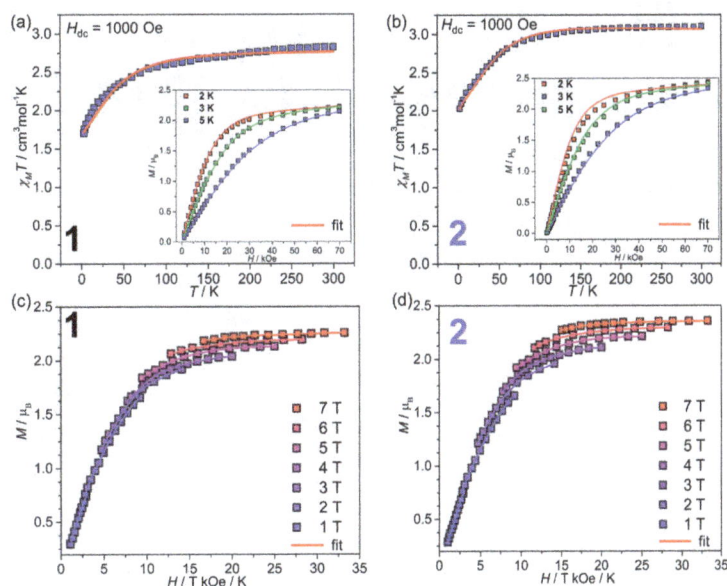

Figure 3. Temperature-dependent magnetic susceptibility data for **1** (**a**) and **2** (**b**) measured under 1 kOe. Inset: the magnetisation curve for **1** measured at 2, 3, and 5 K. The solid lines represent the best fits by PHI program; Reduced magnetisation data of **1** (**c**) and **2** (**d**) collected in the temperature range of 2–10 K under dc fields of 1–7 T. The solid lines correspond to the best fits obtained with Anisofit2.0.

To evaluate the magnetic anisotropy of the Co^{II} centres in **1** and **2**, the magnetic susceptibilities over the whole temperature range and the magnetisation data at 2, 3, and 5 K of **1** and **2** were fitted simultaneously using the PHI [26] program with the following spin Hamiltonian:

$$H = D[S_z^2 - S(S + 1)/3] + E(S_x^2 - S_y^2) + \mu_B\cdot g\cdot S\cdot B \qquad (1)$$

where the D, E, S, B, μ_B represent the axial and rhombic zero-field splitting (ZFS) parameters, the spin operator, magnetic field vectors, and the Bohr magneton, respectively. The best fit parameters are $D = 47.7$ cm^{-1}, $E = 0$ cm^{-1}, $g_z = 2.042$, $g_{x,y} = 2.441$ for **1**, and $D = 62.1$ cm^{-1}, $E = 0.04$ cm^{-1}, $g_z = 1.989$, $g_{x,y} = 2.748$ for **2**. No acceptable fitting can be achieved with the easy-axial type of initially assigned variables. Thus, these values indicate the easy-plane magnetic anisotropy of the octahedral Co^{II} ions in **1** and **2**, which are consistent with the reported results in **CoNCS** [18] and Co(AcO)$_2$(py)$_2$(H$_2$O)$_2$ [27]. Furthermore, the reduced magnetisation data for **1** and **2** collected at different magnetic fields of 1–7 T in the temperature range of 2–10 K (Figure 3) show significant separation between the isofield curves, indicating the large magnetic anisotropy of both complexes. The best fit using Anisofit2.0 [28] gave $D = 36.5$ cm^{-1}, $E = 0.01$ cm^{-1}, and $g = 2.351$ for **1**, and $D = 56.8$ cm^{-1}, $E = 0.1$ cm^{-1}, and $g = 2.34$ for **2**, which agrees well with the fitting results by PHI. For mononuclear octahedral Co^{II} SIMs, positive and negative D values were both observed experimentally and confirmed by theoretical calculations [29]. In addition, to gain a clear scope on reported non-isolated Co^{II} SIMs, the magnetic properties of these systems, including magnetic anisotropy parameters and energy barriers, are compiled in Table S3. From this table, we noticed that the single-ion magnetism in 1D, 2D, and 3D Co^{II} coordination polymers with local O_h symmetry only exhibit an easy-plane type of magnetic anisotropy [11–18]. The situation in the present compounds further support this phenomenon. The magnitude of the ZFS parameters are influenced by the electronic structures of the ligands, charge of the ligands, crystal field distortions, etc. Thus, the larger magnetic anisotropy of **2** could be attributed to the overall change of coordination environment around the Co^{II} ions.

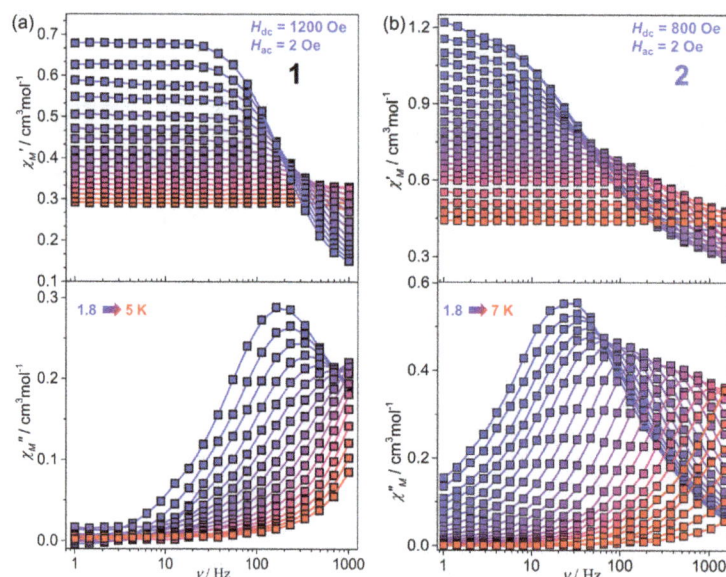

Figure 4. Frequency dependence of the in-phase (χ') and out-of-phase (χ'') part of the alternating current (ac) magnetic susceptibilities for **1** (**a**) and **2** (**b**) collected under a 1200 and 800 Oe dc field, respectively.

To investigate the magnetic dynamics in **1** and **2**, alternating current (ac) magnetic susceptibility was measured at 2.0 K under external dc fields of 0–4000 Oe (Figure S3, Supplementary Materials). No out-of-phase signals (χ'') were observed under a zero dc field due to the occurrence of the fast quantum tunneling process (QTM) induced by intermolecular and/or strong hyperfine interactions, which have been observed in most Co^{II} SIMs [6]. However, after application of dc fields to suppress the QTM effect, both systems showed clear frequency dependence of the in-phase and out-of-phase signals (Figures 4 and 5, Figures S5 and S6, Supplementary Materials), indicating the field-induced SIM behaviour of **1** and **2**. From these ac data, Cole-Cole plots of **1** and **2** at various dc fields were constructed (Figure S6, Supplementary Materials) and fitted by the generalised Debye model [30] to give the field-dependent relaxation time τ (insets of Figure 5). The diagrams of τ vs H suggest that the slowest relaxation occurs at around 1200 Oe for **1** but 800 Oe for **2**, which were then chosen as the optimum fields for further ac measurements of **1** and **2** (Figure 4 and Figure S7, Supplementary Materials). The Cole-Cole plots of **1** and **2** at different temperatures (1.8–5 K for **1**, 1.8–7 K for **1**, Figure 5) were used to extract the temperature-dependent relaxation times (τ). The isothermal susceptibility (χ_T), adiabatic susceptibility (χ_S), τ, and α parameters are listed in Tables S3 and S4. The obtained α parameters are in the range of 0–0.1 for **1** and 0–0.23 for **2**, suggesting the narrow distribution of the relaxation time. As depicted in Figure 5, the $\ln\tau$ vs. $1/T$ plots of **1** and **2** are quite different; it exhibits an almost linear dependence in the temperature range of 1.8–4.0 K for compound **1**, while it shows a clear curvature at the range of 1.8–5.0 K for compound **2**, implying their different relaxation mechanisms. The linear part of the $\ln\tau$ vs. $1/T$ plots can be fitted by the Arrhenius law $\tau = \tau_0 \exp(U_{eff}/k_B T)$ firstly, affording the effective energy barriers (U_{eff}) of 4.8 cm^{-1} and 15.5 cm^{-1} with the pre-exponential factors τ_0 being 6×10^{-6} s and 1×10^{-6} s for **1** and **2**, respectively.

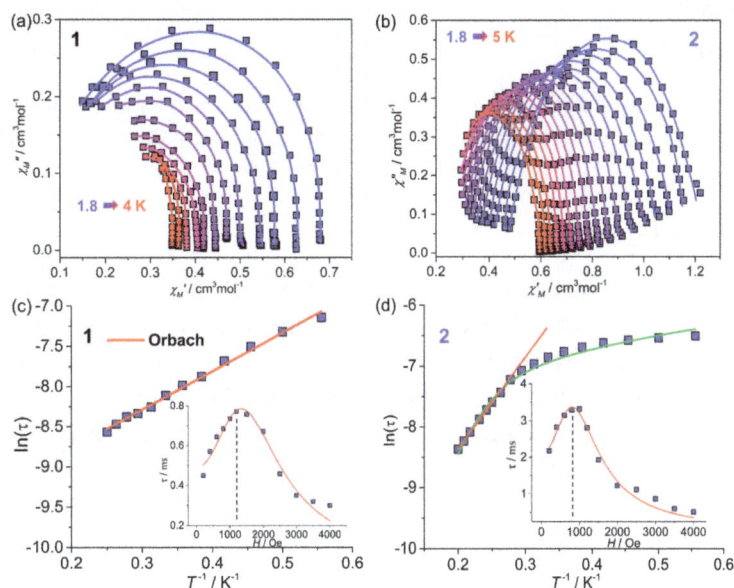

Figure 5. Cole-Cole plots for **1** (**a**) and **2** (**b**) under 1200 and 800 Oe dc fields, respectively. The solid lines are the best fits to the experiments with the generalised Debye model; Arrhenius plot showing $\ln(\tau)$ vs T^{-1} for **1** (**c**) and **2** (**d**). Red lines show fit of the data to the Arrhenius law $\tau = \tau_0 \cdot \exp(U_{eff}/k_B T)$. Green line represents the fit to the data using Equation (2). Inset: Field dependence of the magnetic relaxation time at 2 K for **2** and its approximation by $\tau^{-1} = AH^2 T + B_1/(1 + B_2 H^2)$.

In general, the magnetic relaxation mechanism for a magnetic system is usually quite complex and involves multiple relaxation pathways, such as the Direct, QTM, Raman, and Orbach relaxation processes [31]. These processes can be described by the following equation including the sum of four terms:

$$\tau^{-1} = AH^2 \mathrm{T} + \frac{B_1}{1 + B_2 H^2} + CT^n + \tau_0^{-1} \exp\left(-U_{eff}/k_B T\right) \qquad (2)$$

where A, B_1, B_2, C, n are coefficients and H, T, τ_0, U_{eff}, and k_B have their usual meanings as mentioned before. The first two terms correspond to the field-dependent Direct and QTM processes, while the latter two terms represent the contributions of the two-phonon Raman and thermally activated Orbach mechanisms. Generally, thermally activated behaviour observed in high-temperature region is mainly attributed to the Orbach relaxation process through the excited M_S levels. For **1**, the linear variation of the logarithm of relaxation times at the whole temperature region indicates a dominating Orbach relaxation. It is worth noting that such phenomena have also been observed in some other 2D Co^{II} based SIMs with octahedral local geometry [9–16]. However, multiple relaxation pathways are strongly suggested by the curvature of the Arrhenius plot of **2**. Although a dc field of 800 Oe was applied, there is a temperature-independent region in the temperature range of 1.8–2.4 K, which can be attributed to an unquenched QTM process. To avoid over-parameterisation, the τ vs. H data of **2** at 2 K was fitted with the first two terms of Equation (2) to determine the coefficients related to the Direct and QTM relaxation processes (inset of Figure 5d). This gives $A = 17.3$ s$^{-1}\cdot$K^{-1}, $B_1 = 478$ s^{-1}, $B_2 = 2.5$ kOe^{-2}. These parameters were then fixed, and the $\ln\tau$ vs. $1/T$ plot of **2** was fitted using Equation (2). The obtained values are $U_{eff} = 33.2$ cm^{-1}, $\tau = 4.3 \times 10^{-6}$ s, n = 5.8, C = 0.03 s$^{-1}\cdot$K$^{-5.8}$.

Normally, n equals 7 and 9 for non-Kramers and Kramers ions [32], respectively. When optical and acoustic phonons are taken into consideration depending on the structure of energy levels, n values between 1 and 6 are also reasonable. Thus, the obtained n value suggests an optical acoustic Raman process involving a virtual state, where both acoustic and optical phonons are considered. From these results, we can see that the effective energy barrier of **2** is much larger than that of **1**. A larger easy-plane magnetic anisotropy of **2** might be responsible to this difference. However, the specific reason for the discrepancy of magnetic dynamics of **1** and **2** are currently unknown.

3. Materials and Methods

All reagents were obtained from commercially available sources and used as received unless otherwise noted. 2,4,6-Tris(4-pyridyl)-1,3,5-triazine (TPT) was synthesised according to a previously reported method [33].

3.1. Syntheses of Compounds **1** and **2**

$\{[\text{Co}^{\text{II}}(\text{TPT})_{2/3}(\text{H}_2\text{O})_4][\text{CH}_3\text{COO}]_2 \cdot (\text{H}_2\text{O})_4\}_n$ (**1**): 4 mL ethanol solution of cobalt(II) acetate tetrahydrate (50 mg, 0.2 mmol) was added to 6 mL chloroform solution of TPT ligand (62 mg, 0.2 mmol) and dibenzo-15-crown-5-ether (64 mg, 0.2 mmol) ligand. The mixed solution was then stirred for several hours until all materials dissolved. The solution was then filtered and left to stand at room temperature for weeks. Light orange block crystals were formed, which were collected, washed with water, and dried in air. Yield: about 60% based on Co(ac)$_2 \cdot$4H$_2$O. Elemental analysis (%) for C$_{16}$H$_{30}$CoN$_4$O$_{12}$: C, 36.30; H, 5.71; N, 10.58. Found: C, 36.12, H, 5.26; N, 10.55. IR (KBr, cm^{-1}): 3277(vs, wide), 2433(s), 1976(w), 1617(s), 1558(vs), 1413(vs), 1375(vs), 1231(w), 1214(s), 1160(w), 1056(s), 1021(s), 869(w), 806(vs), 759(s), 736(s), 668(s), 653(w), 518(s), 482(s).

$\{[\text{Co}^{\text{II}}(\text{TPT})_2(\text{CHOO})_2(\text{H}_2\text{O})_2]\}_n$ (**2**): 4 mL aqueous solution of Co(ClO$_4$)$_2 \cdot$6H$_2$O (36 mg, 0.1 mmol) was mixed with 4 mL DMF solution containing TPT (31 mg, 0.1 mmol) and dibenzo-15-crown-5-ether (32 mg, 0.1 mmol) ligands. The solution was sealed in a Teflon-lined stainless steel autoclave and heated at 120 °C for one day. After the autoclave was cooled to room temperature, light orange single crystals of **2** were obtained. Yield: about 50% based on Co(ClO$_4$)$_2 \cdot$6H$_2$O. Elemental analysis (%) for C$_{20}$H$_{21}$CoN$_6$O$_6$: C, 48.00; H, 4.23; N, 16.79. Found: C, 39.83, H, 4.26; N, 16.65. IR (KBr, cm^{-1}): 3399(vs), 3277(vs), 1696(w), 1606(s), 1550(s),1370(s), 1301(vs), 1242(w), 1079(w), 1040(w), 1012(s), 934(w), 848(vs), 762(s), 698(s), 655(w), 642(w), 516(w), 480(s).

Thermal gravimetric analyses (TGA) for the crystal samples of **1** and **2** were performed in the temperature range of 25–700 °C (Figure S8, ESI). The weight loss of 27.6% and 7.3% revealed that the lattice and coordinated water molecules could be removed in the temperature range from 25 to 100 °C, which is consistent with the single crystal data. The purity of the bulk samples of **1** and **2** were confirmed by powder X-ray diffraction (PXRD) spectra and elemental analyses (Figure S9, ESI).

3.2. Physical Measurements

Infrared spectra (IR) data were measured on KBr pellets using a Nexus 870 FT-IR spectrometer (Thermo Fisher Scientific, Waltham, MA, USA) in the range of 4000–400 cm^{-1}. Elemental analyses were performed using an Elementar Vario MICRO analyser (Heraeus, Hanau, Germany). Powder X-ray diffraction data (PXRD) were recorded at 150 K on a Bruker D8 Advanced diffractometer (Bruker, Karlsruhe, Germany) with a Cu Kα X-ray source (λ = 1.54056 Å) operated at 40 kV and 40 mA. Thermal gravimetric analysis (TGA) was conducted in Al$_2$O$_3$ crucibles using PerkinElmer Thermal Analysis (PerkinElmer, Wellesley, MA, USA) in the temperature range of 25–700 °C under a nitrogen atmosphere. All magnetic data were collected using a Quantum Design SQUID VSM magnetometer (Quantum Design, San Diego, CA, USA) on the ground single crystal samples from 2 to 300 K at applied dc fields ranging from 0 to 7 T. Alternative current (ac) magnetic susceptibility data were collected in the temperature range of 2–8 K, under an ac field of 2 Oe, oscillating at frequencies in the range of 1–1500 Hz. All magnetic data were corrected for the diamagnetic contributions of the sample holder and of core diamagnetism of the sample using Pascal's constants.

3.3. X-ray Crystallography

Single crystal X-ray diffraction data were collected on a Bruker APEX Duo diffractometer (Bruker, Karlsruhe, Germany) with a Charge Coupled Device (CCD) area detector (Mo Kα radiation, $\lambda = 0.71073$ Å) at 150 K. The APEX II program (Bruker, Karlsruhe, Germany) was used to determine the unit cell parameters and for data collection. The data were integrated and corrected for Lorentz and polarisation effects using *SAINT* [34]. Absorption corrections were applied with *SADABS* [35]. The structures were solved by direct methods and refined by the full-matrix least-squares method on F^2 using the *SHELXTL* crystallographic software package [36]. All the non-hydrogen atoms were refined anisotropically. Hydrogen atoms of the organic ligands were refined as riding on the corresponding non-hydrogen atoms. Additional details of the data collections and structural refinement parameters are provided in Table S1. Selected bond lengths and angles of **1** and **2** are listed in Table S2 (Supplementary Materials). CCDC 1587443 and 1587444 are the supplementary crystallographic data for this paper. They can be obtained freely from the Cambridge Crystallographic Data Centre via www.ccdc.cam.ac.uk/data_request/cif.

4. Conclusions

In summary, we have synthesised a 2D cationic Co^{II} framework and a neutral 1D zigzag Co^{II} chain using a tritopic ligand 2,4,6-Tris(4-pyridyl)-1,3,5-triazine. The Co^{II} centres are both in a slightly distorted octahedral coordination geometry. The slight discrepancy around Co^{II} ions caused remarkably distinct magnetic anisotropy and dynamic magnetic properties of these two compounds. The large easy-plane magnetic anisotropy and field-induced slow magnetic relaxation have been verified in these two coordination polymers. The present results enlarge the family of framework-based SIMs and highlight the tuneable single-ion magnetism of metal ions via structural variation. Further efforts along this line to realise switchable single-ion magnet behaviour of Co^{II} ions by structural transformation are ongoing in our lab.

Acknowledgments: We thank NSFC (21522103, 21471077, 91622110), the NSF of Jiangsu province (BK20150017), and the Fundamental Research Funds for the Central Universities (020514380006). This work was also supported by the Nanjing University Innovation and Creative Program for Ph.D. candidates.

Author Contributions: Xin-Yi Wang and Hai-Yan Wei conceived and designed this study. Xin-Yi Wang and Dong Shao wrote the manuscript; Dong Shao performed the experiments and analysed the data; Le Shi contributed reagents/materials/analysis tools.

References

1. Thomas, L.; Lionti, F.; Ballou, R.; Gatteschi, D.; Sessoli, R.; Barbara, B. Macroscopic quantum tunneling of magnetization in a single crystal of nanomagnets. *Nature* **1996**, *383*, 145–147. [CrossRef]
2. Wernsdorfer, W.; Aliaga-Alcalde, N.; Hendrickson, D.N.; Christou, G. Exchange-biased quantum tunnelling in a supramolecular dimer of single-molecule magnets. *Nature* **2002**, *416*, 406–408. [CrossRef] [PubMed]
3. Meng, Y.-S.; Jiang, S.-D.; Wang, B.-W.; Gao, S. Understanding the magnetic anisotropy toward single-ion magnets. *Acc. Chem. Res.* **2016**, *49*, 2381–2389. [CrossRef] [PubMed]
4. Craig, G.A.; Murrie, M. 3D single-ion magnets. *Chem. Soc. Rev.* **2015**, *44*, 2135–2147. [CrossRef] [PubMed]
5. Freedman, D.E.; Harman, W.H.; Harris, T.D.; Long, G.J.; Chang, C.J.; Long, J.R. Slow magnetic relaxation in a high-spin iron(II) complex. *J. Am. Chem. Soc.* **2010**, *132*, 1224–1225. [CrossRef] [PubMed]

6.   Bar, A.-K.; Pichon, C.; Sutter, J.-P. Magnetic anisotropy in two- to eight-coordinated transition–metal complexes: Recent developments in molecular magnetism. *Coord. Chem. Rev.* **2016**, *308*, 346–380. [CrossRef]

7.   Zhu, Y.-Y.; Zhu, M.-S.; Yin, T.-T.; Meng, Y.-S.; Wu, Z.-Q.; Zhang, Y.-Q.; Gao, S. Cobalt(II) coordination polymer exhibiting single-ion-magnet-type field-induced slow relaxation behavior. *Inorg. Chem.* **2015**, *54*, 3716–3718. [CrossRef] [PubMed]

8.   Shao, D.; Shi, L.; Shen, F.-X.; Wang, X.-Y. A cyano-bridged coordination polymer nanotube showing field-induced slow magnetic relaxation. *CrystEngComm* **2017**, *19*, 5707–5711. [CrossRef]

9.   Shao, D.; Shi, L.; Zhang, S.-L.; Zhao, X.-H.; Wu, D.-Q.; Wei, X.-Q.; Wang, X.-Y. Syntheses, structures, and magnetic properties of three new chain compounds based on the pentagonal bipyramidal Co(II) building block. *CrystEngComm* **2016**, *18*, 4150–4157. [CrossRef]

10.   Shao, D.; Zhou, Y.; Pi, Q.; Shen, F.-X.; Yang, S.-R.; Zhang, S.-L.; Wang, X.-Y. Two-dimensional frameworks formed by pentagonal bipyramidal cobalt(II) ions and hexacyanometallates: Antiferromagnetic ordering, metamagnetism and slow magnetic relaxation. *Dalton Trans.* **2017**, *46*, 9088–9096. [CrossRef] [PubMed]

11.   Ma, R.-R.; Chen, Z.-W.; Cao, F.; Wang, S.; Huang, X.-Q.; Li, Y.-W.; Lu, J.; Lia, D.-C.; Dou, J.-N. Two 2-D multifunctional cobalt(II) compounds: Field-induced single-ion magnetismand catalytic oxidation of benzylic C–H bonds. *Dalton Trans.* **2017**, *46*, 2137–2145. [CrossRef] [PubMed]

12.   Vallejo, J.; Fortea-Perez, F.R.; Pardo, E.; Benmansour, S.; Castro, I.; Krzystek, J.; Armentano, D.; Cano, J. Guest-dependent single-ion magnet behaviour in a cobalt(II) metal-organic framework. *Chem. Sci.* **2016**, *7*, 2286–2293. [CrossRef]

13.   Świtlicka-Olszewska, A.; Palion-Gazda, J.; Klemens, T.; Machura, B.; Vallejo, J.; Cano, J.; Lloret, F.; Julve, M. Single-ion magnet behaviour in mononuclear and two-dimensional dicyanamide-containing cobalt(II) complexes. *Dalton Trans.* **2016**, *45*, 10181–10193.

14.   Mondal, A.K.; Khatua, S.; Tomar, K.; Konar, S. Field-induced single-ion-magnetic behavior of octahedral Co^{II} in a two-dimensional coordination polymer. *Eur. J. Inorg. Chem.* **2016**, 3545–3552. [CrossRef]

15.   Sun, L.; Zhang, S.; Chen, S.-P.; Yin, B.; Sun, Y.-C.; Wang, Z.-X.; Ouyang, Z.-W.; Ren, J.-J.; Wang, W.-Y.; Wei, Q.; et al. A two-dimensional cobalt(II) network with a remarkable positive axial anisotropy parameter exhibiting field-induced single-ion magnet behavior. *J. Mater. Chem. C* **2016**, *4*, 7798–7808. [CrossRef]

16.   Liu, X.-Y.; Sun, L.; Zhou, H.-L.; Cen, P.-P.; Jin, X.-Y.; Xie, G.; Chen, S.-P.; Hu, Q.-L. Single-ion-magnet behaviour in a two-dimensional coordination polymer constructed from Co^{II} nodes and a pyridylhydrazone derivative. *Inorg. Chem.* **2015**, *54*, 8884–8886. [CrossRef] [PubMed]

17.   Brunet, G.; Safin, D.A.; Jover, J.; Ruiz, E.; Murugesu, M. Single-molecule magnetism arising from cobalt(II) nodes of a crystalline sponge. *J. Mater. Chem. C* **2017**, *5*, 835–841. [CrossRef]

18.   Wang, Y.-L.; Chen, L.; Liu, C.-M.; Du, Z.-Y.; Chen, L.-L.; Liu, Q.-Y. 3D chiral and 2D achiral cobalt(II) compounds constructed from a 4-(benzimidazole-1-yl) benzoic ligand exhibiting field-induced single-ion-magnet-type slow magnetic relaxation. *Dalton Trans.* **2016**, *45*, 7768–7775. [CrossRef] [PubMed]

19.   Liu, K.; Zhang, X.-J.; Meng, X.-X.; Shi, W.; Cheng, P.; Powell, A.K. Constraining the coordination geometries of lanthanide centers and magnetic building blocks in frameworks: A new strategy for molecular nanomagnets. *Chem. Soc. Rev.* **2016**, *45*, 2423–2439. [CrossRef] [PubMed]

20.   Huang, X.-C.; Zhou, C.; Shao, D.; Wang, X.-Y. Field-induced slow magnetic relaxation in cobalt(II) compounds with pentagonal bipyramid geometry. *Inorg. Chem.* **2014**, *53*, 12671–12673. [CrossRef] [PubMed]

21.   Shao, D.; Zhang, S.-L.; Shi, L.; Zhang, Y.-Q.; Wang, X.-Y. Probing the effect of axial ligands on easy-plane anisotropy of pentagonal-bipyramidal cobalt(II) single-ion magnets. *Inorg. Chem.* **2016**, *55*, 10859–10869. [CrossRef] [PubMed]

22.   Llunell, M.; Casanova, D.; Cirera, J.; Alemany, P.; Alvarez, S. *SHAPE*, Version 2.1; Universitat de Barcelona: Barcelona, Spain, 2013.

23.   Polunin, R.A.; Burkovskaya, N.P.; Satska, J.A.; Kolotilov, S.V.; Kiskin, M.A.; Aleksandrov, G.G.; Cador, O.; Ouahab, L.; Eremenko, I.L.; Pavlishchuk, V.V. Solvent-induced change of electronic spectra and magnetic susceptibility of CoII coordination polymer with 2,4,6-Tris(4-pyridyl)-1,3,5-triazine. *Inorg. Chem.* **2015**, *54*, 5232–5238. [CrossRef] [PubMed]

24.   Yoshida, J.; Nishikiori, S.; Kuroda, R. Construction of supramolecular complexes by use of planar bis(β-diketonato)cobalt(II) complexes as building blocks. *Chem. Lett.* **2007**, *36*, 678. [CrossRef]

25.   Wang, X.-Y.; Gan, L.; Zhang, S.-W.; Gao, S. Perovskite-like metal formates with weak ferromagnetism and as precursors to amorphous materials. *Inorg. Chem.* **2004**, *43*, 4615–4625. [CrossRef] [PubMed]

26. Chilton, N.F.; Anderson, R.P.; Turner, L.D.; Soncini, A.; Murray, K.S. PHI: A powerful new program for the analysis of anisotropic monomeric and exchange-coupled polynuclear d- and f-block complexes. *J. Comput. Chem.* **2013**, *34*, 1164–1175. [CrossRef] [PubMed]

27. Walsh, J.P.S.; Bowling, G.; Ariciu, A.-M.; Jailani, N.F.M.; Chilton, N.F.; Waddell, P.G.; Collison, D.; Tuna, F.; Higham, L.J. Evidence of slow magnetic relaxation in $Co(AcO)_2(py)_2(H_2O)_2$. *Magnetochemistry* **2016**, *2*, 23. [CrossRef]

28. Shores, M.P.; Sokol, J.J.; Long, J.R. Nickel(II)−molybdenum(III)−cyanide clusters: Synthesis and magnetic behavior of species incorporating $[(Me_3tacn)Mo(CN)_3]$. *J. Am. Chem. Soc.* **2002**, *124*, 2279–2292. [CrossRef] [PubMed]

29. Gomez-Coca, S.; Cremades, E.; Aliaga-Alcalde, N.; Ruiz, E. Mononuclear single-molecule magnets: Tailoring the magnetic anisotropy of first-row transition-metal complexes. *J. Am. Chem. Soc.* **2013**, *135*, 7010–7018. [CrossRef] [PubMed]

30. Cole, K.S.; Cole, R.H. Dispersion and absorption in dielectrics I. Alternating current characteristics. *J. Chem. Phys.* **1941**, *9*, 341–351. [CrossRef]

31. Zadrozny, J.M.; Atanasov, M.; Bryan, A.M.; Lin, C.-Y.; Rekken, B.D.; Power, P.P.; Neese, F.; Long, J.R. Slow magnetization dynamics in a series of two-coordinate iron(II) complexes. *Chem. Sci.* **2013**, *4*, 125–138. [CrossRef]

32. Abragam, A.; Bleaney, B. *Electron Paramagnetic Resonance of Transition Ions*; Oxford University Press: Oxford, UK, 1970.

33. Anderson, H.L.; Anderson, S.; Sanders, J.K.M. Combined NMR spectroscopy and molecular mechanics studies on the stable structures of calix[n]arenes. *J. Chem. Soc. Perkin Trans.* **1995**, 2231–2245. [CrossRef]

34. *SAINT Software Users Guide*, Version 7.0; Bruker Analytical X-ray Systems, Inc.: Madison, WI, USA, 1999.

35. *SADABS*, Version 2.03; Bruker Analytical X-ray Systems, Inc.: Madison, WI, USA, 2000.

36. *SHELXTL*, Version 6.14; Bruker AXS, Inc.: Madison, WI, USA, 2000–2003.

Permissions

The contributors of this book come from diverse backgrounds, making this book a truly international effort. This book will bring forth new frontiers with its revolutionizing research information and detailed analysis of the nascent developments around the world.

We would like to thank all the contributing authors for lending their expertise to make the book truly unique. They have played a crucial role in the development of this book. Without their invaluable contributions this book wouldn't have been possible. They have made vital efforts to compile up to date information on the varied aspects of this subject to make this book a valuable addition to the collection of many professionals and students.

This book was conceptualized with the vision of imparting up-to-date information and advanced data in this field. To ensure the same, a matchless editorial board was set up. Every individual on the board went through rigorous rounds of assessment to prove their worth. After which they invested a large part of their time researching and compiling the most relevant data for our readers.

The editorial board has been involved in producing this book since its inception. They have spent rigorous hours researching and exploring the diverse topics which have resulted in the successful publishing of this book. They have passed on their knowledge of decades through this book. To expedite this challenging task, the publisher supported the team at every step. A small team of assistant editors was also appointed to further simplify the editing procedure and attain best results for the readers.

Apart from the editorial board, the designing team has also invested a significant amount of their time in understanding the subject and creating the most relevant covers. They scrutinized every image to scout for the most suitable representation of the subject and create an appropriate cover for the book.

The publishing team has been an ardent support to the editorial, designing and production team. Their endless efforts to recruit the best for this project, has resulted in the accomplishment of this book. They are a veteran in the field of academics and their pool of knowledge is as vast as their experience in printing. Their expertise and guidance has proved useful at every step. Their uncompromising quality standards have made this book an exceptional effort. Their encouragement from time to time has been an inspiration for everyone.

The publisher and the editorial board hope that this book will prove to be a valuable piece of knowledge for researchers, students, practitioners and scholars across the globe.

List of Contributors

Christian M. Julien and Henri Groult
PHENIX, UMR 8234, Sorbonne Universités, Univ. Pierre et Marie Curie, Paris-6, 4 Place Jussieu, 75005 Paris, France

Alain Mauger
IMPMC, Sorbonne Universités, Univ. Pierre et Marie Curie, Paris-6, 4 Place Jussieu, 75005 Paris, France

Julie Trottier, Karim Zaghib and Pierre Hovington
Energy Storage and Conversion, Research Institute of Hydro-Québec, Varennes, QC J3X 1S1, Canada

Kornelia Zeckert
Institute of Inorganic Chemistry, University of Leipzig, Johannisallee 29, D-04103 Leipzig, Germany

Fengdong Qu, Bo He and Minghui Yang
Dalian Institute of Chemical Physics, Chinese Academy of Sciences, Dalian 116023, China

Rohiverth Guarecuco
Department of Chemical Engineering, Massachusetts Institute of Technology, Cambridge, MA 02139-4307, USA

Sarote Boonseng, Gavin W. Roffe, Rhiannon N. Jones, John Spencer and Hazel Cox
Department of Chemistry, School of Life Sciences, University of Sussex, Falmer, Brighton, East Sussex, BN1 9QJ, UK

Graham J. Tizzard and Simon J. Coles
UK National Crystallography Service, School of Chemistry, University of Southampton, Highfield, Southampton SO17 1BJ, UK

Peter Höhn and Yurii Prots
Max-Planck-Institut für Chemische Physik fester Stoffe, Nöthnitzer Str. 40, D-01187 Dresden, Germany;

Tanita J. Ball´e, Manuel Fix and Anton Jesche
EP VI, Center for Electronic Correlations and Magnetism, Augsburg University, D-86159 Augsburg, Germany

Antoni Macià Escatllar
Departament de Ciència de Materials i Química Física and Institut de Química Teòrica i Computacional (IQTCUB), Universitat de Barcelona, E-08028 Barcelona, Spain

Piero Ugliengo
Dipartimento di Chimica and NIS Centre, Università degli Studi di Torino, Via P. Giuria 7, I-10125 Torino, Italy

Stefan T. Bromley
Departament de Ciència de Materials i Química Física and Institut de Química Teòrica i Computacional (IQTCUB), Universitat de Barcelona, E-08028 Barcelona, Spain
Institució Catalana de Recerca i Estudis Avançats (ICREA), E-08010 Barcelona, Spain

Yincheng Wang, Andleeb Mehmood, Yanan Zhao, Jingping Qu and Yi Luo
State Key Laboratory of Fine Chemicals, School of Chemical Engineering, Dalian University of Technology, Dalian 116024, China

Takashi Kosone
Department of Creative Technology Engineering Course of Chemical Engineering, Anan College, 265 Aoki, Minobayashi, Anan, Tokushima 774-0017, Japan

Takeshi Kawasaki and Itaru Tomori
Department of Chemistry, Faculty of Science, Toho University, 2-2-1 Miyama, Funabashi, Chiba 274-8510, Japan

Takafumi Kitazawa
Department of Chemistry, Faculty of Science, Toho University, 2-2-1 Miyama, Funabashi, Chiba 274-8510, Japan
Research Centre for Materials with Integrated Properties, Toho University, 2-2-1 Miyama, Funabashi, Chiba 274-8510, Japan

Jun Okabayashi
Research Center for Spectrochemistry, University of Tokyo, Bunkyo-ku, Tokyo 113-0033, Japan

Johannes M. Dieterich
Mechanical and Aerospace Engineering, Princeton University, Princeton, New Jersey, NJ 08544-5263, USA
Institute for Physical Chemistry, Christian-Albrechts-University, 24098 Kiel, Germany

Bernd Hartke
Institute for Physical Chemistry, Christian-Albrechts-University, 24098 Kiel, Germany

Ajay Kumar, Parisa Bashiri, Balaji P. Mandal, Kulwinder S. Dhindsa, Khadije Bazzi, Maryam Nazri, Zhixian Zhou and Ratna Naik
Department of Physics and Astronomy, Wayne State University, Detroit, MI 48202, USA

Gholam-Abbas Nazri
Department of Physics and Astronomy, Wayne State University, Detroit, MI 48202, USA
Electrical and Computer Engineering, Wayne State University, Detroit, MI 48202, USA

Ambesh Dixit
Indian Institute of Technology, Jodhpur 342011, India

Vijayendra K. Garg and Aderbal C. Oliveira
Universidade de Brasilia, Instituto de Fisica, Brasilia, DF 70919-970, Brazil

Prem P. Vaishnava
Department of Physics, Kettering University, Flint, MI 48504, USA

Vaman M. Naik
Department of Natural Sciences, University of Michigan-Dearborn, Dearborn, MI 48128, USA

Tzia Ming Onn, Xinyu Mao, Chao Lin, Cong Wang and Raymond J. Gorte
Department of Chemical and Biomolecular Engineering, University of Pennsylvania, 34th Street, Philadelphia, PA 19104, USA

Yuki Nakagawa, Shigehito Isobe, Takao Ohki and Naoyuki Hashimoto
Graduate School of Engineering, Hokkaido University, N-13, W-8, Sapporo 060-8278, Japan

Norio Nakata, Nanami Kato, Noriko Sekizawa and Akihiko Ishii
Department of Chemistry, Graduate School of Science and Engineering, Saitama University, 255 Shimo-okubo, Sakura-ku, Saitama 338-8570, Japan

Zhixin Zhang, Yehong Wang, Jianmin Lu, Min Wang, Jian Zhang and Feng Wang
State Key Laboratory of Catalysis, Dalian National Laboratory for Clean Energy, Dalian Institute of Chemical Physics, Chinese Academy of Sciences, Dalian 116023, China

Xuebin Liu
Energy Innovation Laboratory, BP Office (Dalian Institute of Chemical Physics), Dalian 116023, China

Linda Iffland, Anette Petuker and Ulf-Peter Apfel
Anorganische Chemie I, Ruhr-Universität Bochum, Universitätsstraße 150, 44801 Bochum, Germany

Maurice van Gastel
Max-Planck-Institut für Chemische Energiekonversion, Stiftstraße 34-36, 45470 Mülheim, Germany

Tomohiro Sugahara and Norihiro Tokitoh
Institute for Chemical Research, Kyoto University, Gokasho, Uji, Kyoto 611-0011, Japan

Takahiro Sasamori
Graduate School of Natural Sciences, Nagoya City University, Yamanohata 1, Mizuho-cho, Mizuho-ku, Nagoya, Aichi 467-8501, Japan

Juan Luis Carrillo-Bucio, Juan Rogelio Tena-García and Karina Suárez-Alcántara
Morelia Unit of the Materials Research Institute of the National Autonomous University of Mexico, Antigua Carretera a Pátzcuaro 8701, Col. Ex Hacienda de San José de la Huerta, Morelia 58190, Michoacán, Mexico

Marina Chong
Pacific Northwest National Laboratory, 902 Battelle Blvd, Richland, WA 99352, USA

Tom Autrey
Pacific Northwest National Laboratory, 902 Battelle Blvd, Richland, WA 99352, USA
Department of Chemistry, University of Hawaii at Manoa, 2545 McCarthy Mall, Honolulu, HI 96822, USA

Craig M. Jensen
Department of Chemistry, University of Hawaii at Manoa, 2545 McCarthy Mall, Honolulu, HI 96822, USA

Dong Shao, Le Shi and Xin-Yi Wang
State Key Laboratory of Coordination Chemistry, Collaborative Innovation Center of Advanced Microstructures, School of Chemistry and Chemical Engineering, Nanjing University, Nanjing 210023, China

Hai-Yan Wei
Jiangsu Key Laboratory of Biofunctional Materials, School of Chemistry and Materials Science, Nanjing Normal University, Nanjing 210023, China

Index